Fundamentals of Environmental Studies

Environmental Studies is a compulsory paper for undergraduate students in almost all disciplines in India. The purpose of teaching this course is to sensitize students towards environmental issues. The course introduces students to the fundamental principles of environmental science, ecology and allied topics including policy, law, pollution control, economics and natural resource management.

This book provides detailed discussion on fundamental concepts and issues – including global warming, acid rain, ozone layer depletion, nuclear accidents and nuclear holocaust – related to the environment. Focussing on the immediate need for public awareness, the book discusses various natural resources such as water, land, forests, food and mineral, and the problems associated with them. Using examples, it introduces and explains different types of ecosystems, biogeochemical cycles and laws of thermodynamics. It also contains chapters on environmental pollution and waste management that visit important topics like air pollution, water pollution, noise pollution, waste water treatment and solid waste management in detail.

Case studies, multiple choice questions, short questions and essay type questions are interspersed throughout the book for assessment of learning achievements.

Mahua Basu is Assistant Professor in the Department of Environmental Studies, St. Xavier's College (Autonomous), Kolkata. She has been teaching courses in environmental studies, renewable and non-renewable resources and solid waste management at the undergraduate level. She has contributed in peer reviewed journals and her areas of interest include environmental impact assessment, toxicology, biodiversity and atmospheric pollution.

S. Xavier is Head of the Department of Environmental Studies, St. Xavier's College (Autonomous), Kolkata. He has been teaching courses in environmental ethics, waste water treatment, environmental management, water pollution and water resource management for more than fifteen years. His research interests include epidemiology, remote sensing, alternative water resource and arsenic mitigation measures.

Fundamentals of
Environmental Studies

Mahua Basu

S. Xavier

CAMBRIDGE
UNIVERSITY PRESS

4843/24, 2nd Floor, Ansari Road, Daryaganj, Delhi - 110002, India

Cambridge University Press is part of the University of Cambridge.

It furthers the University's mission by disseminating knowledge in the pursuit of education, learning and research at the highest international levels of excellence.

www.cambridge.org
Information on this title: www.cambridge.org/9781107536173

First published 2016

Reprint 2017

Printed in India by Thomson Press India Ltd., New Delhi 110001

A catalogue record for this publication is available from the British Library

Library of Congress Cataloging-in-Publication Data
Names: Basu, Mahua. | Xavier, S. (Savarimuthu), 1970-
Title: Fundamentals of environmental studies / Mahua Basu, S. Xavier, SJ.
Description: Delhi, India : Cambridge University Press is part of the
University of Cambridge, [2016] | Includes bibliographical references and index.
Identifiers: LCCN 2015021990 | ISBN 9781107536173 (pbk.)
Subjects: LCSH: Environmental sciences. | Natural resources. | Human ecology.
| Nature--Effect of human beings on.
Classification: LCC GE105 .B37 2016 | DDC 304.2--dc23 LC record available at http://lccn.loc.gov/2015021990

ISBN 978-1-107-53617-3 Paperback

Dedicated to
our Parents and all our Teachers

Contents

Preface

Environmental studies is not just another term paper that students are expected to rush through. On the contrary, it is much more potent in its content and its proper understanding will not only help young learners live their lives in the most sustainable manner, but will also help them make a difference and contribute to the reversal of many of the ailments that afflict Mother Earth.

Our environment, which has been shaped over millennia, is today threatened, primarily because of man's greed and wantonness. Indiscriminate burning of fossil fuels and a mad haste for development over the last three centuries have played havoc with the fragile ecosystems, at places irretrievably. Environmental studies is thus gaining in importance as it not only promises a holistic view of related issues, but is also capable of pointing out human follies and providing answers for a cleaner, greener and more prosperous future world order.

The paper on which this book is printed, the wooden desk on which you are now studying the book, the electricity that is propelling the ceiling fan, the pen that you are using to take down notes – everything around us has an impact on the world that we live in and therefore on our environment. By understanding what starts the domino effect leading to ultimate consequences for the environment – through Environmental Studies – we can not only strive towards a more sustainable lifestyle locally, but can actually make some difference at the planetary level by capping wastages and ensuring the optimal usage of resources.

Each drop of positive action taken by every individual will add up – through families, societies, nations – to encompass entire mankind and will one day make up the ocean that will help us restrain, even reverse the wheels of ecological destruction that we ourselves have set in motion. This is precisely why we have stressed and striven to ensure that the students understand the subject, are able to relate the knowledge they gain to the way they live and finally to empower them to be the change they will profess.

Any textbook can define, dictate and be didactic. Our effort on the contrary is not merely restricted to helping students achieve top grades. That is certainly an important function of the book. However, the core concern and main stress is on engaging young minds and igniting them, so that the students imbibe the values imparted and transform into better human beings.

Being clean, green and sustainable are not mere options. Today, we live in a world, where it is the only option and even as we write this, time is running out on the environment pushing us all towards the brink of destruction. The answer, humanity's response, will depend to a great extent on how conscious and committed future generations are. Only when a million ignited minds attain critical mass and walk the green talk will we be able to make a difference. If this book is able to create an impression on even one such young mind, we will consider having succeeded in our endeavour.

Acknowledgments

It took more than two years to write this book. Prior to writing, it took another few years to assemble all the information that was required to plan and start writing the chapters. The subject is so dynamic that by the time we collected information and started writing, the policies, plan and data would often undergo revision. We have tried our level best to provide updated data, latest developments and case studies in an attempt to make the book apposite to the present day's context and of course ensure that it contains every topic in the UGC syllabus. Our additional inputs shall definitely proffer the utility of this book for students of Honours and Post Graduate courses.

First and foremost we owe a lot to our parents for their constant encouragement and suggestions throughout the period of writing. We express our apology for not being able to give them adequate time during the span of writing.

We gratefully acknowledge the humble support and inspiration we received from our Principal Rev. Fr Dr J. Felix Raj, S.J., the Vice Principals, Rev. Fr Dr Dominic Savio, S.J., Professor Bertram Da Silva, Professor M. M. Rahman, Deans of all the departments in our endeavour.

Our sincere gratitude to all our teachers who imparted knowledge to us since our childhood and encouraged us in our endeavour. Special thanks to Professor Runu Bhattacharya as she would often enrich us with her knowledge and wisdom.

We extend our sincere thanks to Saikat Sen, who believed in us and was the first person to come up with the idea of writing a textbook on Environmental Studies for the Undergraduate level. His support generated the self-confidence required to transform our knowledge into the form of a book. We are grateful to Sabarna Banerjee, Senior Sales Executive, Cambridge University Press. We shall not forget his first telephonic conversation expressing his willingness to refer the manuscript to the editorial sections of Cambridge University Press.

Such confidence was again provided by Rajesh Dey, Commissioning Editor, Cambridge University Press, who after a glance at the initial chapters of the manuscript expressed interest in considering it for publication. We shall remember his guidance and expertise at each step in giving a final shape to this book. Our sincere thanks to Debjani Majumder, Vice President, Cambridge University Press, who, in spite of her busy schedule, would often take out time in order to share her experiences and expectations regarding this book whenever she would get a chance to visit Calcutta.

Our thanks to Professor Bipra Kumar Das, Dr Ankur Ray, St. Xavier's College (Autonomous), Kolkata, Professor Ilora Sen of Shri Shikshayatan College, Kolkata and Late Sri Shyamal Kumar

Bose for their patience in reading the text and for their valuable suggestions for improvements in specific topics. We are grateful to the reviewers who undertook the painstaking job of scrutinizing the chapters and providing valuable comments for enrichment of their contents. Incorporation of these suggestions has taken the book a step ahead from being a text book to becoming a reference book that will be useful both at undergraduate and post graduate levels.

We earnestly thank Suvobrata Ganguly, Editor, Core Sector Communique, who helped in writing a few topics in Chapters 1, 2 and 9. His multifaceted knowledge served as valuable input to this book.

Our sincere thanks to Professor Fr Dejus J.R., St Xavier's College (Autonomous), Kolkata for his contributions in the form of flow charts and illustrations in some chapters. His expertise in this field has simplified the complicated figures into simple, student friendly illustrations. We are indeed very grateful for his support.

We will be failing in our duties if we do not thank our students for whom this book has been written. Our students, both past and present, form our inner strength and are important motivators. It was our students who made us realize the need for such a book and have, since we started working on it, helped us all the way with their valuable inputs, suggestions and carefully thought out intellectual contributions. If they accept this book with the enthusiasm with which they helped us work on it, our purpose will be served.

Our whole hearted thanks to Dr Arijit Basu and Dr Nibha Mishra, Assistant Professor, Department of Pharmaceutical Sciences, BITS, Mesra, who guided us in writing the content in a proper format, and also provided valuable technical inputs and critical comments. They were present right from its inception, up to the proof reading stage. Their constant moral support and guidance enabled us reach our final destination. The book would not have been possible without their encouragement and support.

Lastly, thanks to our priceless friends Saikat Mitra, Anindya Ray, Sudipta Lahiri, Anita Mamidi and Paramita Paul for their constant drive towards completion of this book. In this regard we should not forget to mention Partha Pratim Boral who is not amongst us today, but is present very much in our spirit. His enthusiasm regarding the book release was unbelievable. As friends they were eager to do everything that would help in completion of the book.

Multidisciplinary Nature of Environmental Studies

1

Learning objectives

- To develop a comprehensive understanding of the concept and scope of environment studies.
- To know about the immense importance of environment as a subject.
- To develop public awareness about our environment and elicit collective response for its protection.
- To gather information about organizations and people relentlessly working in this field.
- To know and analyse the types of environment and environmental components and how they affect our survival.

1.1 Definition and Concept of Environment

The word environment is derived from the French word *environ*, meaning external conditions or surroundings that favour the growth of flora and fauna, human beings and their properties and protect them from the effects of pollution.

According to Douglas and Holland (1947), environment is '*a word which describes, in the aggregate, all of the extrinsic (external) forces influences and conditions, which affect the life, nature, behaviour and the growth, development and maturation of living organisms*'.

'*Environment covers all the outside factors that have acted on the individual since he began life*'. (Woodworth and Marques, 1948)

Environment means the aggregate of a complex set of physical, geographical, biological, social, cultural and political conditions that surrounds an individual or organism and eventually determines its appearance as well as nature of its survival.

1.2 Types of Environment

Environment is practically everything that embraces an organism. Out of all the planets comprising the solar system, the only habitable planet to provide all the necessary conditions for existence of life is the Earth. The physical and chemical environments however varies at places and provides unique conditions for living beings to adapt and survive.

On the basis of human interference, environment can be categorized as natural, semi-synthetic or artificial.

- A **natural environment** is inherent, unaltered and not manipulated by man. Life processes and evolution progresses are unhindered in such an environment. However, one does not often find such places in the present day. The core areas of the biosphere reserve are examples of natural ecosystems.

- A **semi-synthetic environment** is the natural environment that is modified partially by human intervention, namely development of lakes, aquaculture tanks and so on.

- An **artificial or man-made or synthetic environment** is when the natural environment is deliberately controlled and converted by mankind. For example, aquariums, cities, community parks, paddy fields or the tissue culture laboratories.

Kurt Lewin, a German-American psychologist, emphasized three types of environment that manipulate the persona of an individual.

- **Physical environment** refers to the physical space, the weather and climatic conditions that influence the organism. The physique and working efficiency of an individual depends much on the climatic conditions. Short and sturdy build-ups are features of humans in cold climates; their reduced body surface area allows more heat to be retained. In hot regions, a thinner and long limbed structure allows more heat to be lost easily. Races such as Ethiopian or Negroids of Africa, the Caucasians of Europe, Western Asia, Australia and major part of America or Mongolians of East Indies, China, Japan, shows variation in skin colour owing to variation in the level of melanin synthesis. Lighter skin allows more penetration of UV rays to facilitate vitamin D synthesis whereas darker skin prevents the penetration of UV rays. The blacks having more dense bones, hence less buoyant but loose less than 1 percent of the bone mass annually after mid-thirties; Whites with less dense bones lose about 2.5 per cent of bone mass annually and is more prone to aging.

- **Social and cultural environment** is made up of moral values, cultural background and emotional drives that modify life and nature of an individual. This in turn is dependent on the social, economic and political conditions surrounding an individual.

Man seems to be the most civilized and skillful of all the organisms. This contributes to a highly systemic social organization.

- **Psychological environment** is the physical, social and cultural environment that limits one's activities. This sets boundaries for the individual, triggering thought processes and changing behaviours of an individual.

1.3 Multidisciplinary Nature of Environmental Studies

Environmental studies cover every aspect that affect a living organism, as it interacts with the surroundings in its quest to live. Environmental studies are integrative, but the core of the subject comprises biological sciences like zoology, botany, microbiology and physiology. Many environmental concerns can be resolved through application of biotechnology and molecular biology, while bioinformatics can serve as a database at molecular level. Environmental studies is therefore multidisciplinary and aims at unraveling the ways in which human beings and nature correlate, sustaining life and man's unquenchable thirst for development with limited and finite resources.

Physics, chemistry, biology, anthropology, geology, engineering, archaeology, sociology, economics, statistics, political science, law, anthropology, management, technology and health sciences are all its components. Among these physics, chemistry, geography, geology and atmospheric science help us understand the basic concepts of structural and functional organization, as well as the physical characteristics of our environment.

Data simulation and interpretation needs the application of statistics and computer application, while mathematical science is often used in environmental modeling. The technical solutions for pollution management, waste management, green building and green energy can be found with expertise from the fields of engineering and architecture. The achievement of sustainability at all levels is interwoven with and dependant on international cooperation which in turn rests on international relations. Principles of sustainable development determine the drafts and negotiation of international accords and security issues. International cooperation is an indispensible factor in dealing with global environmental issues like climate change, trans-boundary pollution, trade in hazardous substances, ozone layer depletion, biodiversity loss, etc. Economics enables us to gain a better understanding of the social background needed to achieve growth and development.

Keeping all these in mind, management studies will enable us to formulate policies, followed by legislation for their implementation. The study and treatment of environment is very much connected with philosophy, ethics and cultural traditions that help us achieve our goal sustainably. The air that we breathe, the water that sustains our lives, the food that gives us energy, the towns and the cities that we live in, in fact everything around us constitute the environment. It is the sum total of all life support systems.

The elements that constitute the environment have been revered and worshipped by our ancestors. Our forefathers, in almost all the major civilization around the world, understood the fragile nature of the environmental system. They also discovered the need to lead a lifestyle that was in sync with the environment. It was this basic understanding, profound as it may sound now, led to their worshipping of nature in its various forms.

Box 1.1: Multidisciplinary nature of environmental studies

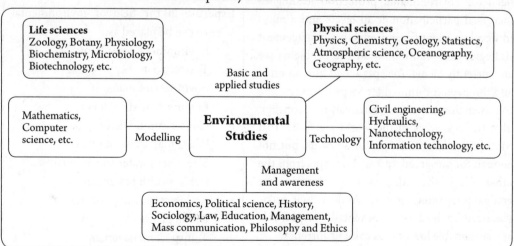

However, the industrial revolution beginning in 1760 introduced a paradigm shift in man's interaction with the environment. Rapid industrialization needed huge amounts of resources to feed the wheels of progress. Europe and then America rode the tidal wave of economic development, discarding the frugal and stoic ways of our forefathers and speeding towards hedonistic lifestyles based heavily on a culture of consumption.

Dams were built that harnessed river water to generate electricity and provide water for agriculture. Traditional farming methods were replaced by ones that depended on massive infusion of chemicals in the form of fertilizers and pesticides. Agricultural production boomed, but entire riverine systems were destroyed irretrievably and ecosystems were devastated.

Factories needed to go full throttle to match the increasing demands, as a result nature's womb was pillaged – fossil fuels were extracted and burnt to power the surge of progress. However, in doing so, humans not only pushed the availability of resources towards exhaustion, but also sacrificed myriad life forms by forcing them towards untimely extinction.

A growing population and rapidly incremental demand needed more and more. This need led to the destruction of Earth's forest-cover, which in turn led to the loss of habitats for organisms that lived in these forests. Finally, it also added to the process of climate change and global warming.

1.4 Scope of Environmental Studies

The last two and a half centuries were most important since the beginning of history in terms of human development. In an all round attempt to control and exploit nature and its services, man has literally whacked up everything in the name of development. However, in doing so we set into motion complex changes that are changing the vary basics of nature and are promising to unleash furies the kind of which have not been witnessed before.

The scope of Environmental Studies is not only limited to studying mere concepts, philosophy, ethics, components and the problems associated with resource depletion, pollution and population explosion, but also to find out a practical global solution in the form of raising public awareness – a heart to feel the immediate need for environmental protection leading to increased participation at all levels and a mind to develop scientific and effective management strategies and solution to all the problems we are currently facing. Acceptance of human error and the need for immediate steps to reverse the trends, in this context, is a relatively new concept. That the resources of the world are finite, that Mother Earth has enough for all of us, but not enough for our greed, too are concepts from the other side of the industrial revolution that are gradually regaining ground. And in their wake, the need to lead environmentally responsible and sustainable lifestyles is gaining currency.

Box 1.2: Scope of environmental studies

Expertise in the field of environmental science can be placed as:
- Environment consultant
- Toxicologist
- Environment manager
- Environmental engineer
- Conservation officer
- Waste management officer
- Scientist in water and air quality
- Public health practitioner
- Architect of landscape design
- Urban planner
- Transport management

1.4.1 Few applications

1.4.1.1 Green marketing

The term 'green marketing' signifies an all-inclusive marketing notion in which the manufacturing, marketing, consumption and disposal of goods and services take place in a manner that is less detrimental to the environment with increasing awareness and realization about the bearings of global warming, non-biodegradable solid leftover, dangerous impact of pollutants etc. This transference to 'green' sounds costly in the short run; but ultimately in the long term it shall certainly prove to be obligatory, valuable and cost effective. The concept of green marketing received importance in the late 1980s and1990s after the first workshop was organized in Austin, Texas in 1975 on the concept of 'ecological marketing'.

According to Peattie (2001) director, BRASS Research Center, Cardiff, UK, green marketing has three segments of growth.

- **'Ecological' green marketing** whereby all marketing activities were centered on removal of environment glitches and to put forward explanation for ecological apprehensions.
- **'Environmental' green marketing** whereby importance is given to clean technology which will address the issues of pollution and waste.
- **'Sustainable' green marketing** which gained importance in the late 1990s and early 2000 lays stress on sustainability.

Box 1.3: Characteristics of green products

- Products which are originally grown.
- Products which are recyclable, reusable and biodegradable.
- Products constituted with natural components.
- Products comprising recycled and non-toxic chemical substances.
- Products containing permitted chemicals.
- Products that will not contaminate the environment.
- Products that are tested on animals.
- Products that are packaged in an eco-friendly way i.e., with reusable and refillable containers etc.

Examples of green marketing in India: case studies

Case Study 1.1: State Bank of India (SBI)

The SBI is currently utilizing eco-friendly and energy-friendly tools in almost 10,000 new automated teller machines (ATMs). This enables SBI to save on power costs and earn carbon credits towards commitment of reducing its carbon footprint. SBI has also launched 'Green Channel Counter' towards green service. It emphasizes on banking without use of paper which means banking without any deposit or withdrawal slip, cheques or money dealing forms. All these dealings can be completed through SBI shopping and ATM cards. In addition to this, the SBI is the first in India to employ a wind farm of 15 MW capacity that has been developed by Suzlon for the generation of power.

1.4.1.2 Environmental management system and ISO14000

The ISO 14000 series provides the requirements and addresses various aspects of environmental management system. It is one of more than 15,000 voluntary International Standards published by the International Organization for Standardizations (ISO). It provides organizations with practical tools, particularly those looking forward to identify and control their environmental impact, and constantly improve their environmental performance.

ISO 1400 has the following benefits:
- low raw material use;
- reduced energy expenditure;
- enhanced process efficiency;
- decreased waste generation and disposal costs; and
- use of recoverable resources.

ISO 14000 standards series
- ISO 14001: Specifies the actual requirements for an environmental management system.
- ISO 14004: General guidelines on principles and on the development and implementation. of environmental management systems and also their co-ordination with other management systems.
- ISO 14010, 14011 and 14012: Guidelines for environmental auditing.
- ISO 14020, 14021, and 14024: Environmental labeling and declarations.
- ISO 14031 and 14032: Environmental Performance Evaluation (EPE).
- ISO 14040, 14041, 14042, and 14043: Life Cycle Assessment (LCA).
- ISO 14050: Terms and definitions.
- ISO/TR 14061: Information to assist forestry organizations in the use of environmental management system standards.
- ISO 14062: Discusses making improvements to environmental impact goals.
- ISO 14063: Environmental communication – guidelines and examples.
- ISO 14064: Measuring, quantifying, and reducing greenhouse gas emissions (GHGs).
- ISO 19011: Specifies one audit protocol for both 14000 and 9000 series standards together.

Figure 1.1: Eco mark logo in India

1.4.1.3 Eco-mark scheme in India

In order to boost customer consciousness or awareness, the Indian government have launched the eco-labeling plan known as 'Eco-mark' in 1991 to facilitate simple recognition of eco-friendly goods.

These standards follow a cradle-to-grave approach, i.e. starting from the extraction of raw materials to production and finally up to disposal. An earthen pot is the logo of Eco-mark scheme in India.

1.5 Components of Environment

The components of environment are atmosphere, hydrosphere, lithosphere and biosphere.

1.5.1 Structure of atmosphere

The layer of air surrounding the Earth is the atmosphere. The atmospheric mass is about 5×10^{18} kg, 75 per cent of which is limited within about 10 km. The atmosphere thins out with increasing altitude, with no distinct boundaries. The Karman line at 100 km is often used as the partition between the atmosphere and outer space. Several layers can be distinguished in the atmosphere, based on composition and temperature variation. Atmospheric science or aerology is the study of Earth's atmosphere and its processes.

Starting from the ground level is the troposphere that varies between 8 and 12 km in thickness. It is thin at the poles (8 km) and thicker at the equator (18 km). Typically air is composed of nitrogen (78.08 per cent), oxygen (20.95 per cent), argon (0.93 per cent), carbon dioxide(0.031 per cent) along with water vapour, neon, helium, krypton, xenon, hydrogen, methane, nitrogen monoxide, ozone, etc. Warm air, being lighter near the surface of the Earth can readily rise above. The molecules can travel to and fro in the troposphere in just a few days. Such vertical movement or convection of air produces clouds and rain and gives rise to most of the weather conditions. Temperature decreases with increase in altitude in the troposphere till people reach the topmost layer or tropopause. Temperature decreases at a rate of 6.5°C for every 1 km and this is known as **environmental lapse rate**. Tropopause may reach a temperature of –55°C at the poles. Sometimes the temperature increases with altitude in the troposphere, giving rise to a situation called **temperature inversion**. Such conditions restrict the vertical mixing of air and result in air pollution incidences at ground level.

The second layer or stratopause extends from tropopause to about 50 km till stratopause. The region is clear and dry with strong and steady winds. Owing to non-turbulence, presence of steady horizontal winds and being located above stormy weather, jet planes route through stratosphere. Temperature is relatively constant up to 25 km and then increases as one goes up the stratosphere. Top of the stratosphere may attain temperatures close to 0°C. Ozone layer is mostly concentrated between 20–30 km. The ozone absorbs the UV (Ultraviolet) B radiation in the wavelength of 290–320 nm. Since ozone is present in this layer it is also called ozonosphere.

Mesosphere lying above the stratosphere extends from above 50 km to about 80 km. It contains almost 0.1 per cent of the atmospheric mass and is rarified. Mesosphere is highly turbulent and experience waves. The excited atoms here absorb a great deal of solar radiation, even then temperature drops to about -100 to -90°C at mesopause. Water vapor freezes into ice clouds which can be seen after sunset if hit by sunlight. They are called Noctilucent Clouds (NLC). This is the stratum in which many meteors burn up while entering the Earth's atmosphere and are perceived as shooting stars from the Earth's surface.

The topmost layer from mesopause is the thermosphere. Temperature increases with increasing altitude and can be more than 1500°C. This is due to the absorption of the UV radiation and X rays by the few molecules present there. Nitrogen and oxygen are found mainly in between 100 to 200 km. The part of the thermosphere (80 km to 500 km) above the Earth's surface is referred to as ionosphere. The atoms exist as ions and hence this layer gets its name. Space Shuttles, the Hubble Telescope and many Earth observing satellites are stationed in this region. The ionization process leads to the creation of beautiful illumination; the Aurora Borealis in the northern hemisphere and Aurora Australis in the southern hemisphere. Auroras are usually observable from within the Arctic or Antarctic circles. The free electrons present in the ionosphere cause the high frequency waves to be refracted and ultimately reflected back to earth. The more the density of the

Figure 1.2: Thermal stratification of atmosphere

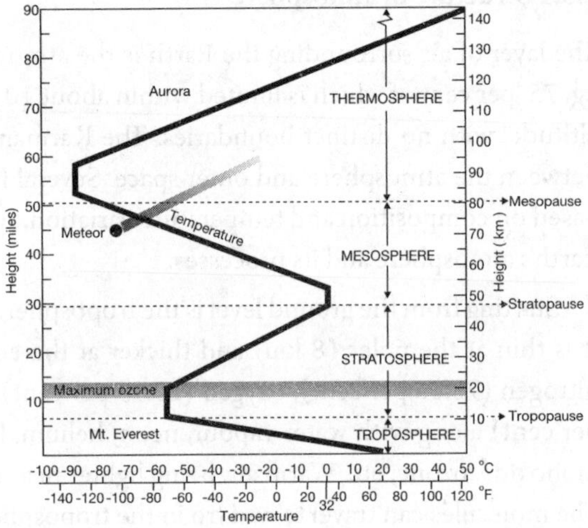

electrons, the higher will be the frequencies that can be reflected. During daytime four regions (D from 50–90 kms, E from 90 to 140 kms, F1 from 140–210 kms and F2 above 210 kms) exist. Sometimes F1 and F2 might merge to form F region. During night time D, E and F1 regions become depleted of free electrons leaving only F2 region available for communications. Hence F2 region is the most important for high frequency radio wave propagation as it allows longest communication pathways on account of its highest altitude and that the lifetime of the free electrons is greatest in this region. The outer most layer above 500 km to about 10,000 km is the exosphere which gradually merges into space. Molecules here have enough kinetic energy to escape the Earth's gravity. Hydrogen and helium are the prime components in this region. The region between 5,000 km to >> 60,000 km is also referred to as magnetosphere as it is strongly influenced by the Earth's magnetic field and the solar wind. This region is occupied by Geosynchronous satellites and comprises the Van Allen radiation.

The atmosphere is a protective blanket of gases, surrounding the Earth that helps in sustaining life on the Earth. It protects us from the hostile environment of outer space, absorbing most of the cosmic rays and harmful UV radiation. It transmits the visible, near infrared radiation (300 to 2,500 nm), part of UV radiation (mainly UV A) and radio waves.

Figure 1.3: Human application of the solar spectrum

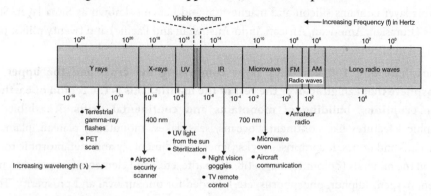

1.5.2 Structure of lithosphere

The diameter of Earth is about 12,700 km. The temperature and pressure increases as one penetrates deeper and deeper. The core temperature is assumed to be 5000–6000°K.

Earth comprises three concentric regions:

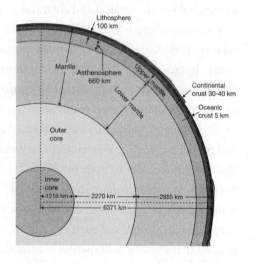

Figure 1.4: A sketch to show the internal structure of the Earth

- The **core** of approximately 7,000 km diameter is divided into inner core and outer core. The solid inner core has a radius of 1,216 km and density of 13 grams/cc. The liquid outer core has an average thickness of 2,270 km and density of 11 grams/cc. The core is supposed to be composed of an iron and nickel alloy. About 10 per cent of the layer is supposed to be composed of sulphur and oxygen as these elements are abundant in the cosmos and dissolve readily in molten iron.

- The middle layer is the **mantle** which is about 2,900 km thick. The upper mantle of 660–670 km thickness from the base of the crust, mostly contain olivine and pyroxene minerals. The asthenosphere, lying at a depth of 100–200 km from the Earth's surface, is a weak and deformed layer, which acts as a lubricant for the plate tectonics to glide and may extend up to 660 km. The lower mantle stretches from 670 km to 2,900 km below the Earth's surface. The lower mantle probably comprises silicon, magnesium and oxygen along with some iron, calcium and aluminium.

- The outer most part of Earth is the **crust**. Outer crust or continental crust may be 30 to 40 km in thickness containing mostly granite rocks with a density of 2.7grams/cc. The predominant elements are silicon and aluminium, hence is known as Sial layer. The inner crust or oceanic crust is only 5 to 10 km thick containing basaltic rocks with an average density of 3.0 grams/cc. This layer contains silicon and magnesium and hence is known as Sima layer. Six major plates (Eurasian, American, African, Indo-Australian and Pacific) and twenty minor plates are involved in plate tectonics.

Lithosphere is a 100 km thick layer comprising the crust and the upper part of asthenosphere that can glide over the rest of the mantle. This is the region of earthquakes, volcanic eruptions, building of mountains and continental drifts. It exhibits various topographical features like continents, oceans, seas, lakes, mountains, plateau, plains, deltas, beaches, cliffs and dunes. It comprises rocks (igneous, sedimentary and metamorphic rocks) that contain all the minerals (dolomite, magnetite, hematite, etc.) and elements (iron, nickel, nitrogen, hydrogen, oxygen, sulphur, phosphorus, etc.) needed for our survival and prosperity. The rocks turn into soil by pedogenesis, which sustains the biota. A tabular representation about earth's interior is uploaded on the website.

1.5.3 Structure of hydrosphere

Water is found in hydrosphere, lithosphere and atmosphere and in almost everything including our body cells. It is an oxide of hydrogen and serves as one of the source to hydrogen and oxygen in metabolism. Its physical and chemical properties make it unique. It is present as water vapour in the atmosphere which takes part in the formation of clouds and fogs thus regulating our weather conditions. In the soil, water is usually present in form of gravitational water, capillary water, hygroscopic water and as combined water. The water that percolates down under gravitational force until it reaches the saturated zone is the gravitational water. The part lying above this saturated zone is water table. Such water is used when it comes out naturally as spring or by digging of wells. The water retained in the soil against the gravitational pull around the soil particles is capillary water. Capillary water is apprehended by cohesion (attraction between the water molecules to each other) and adhesion (attraction of water molecule with the soil particle). Water remaining in the soil after drying of capillary water is known as hygroscopic water which is lost when the soil is subjected to a temperature of 105°C. Water forms very thin film around soil particles and are not available to the plant. After hygroscopic water is lost, all that remains is known as is combined water. Both combined and hygroscopic water is of no use to plants.

Most of the water found on Earth is marine. The oceans and seas contain more than 97 per cent of the hydrosphere. Ice caps and glaciers comprise a slightly more than 2 per cent of all water on Earth. The groundwater and soil water makes up about 0.63 per cent and 0.005 per cent of freshwater, respectively. Soil water refers to the water arrested in spaces amid soil particles. Of the total water on Earth, freshwater streams, rivers, ponds, lakes, and inlands seas comprise less than 0.03 per cent. The water content of the atmosphere is about 0.0001 per cent.

Figure 1.5: Various stores of water in the hydrosphere

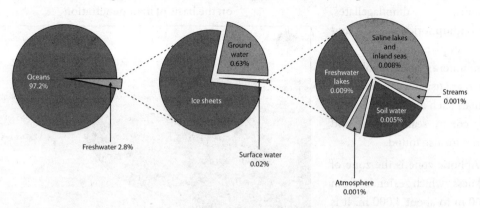

Water in solid form (ice or snow) comprises the 'cryosphere.' The word 'cryosphere' is taken from the Greek word *kryos*, meaning cold. Snow, ice, together with icebergs, glaciers and ice shelves along with permafrost are all the ingredients of cryosphere. Cryosphere exists in the Arctic Ocean, the glaciers, snow and ice covering the nearby land including Greenland, the huge ice sheet covering Antarctica, the shelves of floating ice extending into the ocean and the floating icebergs.

The snow cover on Mt Kilamanjaro in Africa, the highest elevations in Canada, China, Nepal and India, all represent cryosphere. The places that experience snowfall as well as the areas where the soil, lakes, rivers and streams freezes represent the seasonal cryosphere.

Snow is precipitation comprising of ice crystals. It reflects sunlight and serves as habitat for some animals and plants. The principle constituent of the icebergs, glaciers, ice shelves, sea ice and freezing surface is ice. The regions where ice sheets and glaciers budge into the ocean, forming platforms of ice is usually called ice shelves. Such ice shelves are frequent in Antarctica, Alaska, Canada and Greenland. Glaciers refer to thick masses of ice on land. They cover 10 per cent of the global land and store 68.7 per cent of the world's freshwater. 84 per cent of Antarctica and 12 per cent of Greenland is made up of glacial ice that makes a global total of 96 per cent.

There are 9,575 glaciers in the Indian Himalayas as per recent data furnished by the Geological Survey of India, Ministry of Mines, Government of India. These glaciers are dispersed in Jammu and Kashmir, Himachal Pradesh, Sikkim, Arunachal Pradesh and Uttarakhand. The 30 km-long Gangotri glacier covers an area of 148 sq. km.

The permanently frozen ground is the 'permafrost', found mainly in the Arctic, Antarctic and at high elevations. It is a vast storehouse of greenhouse gases like carbon and methane.

1.5.3.1 Vertical stratification of hydrosphere

(a) <u>On the basis of light penetration</u>, there are three zones namely, euphotic, dysphotic and aphotic zone.

Euphotic zone extends up to 200 m up to which light can penetrate. The depth of this zone may alter depending on the turbidity of water. Planktons, kelp forests and sea grasses are found here. Phytoplankton involves diatoms, dinoflagellates, coccolithophorids and the zooplanktons are the foraminifera's and radiolarians which are usually consumed by animals such as copepods. Jelly fish, corals, seals, turtles and sharks are also found.

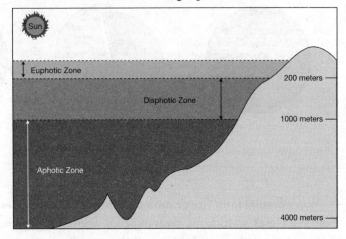

Figure 1.6: Vertical zonation of water bodies on the basis of light penetration

Aphotic zone is the zone of darkness which extends from 1,000 m to about 4,000 m. It is persistently dark and cold. For each 10 m increase in depth, the pressure increases by 1 atmosphere. Several areas are oxygen minimal layers. The organisms found in this areas are red or black in colour. This region is inhabited by angler fish, crabs, lantern fishes, tubeworms etc. Sharp and pointed teeth border the huge mouth of the fishes that are well adapted to feed on large preys. The crude organic dead matter that falls down slowly from the surface waters seems to form the principle food in this region. These are mostly the animal carcasses, faecal matters, etc. Bacteria are also common in this region.

In between the euphotic and aphotic zones is the dysphotic zone or the twilight zone that extends from about 200 m to about 1,000 m. The light intensity is too weak for photosynthesis to ensue along with an oxygen minimum condition. The zone is stable over time in absence of seasonal effects of heating and cooling. Copepods, shrimps, amphipods, ostracods, prawns, squids and fishes are found here. Organisms usually have red to red-orange pigmentation. Predominantly they possess big eyes that exhibit photo-sensing ability. Few organisms have photophores capable of manifesting bioluminescence. Some fishes have bacterial photophores, light being produced by the metabolism of the symbiotic bacteria dwelling in the photophores.

(b) <u>On the basis of depth of the water body</u> there are usually five zones in very deep water bodies such as oceans and seas.

- Epipelagic region refers to the euphotic zone extending from the surface to about 200 m, where most of life forms exist such as planktons, fish, whales, dolphins and sharks.

- Mesopelagic region extends from 200 to about 1,000 m. In very scanty light bioluminescent organisms are usually found.

- Bathypelagic region lies between 1,000 to about 4,000 m where no light reaches. Yet glowing animals, sperm whales are found in this great depth.

- In the abyssopelagic region, the water is too cold, dark and under extreme pressure from 4,000 m to 6,000 m. Some types of squid, worms and sea stars are found here.

Figure 1.7: Vertical zonation of the ocean and seas on the basis of depth

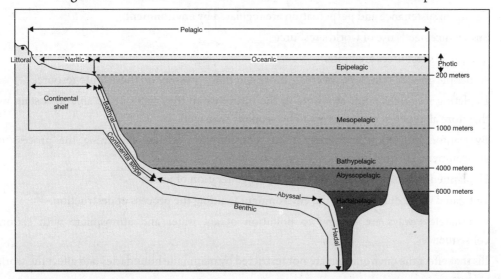

- Hadalpelagic region extends beyond 6,000 m to the deepest trench. Animals like starfish and worms survive even in these places. For example, the Marina trench (11,033 m deep), Tonga Trench (10,882 m deep) and Japan Trench (9,000 m deep) in the Pacific Ocean and Puerto Rico Trench (8,800 m deep) in the Atlantic Ocean.

The physical factors like light, temperature, pressure, salt and other chemicals serves as limiting factors for the distribution of the aquatic organisms.

1.5.4 Biosphere

It comprises all parts of land or air or water where life can exist. It is the realm of living organisms. Life on Earth occupies a very small proportion of these three components. Most living organisms can be found either on the ground or extend up to a few km above and a few m below the Earth's surface. Life is found from Arctic to Antarctic and across the equator. Birds normally hover at elevations of 650 to 1,800 m while Rüppell's vulture can soar above 11,000 m. Although life forms (barophilic microbes) are found even in the trenches of the ocean at more than 10,000 m, their abundance decreases with depth of the water body. Most aquatic life forms can be found within a depth of 200 m. The limits are largely governed by the availability of light, oxygen, moisture, appropriate temperature, pressure, etc.

All the Earth's spheres consisting of the atmosphere, the lithosphere, the hydrosphere and the biosphere make up the **ecosphere**.

1.6 Importance

Environment is not a standalone subject and in order to comprehend its fine nuances, people have to take a holistic view and understand various disciplines like economics, resource studies, biology, population issues, geography, physics, chemistry and the like. Such is the wide scope of environmental studies, that it covers some aspect or the other from every major discipline.

The basic tenets of environmental studies are embedded in simple facts:

- Creation, sustenance and perpetuation are regulated by environment.
- Sustenance needs use of world's resource.
- The world's resources are finite.
- As people grow in numbers, we create more and more pressure on these finite resources.
- Depleting resources point at a time in the future when they will not be able to sustain us; therefore all types of resources are to be properly managed.
- By wanton utilization and conspicuous consumption we are hastening the process of environmental degradation.
- Unplanned and misused resources lead to the generation of waste.
- Untreated waste dumped into the environment is aiding the process of destruction.
- Our careless ways are leading to pollution of air, water and atmosphere with serious consequences.
- Ills that effect the environment are not restricted by manmade boundaries and affect the world at large.
- Reversals of these trends are an immediate must, if we are to stop irreparable damages to the environment.
- Problems are to be identified and appropriate solutions are to be sought.
- Nations and governments alone cannot put a cap on such damages.
- We must, on an individual level, be aware of the issues and lead a life that is sustainable, therefore knowledge and education becomes necessary.
- We need to save humanity from extinction and hence the need for sensible planning of development. Being clean green and sustainable is not a choice; it is a necessity to maintain a natural balance. It should be the common goal for all citizens.

1.6.1 Productive value of nature

With advancement of knowledge humans are reinventing the wheel as it were, viewing in new light traditional knowledge that was passed on by our ancestors. And the more we know, the more we are becoming aware of the fact that the web of life is an intricately woven yarn of interdependence. From the lowly ant and the butterfly to man at the height of evolution, sitting at the top of the food chain, everything is inexplicably interlinked and interwoven as if by some unknown grand design called the environment. This environment is today under terrible stress because of our acts of aggression. We cannot recreate what has been, or is being lost due to our actions and each such missing link is pushing us, dangerously closer to the precipice. Therefore, in order for our selfish survival and sustenance, we must look at the larger picture and ensure that we mend our ways before time runs out.

Nature has an unbelievable and uncountable amount of elements, compounds and other materials that serve as raw materials. Nature is a vast treasure of various life forms. All of these can be used for developing new medicines and useful products. The direct use value can be obtained from goods, which can be extracted, used or directly enjoyed. Such as iron, gold, silver etc. Indirect use value refers to those which cannot be extracted. A wetland, for instance, is a water filter improving water quality for downstream users.

1.6.2 Aesthetic/Recreational value of nature

The beauty that is nature is beyond the realm of the written world, not merely because of its sweep and grandeur but also because of the millions of ways in which it manifests itself. The concrete jungles that humans have made their habitat is in stark contrast to the pristine beauty of nature, which not only soothes and revitalizes but also offers endless lessons of life and its meanings. Great men, down the ages have sought enlightenment in the lap of nature – some attaining nirvana by observing and understanding the laws of nature and imbibing their values. Poets, artists, creators, thinkers too have not been exceptions.

Today, as we inhale the toxic fumes generated by the sins of development, the journey back to nature as a viable alternative is becoming more and more important. Eco tourism is gaining currency by the day with more and more people establishing the connection with nature – not only to learn about her simple and sustaining ways, but also to develop into a responsible Earthling.

1.6.3 Option value of nature

We are greatly dependent on nature for our daily consumption, in fact our very existence. The food we eat, the air we breathe, the clothes we wear, the paper we write on, the electricity that gives us power, the steel that builds the sinews of our nations – everything can be traced back to nature. How we use them – as responsible citizens taking only what is essential and taking steps to replenish what we use or as careless conspicuous consumers with not a bother – will leave a huge impact on nature.

'Option value' refers to the still unknown value of the present and future ecosystems, and may range from genes to the entire biosphere, proving to be useful at some point in the future. Today we are faced with the concept of 'option value' – we have the option to either lead a sustainable lifestyle in sync with nature and aimed at preserving for the future, or to continue the way that has brought us here with the prospect of our generation destroying the goose of nature that laid the golden eggs.

1.7 Need for Public Awareness

Mother Earth is bleeding as she cannot withstand the plunder and pillage anymore. Naturally, if the environment that is the mother to us all and every living creature falls terminally ill, the consequences will be devastating for all, including us humans. The need of the planet and the people has become one.

This simple fact has to be communicated, which will, in turn generate awareness about the ailments that plague the Earth and the possible treatments that have to be administered. However, it is not as simple or straight forward as it seems. For, the greed that has brought us here is still the driving force of humans and their petty national interests. If the environment itself goes comatose, it will not do so keeping in mind man-made geographical boundaries but will unleash its devastations regardless of geographical and political boundaries. . Therefore, the response has to be universal with everyone standing up for nature and ensuring that charity begins at home. We have to sacrifice our needs for instant gratification and immediate consumption and move towards a more sustainable lifestyle which is not only conscious about the environment but is also responsible towards it.

1.7.1 Environmental initiatives at international level

At the international level UN serves as the parent body in drafting guidelines and agreements that are to be accepted by the member countries of UN at various summits, conventions, conference, meets, etc. The following are some of the initiatives at international level:

United Nations Conference on Human Environment (UNCHE), 1972 at Stockholm, where human activities were related to the environment. UNEP was launched and 5 June is observed each year as World Environment Day.

The 1985 **World Commission on Environment and Development (WCED)** or Brundtland Commission, marked the dawn of the concept of sustainable development.

Agenda 21, Convention on Biological Diversity (CBD), Commission for Sustainable Development (CSD), United Nations Framework Convention on Climate Change (UNFCCC), and forest principle emerged as an outcome of the **United Nations Conference on Environment and Development (UNCED)**, Rio de Janeiro, 1992.

The **World Summit on Sustainable Development (WSSD)**, 2002 at Johannesburg addressed socio-economic challenges such as terrorism, trafficking, disasters, etc with sustainable development.

Rio+20, 2012, was organized to obtain renewed political obligation for sustainable development, to assess the progress and gaps in the accomplishment of the major summits and outcomes on sustainable development.

1.7.2 History of the origin of environmental awareness

A retrospect to the Indian history and culture would indicate the importance of forests and wildlife in the Indian system was essentially duty based.

- Hindu mythology like *Vedas, Puranas* and *Upanishads* depict a detailed description of the elements of nature and their importance in people's lives. Cutting trees were a punishable offence as per *Yajnavalkya Smriti*. Chanakya's *Arthashastra* laid immense importance on forest administration in the Mauryan Period.

- On the other hand the Mughals established magnificent gardens, orchards, parks etc.

- The colonial period under the British plundered the natural resources with indifference towards environmental protection. Apart from the forest acts nothing significant was achieved. The legislative measures by the British primarily aimed at earning revenue; even though it marked the first attempt towards conservation.

- Rapid economic, scientific and technological advances in India reflected a massive repercussion in the form of degradation and ecological imbalances. This large scale environmental crisis had put people's lives in jeopardy. Issues like sewage, sanitation and public health were treated independently without any co-ordination.

- Stockholm Conference in 1972, also known as UNCHE, was a landmark in the history of environmental management in India. Mrs Indira Gandhi, the then Prime Minister of India, addressed the assembly with the famous statement that 'poverty is the worst form of pollution', later initiated various environmental conservation measures in the country in the form of environmental policies and legislations.

- In February 1972, under the Department of Science and Technology (DST), National Committee on Environmental Planning and Coordination (NCEPC) was established as an apex advisory body, to bring a coherence and synchronization in environmental policies and programmes. The major enactments are the Wildlife Protection Act, 1972 and Water Act, 1974.

- Article 21 of Indian Constitution provides 'Protection of life and personal liberty' that includes right to safe and sufficient water, sanitation and a healthy environment.

- The Indian Constitution was amended for the 42nd time in 1976 and Article 48(A) and 51A(G) was incorporated. Article 48(A) states: 'Protection and improvement of environment and safeguarding of forests and wild life. The State shall endeavor to protect and improve the environment and to safeguard the forests and wild life of the country'. Article 51A (G) states about the Fundamental Duties. 51A: 'It shall be the duty of every citizen of India- (g) to protect and improve the natural environment including forests, lakes, rivers and wild life, and to have compassion for living creatures'. (http://www.constitution.org/cons/india/p04048a.html)

- The Union list, States list and the Concurrent list evidently shows that the matters of environment forms a decisive guiding dimension for plans and programmes in each sector. Environmental issues are categorized as that arising from poverty and under development and those that arises as harmful impacts of developmental processes.

1.7.3 Knowledge from Indian culture and tradition

Nature has always been attributed with vitality and resilience. In Hindu religion, the sun, the land, the water, the plants are worshipped. These entities form the very basis of one's survival. In the same way animals like lion, tiger, elephant, peacock, owl, garuda, snake, mouse and others accompany Gods and Goddesses and forms an element of their cultural beliefs. We certainly get some instructions related to wildlife conservation in *Vishnu Samhita*. The initial instances of species protection in India were reported during Emperor Ashoka around 300 BC. He had a clear-cut policy of exploiting and conserving natural resources. He employed specific people and assigned specific protection duty.

Sacred spaces are referred to those regions which are left mostly unaffected due to some religious ideas and feelings attached with them. They could be an expanse of forests, lakes and ponds, valleys or mountain tops, islets, marine stretches, swamps, grasslands and virtually all other types of ecological systems. Many of these sacred elements persevere even today, for instance the Mawphlang sacred grove in Meghalaya and the sacred river Rathong Chu, etc.

Indigenous knowledge had always contributed to contemporary healthcare and drugs. For ages, the inherent communities were enduring and altering their farming, fishing and hunting practices in a reaction to changes. The Yanadi tribals in Chittoor, Andhra Pradesh are endowed with medical expertise. Their isolation from forest wealth resulted in forfeiture of Yanadi traditional knowledge.

In 1987, the Mendha village in Gadhchiroli district of Maharashtra was unequivocal in favour of terminating commercial exploitation of forests. They exerted self-control on the amount of resources to be extracted and used. They themselves took specific measures to control soil erosion. Their effort in natural resource conservation is worth mentioning.

The Meetei groups in Manipur and Assam meet a substantial amount of their resource necessities from a reasonably unimportant catchment area, located in the area they were inhabiting for a long time. Their diet comprises of fishes and waterfowl, snail, insects and other aquatic organisms. Sacred groves, or Umang Lais, in the Meetei language, form an integral part of nature worship ritual in Manipur. They strongly protect teak, eucalyptus, ginger, bamboo, lemon, etc.

The rich and traditional knowledge of the fishermen of Greater Mumbai and Sindhudurg finds great application in the fisheries.

In Karnataka Bhadra Wild Life Sanctuary, the villagers use the *Centella asiatica* leaves and *Ichnocarpus frutescens* roots in the treatment of diseases like jaundice and diabetes.

It is highly paradoxical that although the West is seeking consolation in indigenous practices be it medicine or meditation (yoga), we still strive for contemporary practices. We owe a merely less than 1.5 per cent in the world herbal share.

1.7.4 Environmental initiatives in India

The two international conferences, UNCHE held at Stockholm (1972) and UNCED at Rio de Janeiro (1992), have significantly left a footprint on the environmental policies across the globe. India is no exception as predisposed in many of the Directive Principles of State Policy in the Indian Constitution. Furthermore, the Five Year Plans of India emphasizes not only on speedy fiscal expansion, but also on importance of generating employment and alleviating poverty coupled with even-handed regional improvement. The Shore Nuisance (Bombay-Kalova) Act, 1893 and Bombay Smoke Nuisance Act, 1912, were amongst the most notable legislations to protect the environment.

To prevent and mitigate pollution in and around Mumbai and Calcutta, Oriental Gas Company Act was enacted in 1857 followed by the Bengal Smoke Nuisance Act in 1905. Cattle Trespass Act and Indian Forest Act were passed in 1871 and 1927 respectively. The Indian Easement Act was enacted in 1882 to guarantee property rights to the riparian owners. The Part XI of the Constitution of India deals with the relations between the Union and the States in terms of legislative and administrative power. Article 246 deals with the subject matter of laws made by the Parliament and by the state legislatures. List I in the Seventh Schedule is known as Union List, List II as State List while List III is the Concurrent List. Environment features in the Concurrent List.

The laws can be enacted by both the Parliament and state legislatures for subjects under the Concurrent List.

In 1976, The Indian Constitution was amended for the 42nd time to incorporate specific provisions to protect the environment in the form of the Directive Principles of State Policy and Fundamental Duties.

Departments in India bestowed with responsibility for proper implementation of environmental policies:

- The Department of Environment (DoE) formed in 1980 and upgraded to the Ministry of Environment and Forests (MoEF) in 1985
- Department of Non-conventional Energy Sources (DNES), Energy Management Centre
- Central Pollution Control Board (CPCB)
- Department of Environment both at state and union territory levels
- State Pollution Control Board (SPCB)

1.8 Institutions and People

There are several entities that are tirelessly working towards the preservation of the environment and the reversal of trends that have manifested because of man's past misdemeanors. While the government, primarily through the Ministry of Environment and Forests has become extremely stringent about the laying and the implementation of various laws aimed at the preservation and well being of the Nation's environment, various non-government organizations (NGOs) too have been performing a Herculean task, some at the very grassroots.

Globally, there are various organizations that are spreading awareness and stretching the horizons of one's knowledge.

We will fail in our duties if we do not learn about people who are leading this crusade against climate change and are living examples of sustainability. Not only have they tirelessly worked towards bringing these burning issues before the nation but have actually been the catalysts of change, making a difference where it matters the most – on the ground.

- **United Nations: United Nations Educational, Scientific and Cultural Organization (UNESCO), United Nations Development Programme (UNDP), United Nations Environment Programme (UNEP), United Nations Framework Convention on Climate Change (UNFCCC), International Union for Conservation of Nature and Natural Resources (IUCN) and World Wide Fund for Nature (WWF),** etc. endowed with organizing conferences, conventions, summits, formulation of international policies and protocols and mediate treaties and negotiations.
- **GREENPEACE:** global campaigning organization to protect and conserve the environment.

- **Bombay Natural History Society**, BNHS, Mumbai: Publications include HORNBILL, journal on natural history, S. H. Prater's book on Mammals in India, Handbook on Birds by Salim Ali and book on Indian reptiles by J. C. Daniel.

- **Centre for Science and Environment** (CSE), New Delhi, publishes popular magazine *Down to Earth*, a Science and Environment fortnightly.

- **CPR Environmental Education Centre** (CPREEC), Madras is involved in spreading environmental awareness and generating interest in conservation amongst the general mass.

- **Centre for Environment Education** (CEE), Ahmedabad, produces a variety of education materials.

- **Bharati Vidyapeeth Institute of Environment Education and Research** (BVIEER), Pune.

- **Kalpavriksh**, Pune: education and awareness; investigation and research; involved in street demonstrations, creating consumer awareness on the subject of organic food, press statements.

- **Salim Ali Center for Ornithology and Natural History** (SACON), Coimbatore promotes biodiversity conservation.

- **Wildlife Institute of India** (WII), Dehradun organizes training for the forest officials and provides research studies in Wildlife Management.

- **Zoological Survey of India** (ZSI), undertakes systematic survey of fauna in India, a repository of type specimens.

- **Botanical Survey of India** (BSI) is responsible for conducting survey of plants in various bio-geographical regions.

- **Mrs Indira Gandhi**, Prime Minister of India, participated in UNCHE (1972) at Stockholm and delivered a speech to relate future with environment security. She said, 'Are not poverty and need the greatest polluters?' Her government formed the DoE in 1980, enforced several legislations and have supported Chipko and Silent valley movements.

- **Sundarlal Bahuguna**, environmentalist, philosopher and activist, was in the forefront of Chipko movement in 1970s and the Anti-Tehri Dam movement.

- **Salim Ali**, an ornithologist, is also the author of *Book of Indian Birds* and *Fall of a Sparrow*.

- **M. S. Swaminathan**, one of India's foremost agricultural scientists, father of Green Revolution and founder of M. S. Swaminathan Research Foundation in Chennai.

- **Madhav Gadgil**, ecologist in India, keen for developing Community Biodiversity Registers.

- **Anil Agarwal**, a journalist to write the first report on the 'State of India's Environment' in 1982, founder of the Center for Science and Environment.

- **M. C. Mehta**, an environmental lawyer, involved in filing several Public Interest Litigation (PIL) for supporting environmental conservation, protection of Taj Mahal and the cleaning up operations of river Ganga.

- **Vandana Shiva**, active campaigner against Intellectual property rights, Genetically Modified

Organism (GMO); founded Navdanya in 1991, to protect the diversity and integrity of native seeds.

■ **Rajendra Singh**, pioneer in the field of community-based water harvesting and water management, founder of Tarun Bharat Sangh; active campaigner against the destructive mining in the Aravali hills.

1.9 Raising Environmental Awareness in India

In 1983–84, a comprehensive plan known as 'Environmental Education, Awareness and Training Scheme' was established with the following objectives:

■ to encourage environmental awareness amidst all sections of the society;

■ to disseminate environment education (EE), especially through non-formal system;

■ to assist progress of education and instruction materials in the education division formally;

■ to sanction environmental education in the running institutions, scientific and research centers;

■ to guarantee the development of human resource in such education and spread of consciousness;

■ to persuade the NGOs and media in promoting awareness about various environmental issues; the media can be theatre, dramas, music, poetry, films, audio, visuals, print, advertisements, posters, hoardings; for promotions competitions, workshops and seminars may be organized; and

■ to incorporate active involvement of men and women for the purpose of preservation and conservation of environment.

The central government of India have also decided in establishing 'Centre of Excellence' in order to develop resource materials, to organize training programmes, to spread awareness and also capacity building in the area of sustainable development. Notable centers are:

■ CPR Environmental Education Centre (CPREEC) established jointly by the MoEF, Government of India (GoI) and C. P. Ramaswami Aiyar Foundation

■ CEE, Ahmadabad, MoEF

■ Centre for Ecological Sciences, Indian Institute of Science (IIS), Bangalore

■ Centre for Mining Environment (CME), Indian School of Mines, Dhanbad

■ Salim Ali Center for Ornithology and Natural History (SACON)

■ Centre for Environmental Management of Degraded Ecosystem (CEMDE) under the School of Environmental Studies, Delhi University

■ Tropical Botanic Garden and Research Institute (TBGRI), Thiruvananthapuram, Kerala

■ Foundation for Revitalsation of Local Health Traditions (FRLHT), Bangalore

■ Centre for Animals and Environment, MoEF, Bangalore

■ Centre of Excellence in Environmental Economics, Chennai

The principal activities taken under this scheme are:

■ **National Environment Awareness Campaign (NEAC):** Launched in 1986 by the MoEF, the objective was to create awareness at national level through the creation of 34 Regional Resource Agencies (RRA).

- **Eco clubs and National Green Corps (NGC):** Eco clubs provide knowledge about the environment, ecosystems and human interdependence on environment in a non-formal proactive system to the school children. NGC is a national programme started in 2001 to sensitize school children. It comprises 30–50 school children supervised by the teacher in charge selected from the school.

- **Global Learning and Observation to Benefit the Environment (GLOBE):** A worldwide primary and secondary school based science and education programme to endorse inquisition and investigation programme. It works in collaboration with National Aeronautics and Space Administration (NASA), National Oceanic and Atmospheric Administration (NOAA) and serves to connect students, teachers and scientists globally.

- **Formal EE Programme:** The National Policy on Education (NPE), 1986 envisioned environmental protection of utmost importance and that it must be integrated as an integral part of the curriculum at various stages of education. The policy obliged the Ministry of Human Resource Development (MHRD) to pursue necessary steps in imparting of EE in India at school levels. In 1991, under the directives of the Supreme Court, especially in response to PIL filed by *M. C. Mehta vs Union of India*, all curriculums were to be environment oriented. The state of Maharashtra was the first to do so.

The NPE states: 'There is a paramount need to create a consciousness of the environment. It must permeate all ages and all sections of society, beginning with the child. Environmental consciousness should inform teaching in schools and colleges. This aspect will be integrated in the entire educational process'.

- **Environmental Appreciation Course – Distance Education:** Since environment is everybody's business, a three-month course called Appreciation Course on Environment (ACE) was provided by the Indira Gandhi National Open University (IGNOU) to the individuals interested in availing the chance to educate themselves about the relevant environmental issues through distance education.

Human society has achieved materialistic growth but failed miserably to attain value integrated development. Environmental goods were not evaluated in terms of money in the past and were treated as valueless and taken for granted. The result was of course misuse and overuse. The valuation of all environmental goods and services have been estimated to be \$33 trillion annually against the economic value of \$29 trillion annually (Costanza, *et al.*, 1997). The environmental problems people confront today must be handled with appropriate environmental governance and participatory humane approaches. An assault on nature has always led to nature's revenge by the way of robbing our means of sustenance. We can only conquer nature by obeying it and not by waging war against it.

Summary

- Environment is practically everything that surrounds an organism. Out of all the planets in the solar system, Earth is the only habitable planet owing to its environment. On the basis of human interference, environment can be natural, semi-synthetic or artificial.

- Environmental studies cover every aspect that affects a living organism. Environmental studies are integrative. The scope of the subject is not only limited to philosophy, ethics, components and the allied problems but it also enables one to find out a practical and global solutions to problems facing us.

- Mother Earth has enough for all of us, but not enough for human greed. The resources of the world are finite and there is pressing need for environmental protection leading to increased participation at all levels and a mind to develop scientific and effective management strategies as a solution to all problems confronting us.

- The knowledge of environment finds applications in green marketing for effective environmental management system. The Government of India launched 'Eco-mark' in 1991 for easy identification of environment-friendly products.

- Environment comprises atmosphere, hydrosphere, lithosphere and biosphere. The air around the Earth is the atmosphere. Lithosphere is the land mass. Water or hydrosphere is found in hydrosphere, lithosphere and atmosphere and in almost everything including our body cells. Biosphere comprises all parts of land or air or water where life can exist. It is the realm of living organisms.

- Creation, sustenance and perpetuation are all regulated by environment. Nature has an uncountable amount of elements, compounds and other materials that serve as raw material. Nature is a vast treasure of various life forms. The increase in numbers creates more and more pressure on nature's resources. Our means of survival can all be traced back to nature.

- Mother Earth is bleeding and cannot withstand the plunder and pillage anymore. Hence the need for public awareness. There should be a unified response with each one of us standing up for nature.

- At International level, UN serves as the parent body in drafting guidelines and agreements that is enforced by national legislation. The Constitution of India was amended for the 42nd time in 1976 to insert a new Article 48(A) and 51A (G) to include environment.

- A retrospect to the Indian history and culture indicates the importance of forests and wildlife. Hindu mythology like *Vedas*, *Puranas* and *Upanishads* depict a detailed description of the elements of nature and their importance in people's lives.

- Rapid economic, scientific and technological advances in India reflected a massive repercussion in the form of degradation and ecological imbalances. Environmental protection thus became indispensible.

- Several individuals and NGOs are working tirelessly towards the preservation of the environment along with the government. Environmental goods were treated as valueless and taken for granted. The valuation of all environmental goods and services was estimated to be

$33 trillions annually against the economic value of $29 trillions annually. The importance, value and beauty of nature is beyond the realm of the written world.

Exercise

MCQs

Encircle the right option:

1. Match the following:

					Ans:
i.	ISO 14020	a.	Environmental auditing	A.	ic, iia, iiid, ivb
ii.	ISO 14010	b.	Reducing GHGs	B.	ia, iid, iiib, ivc
iii.	ISO 14040	c.	Environmental Labeling	C.	id, iic, iiib, iva
iv.	ISO 14064	d.	Life Cycle Assessment	D.	ic, iia, iiib, ivd

2. Eco-mark of India is:

   A. Swan   B. Earthen pot   C. Earthen saucer   D. Potted plant

3. Match the following:

					Ans:
i.	Oxygen	a.	0.9 per cent	A.	ic, iia, iiid, ivb
ii.	Carbon dioxide	b.	79 per cent	B.	ia, iid, iiib, ivc
iii.	Nitrogen	c.	0.03 per cent	C.	id, iic, iiib, iva
iv.	Argon	d.	21 per cent	D.	ic, iia, iiib, ivd

4. The temperature of the troposphere:

   A. Increases at a rate of 6.5°C for every 1 km decrease
   B. Decreases at a rate of 6.5°C for every 1 km increase
   C. Decreases at a rate of 1°C for every 6.5 km increase
   D. Increases at a rate of 1°C for every 6.5 km decrease

5. Noctilucent clouds are seen in the:

   A. Ionosphere   B. Troposphere   C. Stratosphere   D. Mesosphere

6. Geosynchronous satellites is placed at:

   A. 30000 – 60000 kms   B. 100 – 200 kms   C. 50 – 80 kms   D. 200 – 5000 kms

7. The temperature at the core of the Earth is assumed to be:

   A. 1000 – 2000°K   B. 5000 – 6000°K   C. 50000 – 60000°K   D. 2000 – 3000°K

8. Which form of radiation is used in microwaves?

   A. Infra-red   B. Ultraviolet   C. X rays   D. Visible rays

9. Sial layer mainly comprises:

   A. Pyroxene   B. Basalt   C. Granite   D. Olivine

10. Icecaps and glaciers cover approximately _____ of hydrosphere:

    A. 2 per cent   B. 5 per cent   C. 0.6 per cent   D. 0.001 per cent

Fill up the blanks

1. _____ component of the environment has the least pollutant receptor capacity.

2. The 'Eco-mark' in India is _____.

3. Wangari Mathai won the Nobel Peace for her work on _____.

4. ISO _____ is a part of EMS.

5. The Constitution of India was amended in 1976 to include the new Articles _____ and _____.

State whether the statements are true or false

1. Environmental studies as a subject comprise all branches of science, arts and social science. (T/ F)

2. Life on Earth interacts with its surroundings that essentially comprise abiotic factors. (T/ F)

3. Rio Earth Summit was held in 1993. (T/ F)

4. Rajender Singh is popularly known as the 'water man of India'. (T/ F)

5. Umang Lais form an integral part nature worship ritual in Manipur. (T/ F)

Short questions

1. Define environment.

2. Classify the types of environment on the basis of human interference and give suitable examples.

3. What do you mean by socio-cultural environment?

4. What is green marketing? Give examples.

5. What are the advantages of ISO 14000 certification audit?

6. State the significance of Eco-mark in business.

7. State the tropospheric composition of atmosphere.

8. Define environmental lapse rate.

9. Why is stratosphere important?

10. What is lithosphere?

Essay type questions

1. Explain the multidisciplinary nature of environment as a subject.

2. What are the various scopes of environmental studies?

3. What are the various applications of environmental studies in the context of the present world?

4. What are the salient features of ISO 14000?

5. Explain the components of environment and discuss the significance of each.

Natural Resources – Energy _____ 2

Learning objectives

- *To develop a concept of natural resource.*
- *To be able to classify the different types of natural resource.*
- *To know about the immense importance of renewable and non-renewable energy, their applications, advantages and disadvantages.*
- *To know about such resource in the Indian perspective.*
- *To explore the future possibilities of sustainable energy production.*

2.1 Introduction

Materials that occur in the nature under different environmental conditions are termed as **natural resources**. They are valuable in their natural (unmodified) form and their value is determined by the amount of material that can be extracted from them and the demand for the same. A **commodity** is usually considered a natural resource when the key activities associated with it are extraction and purification in contrast to manufacture. Thus, quarrying, extraction, fishing and forestry are referred to as natural-resource industries, whereas agronomy is not.

2.2 Classification of Natural Resources

Natural resources can be classified on the basis of the following parameters.

A. On the basis of origin

i. Biotic natural resources

Biotic natural resources are obtained from the biosphere either in the raw form or through cultivation. For example, fossil fuels, agricultural products, fruits, wax etc. Petroleum having an organic origin is a biotic resource.

ii. Abiotic natural resources

These resources are procured from land, water and minerals and are non-living. Air, water, land, gold, diamond, silver, bauxite, nickel, copper, iron ore, zinc, lead, sulphur, chromites, talc, marble, limestone, platinum, vanadium, salt, sand, gravel etc are all abiotic resources.

B. Based on the degree to which they are developing/processing

i. Currently used resources

Those resources which are presently used for human use. For example, coal and petroleum.

ii. Potential resources

These include untouched and untapped resources for future use. Hydrogen is one such resource.

C. On the basis of regeneration ability or continual supply

i. Renewable resources

They are resources, which can be recycled, or resources that can be replenished quickly through natural cycle. For example, solar radiation, wind energy, water energy, biomass energy (solar energy stored in wood), agricultural products, forests, wildlife, etc. If they are consumed at a rate exceeding their natural rate of replacement, the stock will eventually run out. Non-living renewable natural resources are soil and water.

ii. Non-renewable resources

The resources, which cannot be replenished or are replenished very slowly, are non-renewable resources. They can be:

- **Recyclable:** These resources can be collected after use and recycled. For example, aluminium and other metals after being used are collected and recycled.

- **Non-recyclable:** These resources cannot be recycled in any way. For example, coal, oil and natural gas.

Natural resources are **natural capital** that can be made in to commodity inputs to infrastructural capital or wealth-creating procedures. They can be soil, timber, oil, minerals and other materials derived from the Earth. Natural resource activities incorporate both the extraction of the elementary resource as well as refining and purifying it into a usable form.

D. Based on physical existence – tangible and non-tangible resources

A tangible resource is something that is physical in as much that we can touch or feel it. A non-tangible resource, on the other hand, is something that cannot be felt. Coal and iron ore for example are tangible resources, while the goodwill of a company or its brand value is an example of non-tangible resource.

Box 2.1: Schematic classification of natural resources

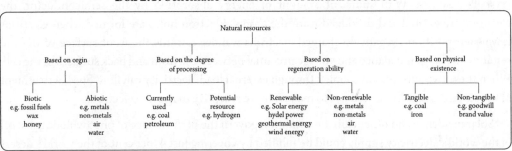

2.3 Renewable and Non-renewable Resources

The sun alone can offer sufficient energy for the world in just 40 minutes, if we had the appropriate technologies to harness it. Before the early twenty-first century, non-renewable resources were somewhat inexpensive to use. That is becoming less true as of 2013 due to their scarcity and high demand.

2.3.1 Renewable energy resources

A. Solar energy

The Earth's most ubiquitous and potent energy source is the sun, located 152 million km away. Solar energy has been reaching Earth in the form of solar radiation for more than billions of years. This energy then gets transformed into other forms of energy. Solar energy has been powering life for many many years. A small fraction of this solar energy that strikes the Earth every minute is enough to assure all the energy needs for the whole year provided it is harnessed properly. Since it is received discontinuously, it can be supplemented with other source such as hydropower, thermal energy etc. The solar thermal collector box was utilized by the British astronomer John Herschel in the 1830s.

Electricity from solar energy can be obtained in two ways:

i. Photovoltaic (PV devices) or 'solar cells'

This technology converts the solar energy directly into electrical power. PV cell is the fundamental unit of the photovoltaic system. The size of each cell may vary from 0.5 to 4 inches in size and is able to generate 1 or 2 watts of power. Cells are electrically coupled and packed into a module which in turn is further linked to form an array of panels. An array may comprise one to thousand modules which in turn depend on the amount of required power output. The front surface of the panel is covered by glass while the back surface can be of plastic, wood, metal, glass or amalgamated materials.

Solar energy hits the earth surface as packets of energy called photons. Photons have variable amounts of energy that correspond to different wavelengths of the solar spectrum. Photons striking a photovoltaic cell may have three consequences – they are either reflected back or they can go through or they can be absorbed. Only the absorbed photons have the potential to generate electricity out of the three options. PV cells are prepared of semiconductors, for instance crystalline silicon. With sufficient amount of energy absorption by the semiconductor, the electrons are excited and dislodged from the atoms. Electron holes are formed when electrons leave their position. Negatively charged electrons wander towards the front surface of the cell resulting in charge imbalance and voltage potential between the front and back surfaces of the cell. When these two surfaces are coupled through external load electricity can flow. Special treatment of the surface is usually carried out to make it more receptive to the free electrons.

Solar incidence and other climatic condition govern the performance of photovoltaic cells. All of the world's electricity supply could be fulfilled by covering just 4 per cent of the world's desert with photovoltaics. Almost all of the world's total electricity demand could be supplied by the Gobi Desert alone. Commercially, the efficiency of the available modules varies from 5 to 15 per cent, though there is a constant effort to increase it by 30 per cent.

Several buildings and houses have installed solar panels on their roofs. There are examples of photovoltaic power plants of 200 MW capacities in China, 48 MW capacities in Nevada, USA and 97 MW capacities in Canada. Numerous such plants are in construction stage throughout the world.

Commercial applications with examples: The application of solar cells includes calculators, wristwatches, solar streetlamps, solar lanterns, solar-power-driven water pumps, solar geyser, solar batteries, power communication equipments and domestic electricity supply.

Advantages of photovoltaic systems are:

- There is no need for heavy mechanical generators as sunlight is directly converted to electricity.
- Installation of PV arrays of any size can be installed fast.
- It has minimal environmental impact and no water is required for system cooling.
- Generates no byproducts.
- Photovoltaic cells, like batteries, generate direct current (DC), usually used for small loads like electronic equipment. For commercial applications the direct current must be converted to alternating current (AC) by using inverters, solid state devices etc.

ii. Solar thermal/electric power plants

In this system, the energy is recovered from the heat of the solar radiation. It can be used both for power production and for heating or cooling applications. The solar energy is concentrated to heat and produce steam connected with a turbine coupled with a generator to produce electricity at variable scales.

California's, **Solar One power station** looks almost like the little Odeillo solar furnace. The exceptions in this case are the mirrors that are organized in rings around the 'power tower'. With the movement of the sun, the mirrors also turn to keep the sun rays focused on the tower. Oil is heated to 3,000°C and this heated oil is utilized to produce steam. The steam drives a turbine while the turbine drives a generator with an ability to provide 10 kW of electrical power.

Solar water heating: The roof is fitted with glass panels. Water is pumped through the pipes in the glass panels at the bottom. The pipes are painted with black colour so that they can absorb more solar radiation. Convection in water will drive the hot water flow from the top. The water should be driven out of the panel to prevent the panels from freezing. Such heating save the electricity bills and is worthwhile in the places with abundant sunshine like Arizona and California.

A more advanced type is the 'Thermomax' panel that comprises a set of glass tubes. Each tube contains a metal plate coated blue which aid in absorbing infrared and UV radiation. The glass tubes should be vacuumed to minimize heat loss. The output is decent even on diffused sunlight.

Solar power driven satellites orbiting the Earth, provides telephones, navigation facilities, satellite TV, weather forcasting and internet. NASA Vanguard satellite comprising monocrystalline backup array was launched into Earth's orbit in 1958.

Figure 2.1: Schematic diagram to show solar water heating

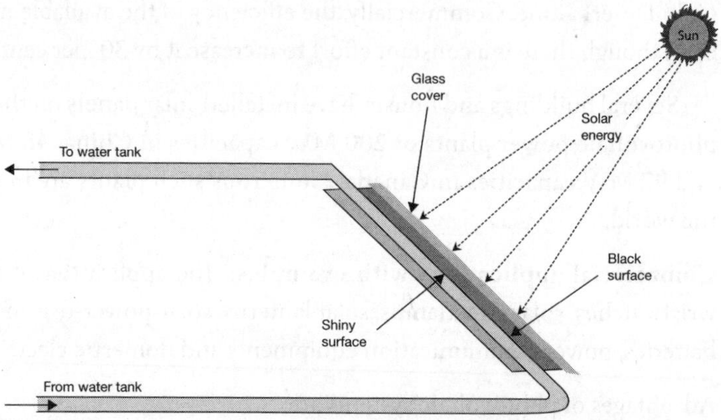

Solar thermal power generating plants served as the major source of electricity in 13 power plants of USA in 2011. Of them 11 plants are in California, 1 in Nevada and 1 in Arizona. Three solar power plants are located in the Mojave Desert of California comprising Solar Energy Generating Systems (SEGS). The two SEGS plants at Harper Lake were the world's largest solar thermal power generating plants in 2011.

The Indian Scenario: India receives around 5,000 trillion kWh of solar radiation per year with the daily solar incidence of 4–7 kWh/m^2. Many of the Indian regions have 250– 300 sunny days yearly. The GoI approved the **Jawaharlal Nehru National Solar Mission (JNNSM)** in January 2010 as a part of National Action Plan on Climate Change and to serve long-term energy and ecological security. The target of this mission is 20,000 MW of solar generating capacity by the Thirteenth Five Year Plan to be achieved by 2022.

During the last three years, it has been decided that 185 grid linked solar power plants with a total capacity of 1.172 gigawatt (GW) are to be set up by the Ministry of New and Renewable Energy (MNRE) under different schemes. Of these, 131 power plants are of about 366 MW aggregate capacities and 1 solar thermal plant with 2.5 MW potential. 10 of the solar thermal power plants will be using foreign technology while some of these will be using local solar cell or module technology.

Tamil Nadu came up with the solar energy policy in 2012 with a target of 3,000 MW of solar power by 2015. Under its Solar Powered Green House Scheme, 3 lakh houses will be constructed with solar power lighting systems over a period of 2011–12 to 2015–16. The government has also decided to install 1 lakh street lights in the village panchayat over a period of five years up to 2016.

Gujarat has an installed capacity of 852 MW as on March 2013 and signed about 88 solar power purchase agreements with high tariff and subsequently backtracked with the fall in tariff rates last year. The biggest solar project of 130 MW capacities was launched at Bhagwanpur in Neemuch (**Madhya Pradesh**) at a cost of ₹ 1,100 crore on 305 hectares of land.

Advantages:

- There is no emission of GHGs or other air pollutants.
- They have minimal impact on the

environment even when they are installed on buildings. ▪ Solar energy is for free and requires no fuel. ▪ Produces no waste. ▪ In sunny countries, solar power can be used simply to supply electricity to a distant place. ▪ Convenient for low-power usages such as solar energy driven garden lights and battery chargers. ▪ It is cost effective in long run.

Disadvantages:

▪ Sun does not shine at night and hence it will not work at night. ▪ Quite expensive to erect solar power stations, although with improvement of technology the cost is coming down. ▪ A large surface area is required to collect the energy at a practical rate as the sun does not deliver much energy at one place at any one time or a large area is needed to mount solar panels in order to get a decent amount of electricity. ▪ Can be unreliable at times. ▪ Amount of sunlight reaching the Earth's surface varies with location, time of day, time of year and other weather conditions. ▪ Transmission remains a barrier that has to be breached.

B. Wind energy

For years, people have been using wind to grind grains, sail ships and for irrigation. This kinetic energy can be converted to more usable forms of power in wind energy systems though wind energy is one of the least used resources. Winds results from uneven heating of earth's atmosphere, which again is due to the surface irregularities and Earth's rotation. It is the kinetic energy associated with atmospheric movement. The patterns of wind flow are governed by the physical features of the Earth, water bodies and the vegetation. This flow when harvested with the help of wind turbines generates electricity. Wind turbines are used either singly or in clusters. Often small wind turbines called aerogenerators are used to charge generators. Clusters of wind turbines are called 'wind farms'.

Modern wind turbines can be of two types

▪ The **vertical-axis** design or Darrieus model, named after its inventor, operates similar to the eggbeater-style.

Vertical axis wind turbines or VAWTs, have the main rotor and shaft arranged vertically. The wind turbine needs not to be pointed into the wind. It is advantageous in places where the wind direction and speed is highly variable. The generator and other major machineries can be conveniently placed near the ground, enabling maintenance easier. The main problem is that it usually creates drag when turning into the wind.

▪ The **horizontal-axis** variety representing most of the large contemporary wind turbines are known as horizontal-axis wind turbines (HAWTs).

It has blades that seem to be a propeller spinning on the horizontal axis. The main rotor, shaft, gearbox and electrical generator is placed at the peak of a tower. It must face the wind. Small turbines are supported by a simple wind vane placed along with the rotor, whilst large turbines normally use a wind sensor attached with a servo motor to revolve the turbine into the wind.

In simple terms, wind turbine works opposite to a fan, where, the mechanical energy is converted to electrical energy. The output (around 700 V) is directed to the transformer which converts the electricity coming out to the right voltage (around 33,000); appropriate for the distribution system or the grid system that transmits the power then. Devices to locate the wind direction and measure wind speed are fitted on top of the nacelle. With change in wind direction, the motors change the nacelle and the blades to face the wind. The nacelle is also provided with brakes so that the turbine can be switched off at very high wind speeds to prevent damage. The information is recorded by computers and sends to the control centre.

Parts and functions of a wind turbine include the following:

- **Fan blades**: Large fan blades are linked to the hub. Force of wind moves the blades to energy which is transferred to the rotor.
- **Shaft (in the nacelle):** Shaft is joined to the center of the rotor. With the rotor spinning, the shaft also spins. Brakes are applied to control the speed of rotation of fan blades and the shaft at time when wind blows fast.
- **Transmission gearbox**: If the speed of the shaft is very slow, a gearbox is connected to the shaft to increase the output speed.
- **Generator**: The high–speed output shaft from the gearbox is coupled to the generator that produces electricity.

Figure 2.2: HAWT and VAWT

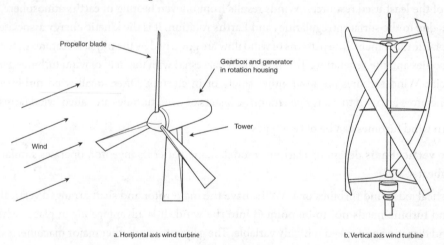

a. Horijontal axis wind turbine   b. Vertical axis wind turbine

Prospects in wind energy: 80 per cent of the world's installed wind energy capacity was held by five nations. During 1990–2000, Germany was the highest producer followed by Spain, USA, Denmark and India.

Table 2.1: World top ten wind energy producers according to Global Wind Energy Council, 2012

Ranks	Country	Total capacity end 2012 (MW)	Ranks	Country	Total capacity end 2012 (MW)
1	China	75,564	6	United Kingdom	8,445
2	United States	60,007	7	Italy	8,144

Ranks	Country	Total capacity end 2012 (MW)	Ranks	Country	Total capacity end 2012 (MW)
3	Germany	31,332	8	France	7,196
4	Spain	22,796	9	Canada	6,200
5	India	19,051	10	Portugal	4,525
	Rest of world	39,852			
	Total	282,482= 282 GW			

India's wind power potential is 45,000 MW. During the Twelfth Five Year Plan period, the government fixed a goal of around 15,000 MW. High installation cost can act as a limitation, for example, the cost of 1 MW power project installation will cost up to ₹ 6 crore. India already has a wind energy set up of 18,551 MW, which comprises 9 per cent of the aggregate installation power capacity.

The optimum sites for wind farm location are the coastal zones, the open plains, the mountain gaps, the rounded hilltops etc. But around 25 km/h of average wind speed is needed.

Wind energy per unit costs only marginally more than that of conventional energy (₹ 4–4.5 crores/MW, compared with thermal power costs of ₹ 3.7 crores/MW). This difference is insignificant, when one considers the environmental costs of thermal energy. Above all, wind is an indigenous energy resource, which one can use in unlimited amounts. Also, it can be produced locally.

To foster private sector investment, the government has announced financial and promotional incentives. The encouragement came in the form of expedite depreciation, discount on import duty on precise parts of the wind electricity generators, release from export duty, etc. Indian Renewable Energy Development Agency (IREDA) together with several financial institutions, provides loans for the installation of windmills. The Chennai based Center of Wind Energy Technology (C-WET) provide the necessary technical backup and is working in tandem with various other organizations to give shape to the government's move towards fulfilling its renewable targets.

Among the Indian states, **Tamil Nadu** and **Gujarat** is ahead in the field of wind energy. Gujarat is followed by Andhra Pradesh. The states of Jharkhand and UP do not have prospective sites for wind energy.

Advantages:

▪ Wind farms require no fuel and hence wind energy is is free. ▪ No waste or greenhouse gas generation occurs. ▪ The land underneath may be used for farming. ▪ Wind farms can attract tourists. ▪ It is a decent method of supplying energy to remote areas.

Disadvantages:

▪ It is unpredictable and intermittent as some days may not be windy. Winds must have a speed above 12 to 14 miles/hour to turn the turbines efficiently to generate electricity. ▪ Optimum areas for wind farms are often the open plains, the coast, where land is expensive. ■ Covering the landscape with these towers is at times unsightly and unaesthetic. ■ Migrating flocks might get killed if the wind farms fall in the migratory route. ■ Can affect television reception. ■ Can be noisy as wind generators have a reputation for producing a persistent, 'swooshing' sound throughout. However, aerodynamic designs in modern **wind farms are much quieter**. The small modern wind generators located on boats and caravans hardly makes any sound. ■ Intention of most Indian companies to set up windmills is for the tax breaks and not for the sake of clean energy.

C. Tidal power

Tidal energy is a type of oceanic energy and has been in use since the eleventh century. Tides are created due to the gravitational pull by sun and moon on the water. The capacity of tidal electricity production is directly proportional to the strength of the tide.

There are two methods of generating tidal power – tidal range and tidal stream. **Tidal range** is the difference between the high tide and low tide. Utilization of tidal range involves the construction of small dams or barrages along estuaries. It is almost like a hydroelectric project where the dam is of much bigger size. 7 m rise is required for economical operation; at least to function efficiently a rise of 16 feet is required between the high tide and low tide. A barrage is built across which traps the approaching tidewater. The sluice gates are kept open to allow the waters to fill in the tidal basin. The water stored at the back of the embankment or barrage is allowed to rush out as soon as the tide falls. Turbines are placed in the path and when the water flows through the tunnels in the barrage with the receding tide, the flow of water could be used to turn a turbine in order to produce power. In some cases, double effect turbines that are able to generate power even in the time of filling in are used.

Tidal streams utilize the kinetic energy of flowing water to drive the turbines. Optionally, offshore turbines could also be used which operates like underwater wind farm. Such type eliminates the possibility of environmental problems, which the tidal barrage could possibly have and is also cheaper to construct. North Orkney, Pentaland Firth, Islay etc are the prospective sites. Only 20 sites all over the world have been identified as potential power stations. Severn, Dee, Soloway and Humber estuaries around Britain were identified as potential sites. Tidal reef almost like a tidal barrage was proposed across the Severn estuary. The design is such that it does not hinder much of the water movement and so lessens the environmental consequences like as storm surges and flooding of low lying land. Migratory fishes could easily get through. Power could be generated for more hours as the mud flats could be exposed at low tide. It could be constructed in parts and so power generation could be earlier. Sections of it could be open during shipping.

Case study 2.1: La Rance Tidal Power Station in France

La Rance Station in France can harness 240 MW from tidal energy, enough to power 2,40,000 homes. It started generating electricity in 1966 and is the world's leading tidal power station and is located in the Rance estuary in northern France, near St. Malo. The barrage extends from Brebis point to Briantais point and is 750 m long. Its capacity is approximately one-fifth of a usual coal-based or nuclear power plant. La Rance Power Station generates ten times more power than 20 MW capacity Annapolis Royal Generating Station located in Annapolis Royal, Nova Scotia, Canada usually generates.

Figure 2.3: Schematic diagrams for harnessing of tidal energy

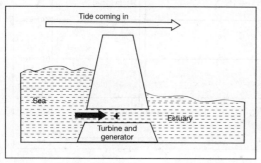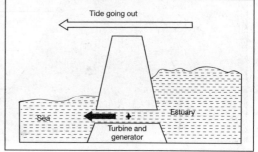

Since 2011 Sihwa Lake Tidal Power Plant in South Korea is the world's biggest tidal power plant of 254 MW capacity followed by Rance tidal power plant of 240 MW capacity at La Rance in France, which was also the world's first tidal power station.

Ocean thermal energy conversion (OTEC): It was in 1881 that Jacques D'Arsonval, a French engineer, conceived the idea of using temperature differences in the ocean. Oceans being the largest solar energy collector, OTEC applies the conversion of solar energy absorbed into electrical energy. The water gets colder and colder as we go deeper and deeper into the ocean. This difference in the temperature can be used to harness energy. A temperature gradient of 38°F is required between the warm surface and cold deep water to produce electricity. The water temperature more or less remains constant around 4°C at a depth of 1,000 m. The difference in temperature is utilized here to vaporize (and even condense) the selected working fluid which can be ammonia, to rotate the turbine generator to create power. Comparable OTEC in Hawaii has been very useful in supplying power for offshore mining. OTEC can also coproduce during water by desalination up to 2 million liters a day for each megawatt of power generated. Other prospective places can be Haiti, Bahamas, Dominican Republic, Trinidad, Cuba, Jamaica, etc.

Tidal energy prospects in India: India with a coastline, gulfs, bays and estuaries is potential enough to harness tidal energy for electric power generation. The Gulf of Cambay with a maximum tidal range of 11m and average tidal range of 6.77 m has a potential of about 7,000 MW. The Gulf of Kutch with a maximum tidal rage of 8 m and average tidal range of 5.23 m has a potential of about 1,200 MW. The Gangetic delta of Sunderbans with a capacity of 100 MW has a maximum tidal range of 5 m with an average tidal range of 2.97 m.

Figure 2.4: Schematic diagram of OTEC plant operation

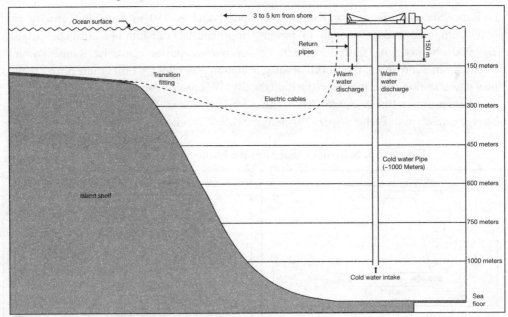

A demonstration tidal power plant of 3.75 MW at Durgaduani Creek in the Sunderbans was sanctioned to the WBREDA, Kolkata. The project is being executed by National Hydro Power Corporation Ltd. (NHPC). A special effort was made by the government of Gujarat to study the possibility of tidal power projects under water without conventional methods.

Advantages:

■ After project completion tidal energy is free. ■ Requires no supply of fuel. ■ No production of wastes or other pollutants. ■ Reliable production of electricity. ■ Maintenance cost is cheap. ■ Tides are totally predictable. ■ OTEC and offshore turbines have minimal environmental impact and OTEC may help in increasing fish production. It is also supposed to have a mitigation role in the intensity of the hurricanes.

Disadvantages:

■ Construction of barrages is usually expensive. Both the upstream and downstream is changed making it difficult for birds to feed and fishes to migrate without fish ladders. ■ Tidal power stations can generate power only for 10 hours a day, i.e., only when the tide is flowing. ■ Few suitable sites are available for tidal barrages. ■ In case of OTEC, the salty marine water is biologically very active and may deposit algae on the pipes and heat exchangers surface to form a coat, thus, reducing the efficiency of heat exchange.

D. Wave power

Waves in motion comprise kinetic energy and are the consequence of winds blowing across the seas and oceans. The motion can be used to drive turbines and generate electricity. Energy can be obtained from the waves in a variety of methods.

One may work like a reverse swimming pool wave machine. In a wave power station, the approaching waves may cause the water inside the chamber to bob up and down. This will subsequently displace air, which is then forced through the turbines to produce electricity. The rushing air can be noisy but fitting a silencer to the turbine can reduce the noise. Other mechanism uses the up and down wave motion to push a piston inside a cylinder up and down which can also revolve a generator. Wave power stations are rare since they generate small amount of power that can be used to power a small lighthouse or a warning buoy.

Examples are as follows:

- **Pelamis wave power:** They used a long, hinged, floating tube called Pelamis, which may rise up and down with the waves. Such movement bends the hinges, which then pump hydraulic fluid to drive the generators.

- **Renewable energy holdings:** They use underwater equipment on the seabed near coastline. Moving waves crossing the top of such unit moves the piston that may drive the generators on land.

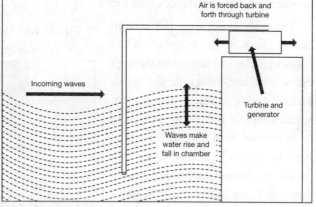

Figure 2.5: Process of wave power generation

Air is forced back and forth through turbine

Incoming waves

Turbine and generator

Waves make water rise and fall in chamber

Global and Indian scenario

The global technological probability of wave energy production is estimated at 11,400 terawatt-hours (TWh) per year. Denmark and Norway and Portugal have already installed wave energy. Scotland and Japan can also be harnessed for wave energy.

The Indian Wave Energy Programme started at the Indian Institute of Technology (IIT), Madras in 1983. A 150 kW pilot plant was set up in 1991 in the Vizinjham Fisheries Harbour breakwater close to Trivandrum in Kerala by the National Institute of Ocean Technology (NIOT), Chennai. Maharashtra government conducted research of the coastal areas and came out with some potential sites such as Vengurla rocks, Malvan rocks, Redi, Pawa Point, Ratnagiri and Girye, Square Rock, etc.

Advantages:

- It is completely free energy, no fuel needed. ▪ No waste is generated. ▪ It also boasts of more or less cheap operational and maintenance costs.

Disadvantages:

- Energy generation depends on the waves – sometimes one can get lot of energy, sometimes almost nothing. ▪ Requires suitable location, where waves are consistent and powerful. ▪ Some designs may be noisy. ▪ Must be able to endure very irregular weather conditions.

E. Geothermal

The temperature increases as one goes inside the earth and the temperature at Earth's centre is about 6,000°C. If the crust is thin, the temperature can even be 250°C a few kilometers down. Temperature increases by about 3°C for every 100 m increase in depth. So it is very obvious that one will find heated rock hot enough to boil water at some distance below the ground. Geothermal energy is the energy obtained from the stored heat inside earth's crust. This form of energy was prevalent since the existence of the earth.

The crust floats over the molten mantle known as magma. When magma gushes and forces out through the cracks and faults in the earth surface during volcanic eruption, it is called lava. If water comes close to or in contact with such hot rocks it starts boiling and quickly changes into steam. The temperature may be more than 300°F. And when this hot water comes out through cracks it is called hot spring, such as Emerald Pool at Yellowstone National Park. Sometimes the hot water explodes in air to form a geyser like Old Faithful Geyser. When holes are drilled, the steam comes up which can then turn turbines to drive electrical generators.

Figure 2.6: Mechanism of harnessing geothermal energy through injection well

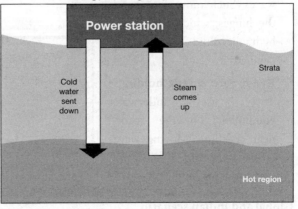

Natural 'groundwater' may be present in the hot rocks or alternatively one has to drill holes to force water downward to them. Water can be forced down an 'injection well', and comes back up the 'recovery well' under high pressure. It bursts into steam reaching the surface.

Geothermal energy scenario – India and the world: The first ever geothermal power station was built at Landrello, Italy and the second plant was at Wairekei, New Zealand. Many geothermal stations are operating in countries like Iceland, USA, Japan and Philippines. Geothermal heat is used to heat houses and to generate electricity in Iceland.

Geothermal energy was limited to regions around tectonic plate boundaries. Globally, geothermal energy production has risen from 5,800 MW to 8,400 MW from 1998 to 1999.

Geothermal power plants worked in nearly 24 countries in 2010 and geothermal energy for heat was in use at least 78 countries with a potential of 67,246 Gigawatt Hour (GWh) of electricity. Presently these nations have geothermal power plants with a total capability of 10.7 GW. 88 per cent of this amount is produced in seven countries: the Iceland, Turkey, Philippines, USA, Indonesia,

Mexico, Italy and New Zealand. Wineagel Developers in California, USA, produces 750 kW. USA leads the world and Philippines is second highest in production, with 1,904 MW capacity. The Yangbajain Geothermal Power Station in Tibet with a capacity of 25 MW ranks tenth in the world.

India has a potential of 10,600 MW though opinions may vary. Thermax, a capital goods manufacturing company has signed an agreement with Icelandic firm Reykjavík Geothermal. India's first geothermal power plant is to be set up at Tattapani in Chattisgarh as per the memorandum of understanding (MoU) between Chhattisgarh Renewable Energy Development Agency and National Thermal Power Corporation (NTPC) Ltd.

The prospective sites in India are:

- Puga Valley (Jammu and Kashmir)
- Tatapani (Chhattisgarh)
- Godavari Basin Manikaran (Himachal Pradesh)
- Bakreshwar (West Bengal)
- Tuwa (Gujarat)
- Unai (Maharashtra)
- Jalgaon (Maharashtra)

Advantages:

■ Does not produce any pollution, and does not add to the GHGs. ■ The power stations do involve much space, so impact on the environment is negligible. ■ No need of fuel. ■ After installation, the energy is almost free. A little energy may be required to run a pump.

Disadvantages:

■ Difficult to find prospective sites. Appropriate hot rocks at a depth to be drilled down should be found. ■ Occasionally such site may 'run out of steam'. ■ Toxic gases and minerals may come out from underground along with steam, may be difficult to handle.

F. Hydroelectricity

The power of water has been benefitting people for more than 2,000 years. Water wheels were used to grind flour and later it was used to generate electricity. At the end of nineteenth century, the water turbines replaced water wheels and storage devices were constructed to regulate the flow of water. Hydropower is a renewable form of energy, economic, apparently non–polluting and environmentally benign. In India such power is over 100 years old. With time the electricity requirements increased, technologies advanced and emphasis was given to the installation of big sized hydro power plants. In 1963, the hydropower had achieved a 50.62 per cent share out of the total capacity of power production in India.

Principle of Hydropower generation: Hydropower can convert the natural flow of water into electricity. Descending water flow that turns the turbine blades coupled with a generator to produce electricity creates the energy. The amount of electricity production depends on the volume of water passing through a turbine and the elevation from which the water falls. The flow and the head are directly proportional to the amount of electricity generated. A dam is built to trap water, much thicker at the base than the top to bear the load of water. Gravitational potential energy is stored in the water and is allowed to flow through passageways in the barrage to turn turbines and drive generators. A station can be built next to a fast-flowing river so that the outgoing water flows normally. So, there can be dams to raise the head and control the water flow and reservoirs that may store water for future; while others generate electricity immediately using the water flow. Once built, water flow is free and power is cheap. More than 50 per cent of the country's energy requirement is met by hydropower in Switzerland and New Zealand. Hoover Dam, constructed on Colorado River, supplied most of the electricity need for Las Vegas city that time.

Table 2.2: Some important hydroelectric power projects in India

Station	River	State	Operator	Capacity MW
Nagarjunasagar	Krishna	Andhra Pradesh	Andhra Pradesh Power Generation Corporation (AP Genco)	965
Sardar Sarovar	Narmada	Gujarat	Sardar Sarovar Narmada Nigam Ltd (SSNNL)	1,450
Tehri Dam	Bhagirathi	Uttarakhand	Tehri Hydro Development Corporation Ltd (THDC India Ltd)	2400
Srisailam Dam	Krishna	Andhra Pradesh	AP Genco	1,670
Bhakra Dam	Sutlej	Punjab	Bhakra Beas Management Board (BBMB)	1,325
Pong	Beas	Himachal Pradesh	Bhakra Beas Management Board (BBMB)	396
Indira Sagar	Narmada	Madhya Pradesh	NHPC	1,000

Advantages:

- Hydropower uses the velocity of flowing water and therefore meets the definition of renewable energy.
- Hydropower possesses unique operational flexibility capable of reacting immediately to fluctuating demands for electricity and is the best source to support the exploitation of wind or solar energy. It is more reliable than wind, solar, tidal or wave power.
- Hydropower reservoirs collect rainfall, thereby, it can store and supply freshwater for drinking,

sanitation and irrigation. This storing of freshwater safeguards aquifers and diminishes our susceptibility to floods and droughts.

- Hydropower is an apparently environmentally friendly, renewable clean source of electricity as it emits a small number of greenhouse gases, no other forms of air pollutants and does not produce any hazardous byproducts.

- It is a clean source of energy. By offsetting carbon discharges from fossil-fuel-driven power plants, hydropower contribute to reducing air pollution and slows down global warming.

- Hydropower development brings about power, roadways, industries and commerce to the public, promoting the economy, ameliorating access to healthiness and education, and improving the quality of life with little or no negative consequences.

- Hydropower guarantees an effective power networking system through its supple, dependable process, where the enactment of thermal plants is boosted and air emissions decreased.

Case Study 2.2: Hydel power in Western Ghats

The Western Ghats extend from the Satpura ranges in the north and includes Goa, parts of Karnataka, Tamil Nadu and Kerala ending up at Kanya kumara, more than 1,500 km. The notable hills are Lonavala-Khandala, Mahabaleshwar, Panchgani, Matheran, etc.. It is also known as Sahyadhri Hills.

It is one of the 25 biodiversity hotspots and offers innumerable ecological services of great economic value to our community. Western Ghats form the source for many major and minor rivers that provide employment for majority of families. Western Ghats harbour very fragile ecosystems with a third of the Indian angiosperms occurring here. About 40 per cent of this is endemic. More than 250 avian species are found along with other fauna.

Hydel power projects require proper infrastructure for proper functioning. Such requirements are dams or reservoirs, power station, transmission lines, offices,

control centers and staff quarters. This leads to social, cultural and environmental impacts and leaves open a number of ethical, constitutional and legal issues. The important impacts that the people face are:

- Large stretches of forest will get diverted or fragmented due to the construction of the power plant or due to the transmission lines.

- Migration and foraging of wild animals will be affected and invasive species might find their entry route thus threatening survival of others.

- Alleviate man-wildlife conflicts.

- The agricultural land fed by the flowing rivers will also be diverted by such construction.

- Residential lands, agricultural fields, forests and grazing land are like to be submerged that will lead to displacement of a large number of people.

- Water from rivers is a renewable source that is outside the scope of fuel price fluctuations or supply constraints; therefore, hydropower fosters energy independence and security.

- With an average lifetime of 50 to 100 years, these projects are long-term investments that can without difficulty be improved to avail gains of the new technologies and normally pay back within a short period of time. Hydropower is a power source with protracted feasibility and very low maneuver and maintenance costs that one generation presents onto the subsequent generations to come.

- Hydropower projects grow and operate in an economically feasible, environmentally sound and socially accountable manner that characterize sustainable development.

Disadvantages:

There are a number of constraints involving hydel power projects.

- Usually it takes a long time to build a hydroelectric power plant as compared to that of a thermal power plant.
- Dams are quite expensive to build. The cost is also three times higher than setting a thermal power plant.
- Usually a large dam will overflow an extensive area upstream.
- Essentially the land should have a slope.
- Perennial rivers must be selected but most rivers in India are monsoon fed. Seasonal failure may affect water flow.
- Temperature of water should be above 4°C or else the water will freeze during winter.
- Land acquiring problems: Hydropower projects face a plethora of land acquisition bottlenecks due to litigation problems, poor keeping of land records, etc.
- Resettlement and rehabilitation problems: As reservoir schemes require evacuation of large extents of lands and results in displacement of entire communities.
- Law and order problems: Projects in some states face problems on account of insurgency, terrorism, etc.
- Difficult/Inaccessible sites: In remote areas, infrastructure such as roadways needs to be first built prior to the onset of work. Power supply in remote areas also requires construction of long transmission lines with natural logistical problems.
- Geological surprises: In the mountains, geological surprises while tunneling are time consuming and leads to cost overruns.
- Postponements in environment and forest clearances: Getting environment and forest clearance is burdensome and involves inputs from concerned department of state and centre which is becoming more and more difficult with rising awareness making way for stringent legislations.
- Inter-state aspects: Inter-state water disputes if any may unnecessarily take away time.
- Funding of hydropower projects: Hydro projects were chiefly funded by government agencies and hence limited number could be taken up.

Small hydro potential in India and strategy for SHP development: SHP method was introduced in India after the installation of the world's first hydroelectric project at Appleton, USA, 1882. River-based projects have a prospect in the Himalayan states whereas other states can set up projects based on irrigation canals. Ministry of New and Renewable energy is responsible for SHP. The first SHP installation in the country was 130 kW plant at Sidrapong in Darjeeling, 1897. Other SHP projects include a 2 MW project in Mysore at Shivasundaram, a 3 MW project at Galgoi in Mussoorie, 1.75 MW in Chaba and 50 MW in Jubbal, both near Shimla. As per Twelfth Five Year Plan, Indian aims to achieve about 7,000 MW out of SHP.

Many of these powerhouses utilize high head accessible at the sites. In the beginning, the development of SHP was limited to the hilly streams of the Himalayan region. Later, SHPs were installed on several canals on the Ganga. The major difficulty in SHP stations was that high voltage transmission lines were not laid that resulted in heavy line losses.

Table 2.3: Classifications of micro, mini and SHP based on capacity as per MNRE

Class	Station capacity in Kilowatt (kW)
Micro hydro	Up to 100
Mini hydro	101 to 2,000
Small hydro	2,001 to 25,000

Advantages of small hydro plants:

- It requires small amount of flow and a height of 2 ft can generate electricity.
- Continuous supply of energy.
- It is a clean process of power generation.
- No reservoir is needed and functions as a runoff river system – water flowing out of generator is directed back to the stream.
- SHP's generate renewable energy apart from catering to the upliftment of the rural masses, more so in remote, inaccessible areas.
- They are a most cost effective option for power generation.
- They provide a stable electricity supply even at remote areas and act as catalyst of economic development.
- SHPs resolve the low voltage problem in the remote regions and help decrease in transmission and distribution losses.
- SHP also helps to provide water for drinking, sanitation and irrigation.
- They also help foster economic development by aiding industrialization.
- SHP calls for minimum rehabilitation and resettlement apart from being environmental friendly.
- They also help in employment generation.

Disadvantages:

- Power expansion is not possible.
- Low power production during the summer due to decreased water flow.

Global scenario: Worldwide, hydropower provides 17 per cent of the overall electricity. This makes hydropower by and large the most important renewable energy for electricity generation. The total installed capacity of SHP is 47,000 MW and the predicted potential is 180,000 MW. The development of small hydro projects is strong in Asia.

The Indian scenario: Earlier, the onus of controlling and operating hydropower production were separated between the Ministry of Power and the Central Electricity Authority (CEA). Since 1989, Ministry of Non-conventional Energy Sources (MNES) is accountable for small and mini hydro projects.

15 states in India namely, HP, UP, Uttaranchal in the North, Haryana, Punjab, Rajasthan, Maharashtra in the West, Karnataka, Kerala, Andhra Pradesh, Tamil Nadu in the South, Orissa, WB in the East, MP, Chhattisgarh in the centre, have publicized their strategies for setting up of commercial SHP projects. Uttarakhand has an SHP potential of 1,710 MW with about 40 under SHP category on the Bhagirathi and Alaknanda basins.

Small hydropower is expected to have a potential of around 15,000 MW. The database for SHP projects generated by MNES comprises 5,415 prospective locations with accumulative capacity of 14,305.47 MW.

Mini and micro hydro power: CDM are the main motives for the development of mini and micro hydro power systems. In general, micro hydro is pegged at less than 100 kW capacity, while Mini hydro ranges from more than 100 kW to less than 10 MW. Pico hydro generates up to 10 kW. The benefit of small hydro plants is that they are cost effective and quite dependable to provide clean electricity. Such systems can be set up in river or streams with negligible harmful environmental impacts and most of them do not need a dam or the diversion of water flows.

Micro hydro power generation provides a decent choice for electrification in the rural areas and several such plants are presently working in developing countries and helping rural populations. The power generation probability depends on the height (head) of water, the rate of flow of water and hydraulic efficacy of the turbine.

Hydroelectric potential in India: The total capacity of India as of June 2011 is 1,76,990 MW and hydropower supplies 38,106 MW which is nearly 21.5 per cent. As per Eleventh Plan, 78,700 MW additional capacities are envisaged from various conventional sources of which 15,627 MW is from large hydro projects. 1,400 MW is expected to come from SHP projects. The Central Electricity Authority plans an addition of 11,897 MW as per the Twelfth Five Year Plan. The total hydroelectric power prospect of the country is expected to be about 1,50,000 MW.

G. Biomass energy

Biomass is organic matter from plants and animals containing the stored energy of the sun. Wood, manure and certain types of garbage are examples of biomass fuels. Biomass energy is reusable −− dead tree parts, branches, grass clippings, leftover crop residue, wood chips and barks, twigs and sawdust. It also includes used tires and livestock manure.

Biomass is a renewable energy because as long as life forms are there people will continue to get biomass energy.

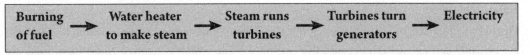

| Burning of fuel | → | Water heater to make steam | → | Steam runs turbines | → | Turbines turn generators | → | Electricity |

Advantages:

- It is sensible to use waste materials.
- It reduces dependence on the fossil fuels.
- The fuel source is cheap.

Disadvantages:

- Gathering fuel in sufficient quantities is difficult.
- It is not available all year round.
- Emission of greenhouse gases.

Converting biomass to other forms of energy: Around 32 per cent of the total primary energy use in India is still obtained from biomass and over 70 per cent of the people depend on biomass for energy. Biomass can be easily transformed into other forms of usable energy, such as methane or other transportation fuels, such as ethanol and biodiesel. However, methane gas is the principal constituent of natural gas. Composting material can be used as manure that can help plants grow. India's potential is around 18,000 MW. Additionally, about 5,000 MW power could be produced through bagasse-based cogeneration in about 550 Sugar mills.

H. Biogas

Biogas is a biofuel, and refers to the **mix of hydrogen and methane** as a result of **bacterial decomposition.** The waste is digested in anaerobic conditions by bacteria, a process called 'fermentation' at about 35–40°C. Some farmers may carry on such process in large tanks called 'digesters' and may cover their manure ponds to capture biogas. Biogas can be utilized to generate electricity or heat. According to the Ministry of New and Renewable Energy, India has the capacity to meet 10 per cent its energy requirements by shifting to biogas technology, which amounts to around 17,000 MW. Nearly 4.30 million biogas units of domestic scale have been installed in 2012, although the prospective capacity was anticipated to reach 12 million. The states of Punjab, Andhra Pradesh, Maharashtra, West Bengal and Madhya Pradesh are among the leading producers in India.

Advantages:

- It is sensible to use waste materials.
- The methane, a GHG can be used for electricity.
- The fuel is cheap.
- Low dependence on fossil fuels.
- Some industry sectors like leather, paper and pulp, sugar, poultry industry, breweries, etc., generating a lot of organic waste, can meet their power demand from their own waste.

Disadvantages:

- Upon combustion it emits greenhouse gases.

I. Biofuel

One way to biomass utilization is ethanol production. Ethanol can be used in vehicles. Biofuels are possibly carbon-neutral, because the carbon dioxide which is emitted when fossil fuels are burnt is also taken in by the plants as they propagate. Vehicles are either powered by bioethanol or biodiesel. Bioethanol is usually mixed with petrol, while biodiesel can be used as it is. The Union Cabinet have permitted a national policy on biofuels that aims to employ 20 per cent amalgamation of biodiesel with diesel and ethanol with petrol by the year 2017. It also plans for setting up of a National Biofuel Fund with a National Biofuel Coordination Committee under the prime minister.

Crops like corn, maize, sugar and cassava can often be used for production of ethanol. Sugar cane leftover pulp, known as 'bagasse' can be used to produce bioethanol. Ethanol may also be made from wastepaper. India presently is the fourth largest producer of ethanol after US, Brazil and China. Common ethanol blends are prefixed E. it varies from low blend like E5, E10, E20 to high E85. E10 is the most commonly used gasohol with 10 per cent ethanol blended with 90 per cent gasoline.

Biodiesel is a ethyl or methyl ester of fatty acids prepared from used or unused vegetable oils and animal fats. The main sources can be plant species such as *Jatropha curcas* (Ratanjyot), *Pongamia pinnata* (Karanj), *Calophyllum inophyllum* (Nagchampa), *Hevea brasiliensis* (Rubber), marine algae etc. It does not contain petroleum and can be mixed at various proportions with diesel to create a biodiesel blend or it can even be used in its pure form it has almost no sulphur content and other aromatic compounds but has about 10 per cent built-in oxygen, which helps in complete combustion. It has a higher cetane number that imparts improved ignition quality.

In Kharagpur of West Bengal, Jatropha plants are cultivated for biodiesel production by South Eastern Railway. About 200 ha of area have over 4 lakh trees. The biotechnology board has planned to set up a Jatropha plant park in Salt Lake. Indian Oil Corporation have acquired 30,000 ha and 2000 ha of land in Chhattisgarh and Madhya Pradesh respectively for biodiesel production. It is looking for 50,000 ha of waste land in Uttar Pradesh to plant Jatropha and Karanjia for the same.

Advantages:

- Less dependence on the fossil fuels.
- No sulphur content; less amount of particulate matter, carbon monoxide, other hydrocarbons, and nitrogen oxide are released.
- Carbon-neutral as compared to other fossil fuels.

Disadvantages:

- Requirement of larger area to grow crops for biofuels.
- Inconsistent supply of the materials.
- Burning does produce carbon dioxide.
- Ethanol with high 'evaporative emissions' from fuel tanks and dispensing apparatus is a limiting factor.

2.3.2 Non-renewable energy resources

Non-renewable energy sources: A non-renewable energy source is a type of energy source, which is somewhat restricted in supply. These energy sources may be copious, but they cannot be created or generated in same speed as they are used up. Apart from them being limited, their mining also has dire consequences on the environment.

A. Fossil fuels:

Fossil fuels principally comprise hydrocarbons. Fossil fuels involve deposits of once living organisms. This may take centuries to form. Fossil fuels for energy provision are of three types; coal, oil and natural gas.

Formation of fossil fuel: Fossil fuels were created several hundreds of million years ago (mya) in the Carboniferous Period, a part of the Paleozoic Era, about 360 to 286 mya. In those periods, swamps, ferns and huge trees reigned the Earth the water bodies were filled with algae. After the plant and trees died, their bodies got submerged into the lowermost level of the swamps and were later transformed into a spongy material called **peat**. Peat remained covered by sand and clay for hundreds of years and gradually got turned into 'sedimentary rock'. Gradually added rocks stacked above the existing rocks and exerted pressure on the peat. The peat was pressed and pressed and ultimately, it formed coal, oil or petroleum and natural gas. The formation of various types of fossil fuels was largely depending on the combination of animal and plant debris, the span of time the material was buried and what settings of temperature and pressure occurred when they were putrefying.

Oil and natural gas were produced from aquatic organisms buried under ocean or river deposits. Extreme heat, pressure and bacteria combined to squeeze the organic material. Thick liquid called oil was formed first, but in hotter regions natural gas was formed. Coal formed in a similar fashion, from the dead residues of trees, ferns and other fauna that existed 300 to 400 mya. Coal was molded from swamps covered by marine water. The seawater contained considerable amount of sulfur; as it dried up, the sulfur was left over in the coal.

i. Coal: Coal is a combustible sedimentary rock comprising hydrocarbon, oxygen, nitrogen and varying amounts of sulphur. Coal can be categorized in three groups – anthracite coal, bituminous coal and lignite coal.

Of these, the toughest variety of coal is the anthracite. More carbon content add to its superiority along with high content of energy.

The softest of all is the lignite coal. It is low in carbon but the amount of oxygen and hydrogen is high. Bituminous coal is the intermediate.

The earliest recognized use of coal was in China and now it occurs throughout the world. There are over 861 billion tons of proven coal reserves, as per estimates, worldwide. This indicates that coal may last around 112 years at present rate of production. But, one has reached the spectre of 'Peak Oil' with the capacity to harvest more pointed off.

Coal reserves are available in almost each and every country, with recoverable reserves in around 70 countries. The largest known reserves are in the USA, Russia, China, Australia and India.

Table 2.4: Coal proved reserves of top ten countries as per BP Statistical Review of World Energy, June 2013

Rank	Countries	Million tonnes	Share of total coal reserve (%)
1	US	237,295	27.6
2	Russian Federation	157,010	18.2
3	China	114,500	13.3
4	Australia	76,400	8.9
5	India	60,600	7.0
6	Germany	40,699	4.7
7	Ukraine	33,873	3.9
8	Kazakhstan	33,600	3.9
9	South Africa	30,156	3.5
10	Colombia	6,746	0.8

Table 2.5: Coal production and coal consumption by top ten countries as on 2012 as per BP Statistical Review of World Energy, June 2013

	Coal production		Coal consumption	
Rank	Countries	Amount (million tonnes oil equivalent)	Countries	Amount (million tonnes oil equivalent)
1	China	1,825	China	1,873.3
2	United States	515.9	United States	437.8
3	Australia	241.1	India	298.3
4	Indonesia	237.4	Japan	124.4
5	India	228.8	Russia	93.9
6	Russia	168.1	South Africa	89.8
7	South Africa	146.6	Germany	79.2
8	Kazakhstan	58.8	South Korea	81.8
9	Poland	58.8	Poland	54
10	Columbia	58.8	Australia	49.3

ii. Petroleum: Oil formed more than 300 mya ago and has been in use for more than 5,000–6,000 years. The Dead Sea, near Israel was referred to as Lake Asphaltites. Lump of viscous oil splashed up on the lake shorelines from underwater leaks, leading to the usage of the term 'asphalt'.

Table 2.6: Oil reserves by top ten countries at the end of 2012 as per BP Statistical Review of World Energy, June 2013

Rank	Countries	Amount (thousand million barrels)
1	Venezuela	297.6
2	Saudi Arabia	265.9
3	Canada	173.9
4	Iran	157
5	Iraq	150
6	Kuwait	101.5
7	United Arab Emirates	97.8
8	Russia	87.2
9	Libya	48
10	Nigeria	37.2

Table 2.7: Oil production and oil consumption by top ten countries at the end of 2012 as per BP Statistical Review of World Energy, June 2013

Rank	Oil production		Oil consumption	
	Countries	Amount (thousand bbl/day)	Countries	Amount (thousand bbl/day)
1	Saudi Arabia	11,530	United States	18,555
2	Russia	10,643	China	10,221
3	USA	8,905	Japan	4,717
4	Iran	3,680	India	3,652
5	China	4,155	Russia	3,174
6	Canada	3,741	Saudi Arabia	2,935
7	United Arab Emirates	3,380	Brazil	2,805
8	Kuwait	3,127	South Korea	2,458
9	Iraq	3,115	Canada	2,412
10	Mexico	2,911	Germany	2,358

iii. Natural Gas: Natural gas is an inflammable mixture, largely of methane and lighter than air. It is produced by the decay of methanogenic organisms in marshland, bog land, and landfills. Low temperatures are expected to produce more petroleum while high temperatures are expected to produce more natural gas. Natural gas is odorless and invisible. Natural gas was added up with mercaptan prior to its storage in tanks or supplied into the pipelines for the purpose of safety. Mercaptan smells of rotten eggs and hence its addition will impart a strong odour facilitating easy identification.

Table 2.8: Natural gas reserves by top ten countries at the end of 2012 as per BP Statistical Review of World Energy, June 2013

Rank	Countries	Amount (trillion cubic metres)
1	Iran	33.6
2	Russia	32.9
3	Qatar	25.0
4	Turkmenistan	17.5
5	USA	8.5
6	Saudi Arabia	8.2
7	UAE	6.1
8	Venezuela	5.6
9	Nigeria	5.2
10	Algeria	4.5

Table 2.9: Natural gas production consumption by top ten countries at the end of 2012 as per BP Statistical Review of World Energy, June 2013.

Rank	Natural gas production		Natural gas consumption	
	Countries	Amount (billion cubic metres)	Countries	Amount (billion cubic metres)
1	USA	681.4	USA	722.1
2	Russia	592.3	Russia	416.2
3	Iran	160.5	Iran	156.1
4	Qatar	157.0	China	143.8
5	Canada	156.5	Japan	116.7
6	Norway	114.9	Saudi Arabia	102.8
7	China	107.2	Canada	100.7
8	Saudi Arabia	102.8	Mexico	83.7
9	Algeria	81.5	United Kingdom	78.3
10	Indonesia	71.1	Germany	75.2

Energy obtained by combusting fossil fuels are converted to electricity. Heat released during the reaction further intensifies the reaction. In most instances, more electricity is produced than is essentially needed, since electricity cannot be stored. Electricity demands vary all through the year and the delivery must satisfy the **peak load**, which means the maximum possible demand within a year. If demands considerably exceed the capacity of the power plant to generate electricity, it may lead to temporary blackouts.

Environmental impacts: Burning fossil fuels is accountable for global **environmental issues** that feature high on the political itinerary in present day's context. Examples are greenhouse gas increase, acidification, pollution of air and water, degradation of land and ground-level ozone accretion. These environmental glitches are mainly by discharge of pollutants that are inherently present in fossil fuel composition, such as sulphur and nitrogen. At present, oil combustion contributes to around 30 per cent of all carbon dioxide emissions in air. Natural gas does not discharge much of carbon dioxide (CO_2) because of its methane (CH_4) configuration in the mixture. The biggest emissions are from coal combustion. Coal also results in unintentional underground fires that are practically unmanageable and impossible to extinguish. Coal dust can also burst which makes coal mining a very risky occupation. Oil may accumulate in soil or water in unrefined form, for instance during oil spills or wars. This has instigated many manmade disasters in the past.

Fossil fuels are used to such extent because it is cheaper than any other type of reasonable alternative known to us.

Effect of oil spill in general: Oceanic mammals and seabirds face great threat from floating oil because they have regular contact with the aquatic surface. Oil smearing of fur coat or plumages reduces the insulating capacity and can cause death from hypothermia, smothering, drowning and consumption of deadly chemicals. Oil harms the wildlife through bodily contact, consumption, breathing and absorption. Oils are persistent in nature and exert long-term impacts on fish and aquatic life, interacting with the environment. Floating oil can affect the planktons, which can be algae, fish eggs and the larvae of numerous invertebrates. Long term destruction to lower nutrient levels is hard to assess, but might pose ecological menaces in the Gulf of Mexico for years, grounded on its intervention with metabolism of thousands of species; birds floating on the water or diving for fish through oil–slicked water are easily exposed. Oiled birds lose their flight ability, but rather ingest that oil while preening. Loggerheads and leatherbacks turtles are affected badly as they swim ashore for nesting. Turtle nest eggs may be spoiled if an oiled adult stays in the nest. Scavengers such as bald eagles, gulls, raccoons, and skunks feeding on bodies and skeletons of polluted fish and wildlife are often exposed to oil.

Uses of fossil fuels:

- Gasoline is highly combustible, is the fuel source for traditional and hybrid vehicles, jet airplanes and racing cars. Diesel, obtained by refining fossil fuel is also used in cars, trucks and trains to power their engines.
- Kerosene, a byproduct of crude oil is also used to drive heaters. Propane is used as fuel in cooking and heating, similar to natural gas is sold as Liquid Petroleum Gas (LPG).

- Benzene or benzol forms ingredient in medicines, synthesized vitamins, synthetic rubber, pesticides, solvents, dyes etc.
- The chief fuel source for thermal power plants is coal.
- Motor oil, petroleum jelly, hydraulic fluid is used to lubricate the machine parts.
- Plastic and polyester, produced in the process of refining is used in making of containers, electronics components and building materials. Polyester is a multipurpose and tough component of modern clothing.

Case Study 2.3: Mumbai oil spill, 2010

In August 2010, Panama flagged MV MSC Chitra and MV Khalijia met with a collision off the coast near Mumbai. This resulted in around 200 cargo containers being thrown into the Arabian Sea. The collision, 10 km off the Mumbai harbour, resulted in overturning of the former ship and leading to discharging of 800 tons of oil into the Arabian Sea. The ship was also transporting containers laden with toxic organophosphate pesticides which fell off the deck and drowned in the sea.

The spill is small as compared to the Gulf of Mexico oil spill but had a great threat impact on the marine ecology along the coast. According to Mumbai Port Trust (MBPT), dolphins were reported to be blackened and also 500 kg of fish samples was found to be contaminated. This resulted in poor revenue for the fishermen, as they cannot sell the contaminated fishes in the market, which depend on the sea for their livelihood. The two most severally affected lands were the Elephanta islands and the Butcher islands which is densely populated with mangrove forests. According to BNHS reports, the mangrove has been affected due to the oil slick and the roots are becoming shorter and weak. The tides along the coast are not strong enough to wash away the oil from the coastline. In the smaller islands, the mangrove population has been affected up to 5ft. The animals on the islands are also being affected specially monkeys.

Advantages:

- They have high calorific value.

- It has a vast potential to power the entire world.
- Infrastructure for fossil fuel energy is entirely developed.
- Easy transportation of liquid or gaseous fossil fuels.
- Electricity can be produced by simple combustion process.

- They are highly stable in nature as compared to other fuels.
- Coal is obtainable in abundance.

- Tried and tested type of fuel.

- Cheaper source than non-conventional forms of energy.

Disadvantages:

- Over exploitation has caused in their considerable depletion.

- Largest emitters of greenhouse gases such as carbon dioxide and methane, responsible for global warming.

- Emission of sulphur dioxide, which causes acid rain.
- Non-renewable in nature.

- Threatens the ecological balance and may be a cause of earthquakes.
- Formation of fossil fuels takes millions of years.

B. Nuclear power

Another most important type of energy is nuclear energy which may be described as the energy trapped inside each and every atom. Matter can be transformed into energy. According to Albert Einstein, E [energy] = m [mass] x c^2 [c is the velocity or the speed of light]. A nuclear power plant generally uses uranium as a 'fuel.'

Nuclear fission: When an atom's nucleus is split apart, tremendous amount of energy is released in form of heat and light energy. This energy, when leased out in controlled manner, can be used to generate electricity. When it is leased out all at once, it makes a tremendous explosion like that of an atomic bomb.

Nuclear power is produced by the process of controlled nuclear fission (splitting of atoms). The word fission means to split apart. In majority of cases, nuclear fission reactions are used to heat water, the steam is further used to produce electricity. Uranium-235 and Plutonium-239 can be easily fissioned. 95 per cent of naturally occurring Uranium is U-238. Thorium is much more abundant; it has better physical and nuclear fuel properties; produces much less nuclear wastes, hence a thorium fuel cycle is potentially more advantageous. It involves high production and processing costs, and lacks weaponization potential.

Uranium 233 can be transmuted from Thorium 232 and used as a fuel. The liquid fluoride thorium reactor (LFTR), is on its way in countries like India, US, Israel, china, Norway and Russia.

Largest uranium deposits are in Australia. Kazakhstan, Canada, Russia, Namibia, Niger, USA are also important in terms of reserves. The primary uranium ore is uraninite or pitchblende.

A nuclear reactor is a device designed to control a chain of fission reaction that produces a stream of neutrons. Uranium is dug out of the ground, organized into small pellets, stacked into long rods and positioned in the reactor of the power plants. Uranium atoms are bombarded. With the onset of the chain reaction the atoms split and release the particles. These particles then go out and bombard another atom of uranium breaking them in turn. Control rods or efficient neutron capturers are also in the core to keep the fission regulated. Silver-indium-cadmium alloy is commonly used control rod in PWR. Moderator is another component to impede the high speed neutrons. Commonly used moderators can be light water, solid graphite, heavy water etc. The heat generated in this reaction is transferred to a liquid known as coolant.

Such a chain reaction releases enormous heat that is used to heat up the water in the reactor core. So instead of combusting a fuel, nuclear power plants practice the sequence reaction of atoms breaking to alter the energy of atoms into thermal energy.

This water from the nuclear core is directed to another sector of the power plant. Here, in the heat exchanger, it heats extra set of water filled tubing to turn it into steam. The vapour in this second set of piping rotates a turbine to produce electrical energy.

Table 2.10: Common types of nuclear reactor

	Type of reactor	Fuel	Coolant	Moderator
1	Boiling water reactor (BWR)	Enriched uranium	Light water	Light water
2	Pressurized water reactor (PWR)	Enriched uranium	Light water	Light water
3	Pressurized heavy water reactor (PHWR)	Natural uranium	Heavy water	Heavy water
4	Magnox, advanced gas-cooled reactor (AGR), UNGG	Natural uranium	CO_2	Graphite
5	RBMK (Reaktor Bolshoy Moshchnosti Kanalniy)	Enriched uranium	Pressurized boiling water	Graphite

Figure 2.7: Schematic diagram to show the cross section of a nuclear reactor of a characteristic nuclear power plant

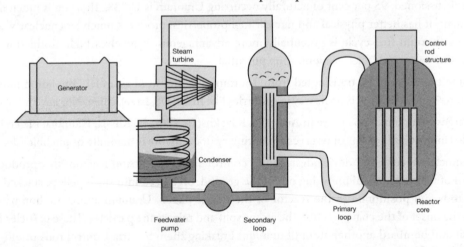

Nuclear fusion: Another form of nuclear energy is known as fusion. Fusion refers to union of smaller nuclei to make a larger nucleus. People get solar energy because of the nuclear union of atoms of hydrogen into helium atoms. This in turn gives out heat and light and a spectrum of radiation. For instance, deuterium and tritium, react to form a helium atom and an extra neutron particle.

Scientists have been attempting to construct a fusion reactor to generate electricity. But they have been facing trouble as they are to regulate the reaction in a confined space. The plus point of nuclear fusion is that it generates a reduced amount of radioactive material than fission, and it has a longer fuel supply than the sun.

At present, the installed nuclear power potential of India is 4780 MW. This is produced from

20 nuclear reactors in six power plants. Seven more nuclear power reactors, with a capacity of 5300 MW, are in the phase of construction. The details are as follows:

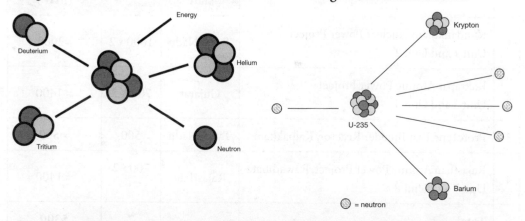

Figure 2.8: Mechanism of fusion reaction

Figure 2.9: Fission reaction

Table 2.11: Current installed nuclear power reactors and power plants

	Name of state	Nuclear power plant	Type of reactor	Operator	Units	Total capacity (MW)
1	Maharashtra, Thane	Tarapur first nuclear reactor called APSARA	BWR, PHWR	NPCIL	160 x 2 and 540 x 2	=1400
2	Gujarat near Surat	Kakrapar	PHWR	NPCIL	220 x 2	= 440
3	Karnataka Uttar Kannada	Kaiga	PHWR	NPCIL	220 x 4	= 880
4	Uttar Pradesh – Bulandshahar	Narora	PHWR	NPCIL	220 x 2	= 440
5	Tamil Nadu, coast of Coromandel	Kalpakkam	PHWR	NPCIL	220 x 2	= 440
6	Rajasthan, Chittorgarh	Rawatbhata	PHWR	NPCIL	100 x 1 200 x 1 220 x 4	= 1180
	Total				20	4780

Table 2.12: Nuclear power reactors under construction

	Nuclear power plant	Name of state	Units	Total capacity (MW)
1	Kundankulam Nuclear Power Project Unit 1 and Unit 2	Tamil Nadu	1000 x 2	=2000
2	Kakrapar Atomic Power Project Unit 3 and Unit 4	Gujarat	700 X 2	= 1400
3	Prototype Fast Breeder Reactor, Kalpakkam	Tamil Nadu	500	= 500
4	Rajasthan Atomic Power Project, Rawatbhata Unit 7 and Unit 8	Rajasthan	700 x 2	= 1400
	Total		7	**5300**

India plans to generate 25 per cent of its electricity from nuclear power by 2050.

Table 2.13: The Twelfth Five Year Plan proposes a further 19 nuclear power reactors of various types with a total capacity of 17,400 MW

	Name of state	Nuclear power plant	Type of reactor	Units	Total capacity (MW)
Indigenous reactors					
1	Haryana– Gorakhpur	Gorakhpur Units 1 and 2	BWR, PHWR	700 x 2	=1400
2	Madhya Pradesh	Chutka Units 1 and 2	PHWR	700 X 2	= 1400
3	Karnataka	Kaiga Units 5 and 6	PHWR	700 x 2	= 1400
4	Rajasthan	Mahi Banswara Units 1 and 2	PHWR	700 X 2	= 1400
5	Tamil Nadu	Kalpakkam Fast Breeder Reactor Units 1 and 2	Fast Breeder Reactor (FBR)	500 X 2	= 1000
6		To be decided	Advanced heavy water reactor	300	= 300
LWRS with international cooperation					

	Name of state	Nuclear power plant	Type of reactor	Units	Total capacity (MW)
7	Tamil Nadu	Kundankulam Units 3 and 4	Light water reactor (LWR)	1000 X 2	= 2000
8	Maharashtra	Jaitapur Units 1 and 2	LWR	1650 X 2	= 3300
9	Gujarat	Chhaya Mithi Virdi Units 1 and 2	LWR	1100 X 2	= 2200
10	Andhra Pradesh	Kovvadda Units 1 and 2	LWR	1500 X 2	= 3000
	Total				17400

Safety is accorded paramount importance in all phases of nuclear power plants right from location, design, construction, commissioning, function and upholding and ultimate decommissioning. The operation of nuclear power plants is carried out by trained and skilled people in accordance with the regulatory body appropriately accredited by the Atomic Energy Regulatory Board (AERB).

Advantages:

■ Emits very little GHGs. Hence does not contribute to global warming. ■ Readily available technology. ■ Generates high amount of electricity from very small use of nuclear fuel. ■ Low operating costs. ■ Is able to meet both industrial and domestic needs. ■ Nuclear waste may be reduced through reprocessing or recycling.

Disadvantages:

■ High installation cost due to radiation containment and procedures. Needs centralized power source with large infrastructure. ■ High known and unknown risks. ■ Requires long construction period. ■ Nuclear fuel is a finite source. Uranium may last for 30 to 60 years. ■ Installation and operation need high expertise and skill. ■ Mining involves health cost and other catastrophe. ■ Requires huge amount of water. ■ Disposing spent fuel is a problem. Wastes may last for 200 to 500 years. ■ Target for terrorist activities. ■ Average life span of nuclear reactors is usually 40–50 years.

Case Study 2.4: The Chernobyl nuclear disaster, 26 April 1986

The Chernobyl Power Complex, located about 130 km north of Kiev, Ukraine, and about 20 km south of the border with Belarus, comprised four nuclear reactors. The power plants were constructed between 1970–80. The water required for cooling the reactors was procured from the manmade lake of about 22 sq. km. This artificial lake was constructed which was fed from the waters from the river Pripyat, a tributary of Dnieper. The population density was low in this region. There were 49,000 inhabitants in the new city of Pripyat situated 3 kms away. The old town of Chernobyl with a population of 12,500 was about 15 km away from the plant. The newly built city of Pripyat close to the power plant had approximately 50,000 inhabitants.

The accident crashed the fourth reactor and killed 30 operators and firemen within three months with quite a few deaths later. Huge amounts of radioactive substances were free, almost for ten days. Iodine-131, Caesium-134 and Caesium-137 were predominantly discharged. Of these radioactive iodine has rather short half life as compared to that of Caesium. Hence, Caesium easily routes in through ingestion.134 plant employees out of a total of 600 that were present in site suffered from radiation ailments. Xenon gas, iodine, caesium and other radioactive substances ware set on loose after the incident. As per estimate a total of 192 tons of radioactive fuel was released that disrupted the lives in Belarus, Ukraine and Russia. The trees died after the accident and the region was then referred to as 'Red Forest'. Though bulk of the radioactive substances got settled as debris, the lighter materials were carried by air over these nations and further away to European and Scandinavian countries.

About 2 lakh people were involved in the clean-up operations between 1986 and 1987. These people were exposed to higher dose of radiation, on an average of 100 millisieverts. Few are said to receive even more. The first symptoms of Acute Radiation Syndrome (ARS) were diagnosed in 237 persons involved in cleaning, later established to be 134 people. 28 deaths occurred immediately after the accident. A large number of children (approximately 4,000) were detected with thyroid cancers in Belarus, Russia and Ukraine, which may be due to consumption of iodine fallout. Incidence of leukemia, heart disease and congenital defects went on a rise. People suffered from a paralyzing pessimism due to myths and misunderstandings about radiation threat. Relocations were very upsetting and painful and one could hardly do anything to reduce radiation exposure. Psycho-social effects were widespread.

2.4 Energy Resources: Present Trend and Future Energy Sources

2.4.1 Present increasing energy needs

The chief energy-consuming nations have declared new actions. China is aiming a 16 per cent decrease in energy power by 2015; while United States has embraced fresh standards on fuel economy; the European Union has devoted to cut 20 per cent in its 2020 energy requirement; and Japan to curtail 10 per cent from current use by 2030. Renewables will turn into second leading source of electricity production globally by 2015.

In spite of development taken place in the last year, roughly, 1,300 million people are left with no access to electricity; 2,600 million lacks basic clean cooking amenities. Out of the top ten countries, four are Asian and six are sub-Saharan African countries that account for two-third of such people who have no access to electricity. In India, Bangladesh and China alone more than half of the population lack hygienic cooking amenities. The Rio summit of 2012 made an obligatory pledge to move towards global contemporary energy installment by the United Nation Year of Sustainable Energy and all have whole heartedly welcomed such pledges in the direction of reaching this goal. But a lot more needs to be done. There is no doubt that the cost of electricity produced from coal is far less expensive compared to other fuels. But the expenditure spent to meet the consequences (like disposal of fly ash) is very high. If the 'carbon cost' – the cost of the damage caused to the environment by burning coal is added, then needless to say, this cost becomes exorbitant.

GoI endeavour towards energy security: As per International Energy Association (IEA), energy security is the incessant availability of energy sources for energy consumption at a reasonable price. Since access to cheap energy is a necessity for proper functioning of modern economy, it is linked with the national security.

India being a developing nation with ever increasing population, faces critical challenge to meet its energy demand. Globally, it ranks sixth in terms of energy demand. The World Energy Outlook of IEA projects that by 2020, the import of oil will rise by 91.6 per cent in India. In spite of its efforts to increase domestic energy production, it still faces an energy and peak shortages of nearly 8–12 per cent respectively. India's per capita consumption of 439 kg of oil equivalent is far behind the global average per capita consumption of 1,688 kg.

India stands third in terms of production and consumption of coal. 77 per cent of India's coal reserves are in Chhattisgarh, Jharkhand, Orissa and West Bengal. A cumulative total of 253.3 billion tonnes of coal resources have been estimated through the exploration carried out by Geological Survey of India (GSI), Central Mine Planning and Design Institute Ltd. (CMPDI) and Mineral Exploration Corporation Limited (MECL). India's low quality coal makes it a small exporter. On the contrary, India imports 20 million tones of coking coal and 17 million tones of steam coal per year from Australia, Indonesia, China and South Africa. The current peak demand for crude oil in India is 110×10^6 tonnes whereas India's yearly crude oil production maximized at 32×10^6 tonnes. To meet such demands, it has to import 70 per cent of crude oil from the Gulf nations. Bulk of Indian oil reserves are in the Bombay High, Assam and Krishna-Godavari basin. India's natural gas reserves are about 0.5 per cent of the world which lies mostly in two places – one lie offshore of Bombay in the Arabian Sea and the other lie onshore in Gujrat.

India has a plan of a nuclear capacity of 63,000 MW by 2032. India was excluded from Nuclear Non-Proliferation Treaty (NPT) as a result of being capable of acquiring nuclear weapons after 1970. As a consequence its strategy for civil nuclear use is directed towards independence and characterized by low capacity and technical difficulties. India has poor uranium reserves that can only supply 10,000 MW of PHWR. Moreover, this uranium is as low as 0.1 per cent, enriching of which make nuclear fuel two to three times costlier than other nations. Again with substantial thorium reserves it requires a conversion of thorium into fissionable material.

Estimates stand out at 84,000 MW in terms of hydropower in India. Overall India is in increasing need of power. There is a 64 per cent increase in the annual electricity production and consumption. As per Ministry of Power, there is an objective to add a production potential of

1,00,000 MW by the year 2002–07. More than 50 per cent of rural households and some fraction of urban households do not have electricity. So, the access to electricity is highly uneven. Currently, coal accounts for 65 per cent of electricity production.

India has a huge potential for renewable energy of over 1,00,000 MW; so far achieving about 10 per cent of installed capacity. In order to march towards sustainable development, people need energy conservation along with energy efficiency.

Energy conservation can be met with the reduction in consumption measurable in physical terms. Whereas, 'energy efficiency' is practically attained through reduction in energy intensity of a product or process keeping its output, consumption and comfort levels unaffected. For example, CFL replacing traditional bulbs uses one quarter of the energy to light a room along with pollution reduction. Energy effectiveness of a product or process can be enhanced right from resource extraction, processing, transportation to consumption. Efficiency in use of energy can leave considerable impact on mining, water pumping, electricity production and transmission, mass transport system, building, lightings, domestic appliances and industrial processes.

The government is keen to focus on low cost power generation, optimization in utilization, optimizing the fuel mix, input cost control, upgradation of technology and increasing non-conventional energy sources.

Regarding transmission, National Grids with interstate connections are to be developed along with cutting down the transmission cost. Regulation should be implemented for the commercial viability of the power sector along with protection of consumers' interest. Emphasis is laid on rural electrification. Some state governments provide electricity at subsidized rates or even free power to some, such as agriculture and backward classes.

Major steps taken by GoI towards energy conservation and energy efficiency are provided below:

- Ministry of New and Renewable Energy (MNRE) is the authority in planning and making policy. It is responsible in the development of alternative fuels, renewable energy products and services to rural, urban, commercial and industrial needs. The MNRE were assigned the research and development of biogas units, improved chulhas, IREDA, SHP, mini and micro hydel projects, tidal energy, geothermal energy, Integrated Rural Energy Programme (IREP), etc.

- IREDA, established in 1997, is a non-banking financial company meant to offer loans in the field of renewable energy ventures.

- To achieve the government's energy goals the Bureau of Energy Efficiency (BEE) was set up in 2002 under Energy Conservation Act, 2001 attained by the integration of Ministry of Power and Energy Management Centre. BEE's objective is to reduce the energy intensity of Indian economy. BEE had been instrumental in the development of energy performance labels in motors and equipments, refrigerators and air conditioners; helped the industries to benchmark their energy use. BEE in addition, also prepared the strategy for Energy Conservation Building Codes (ECBC) and carries out energy audits of government buildings and certifies energy auditors and energy managers, etc.

- Indian industry also played a vital role in promotion of 'energy efficiency'. For capacity building, the responsibility lies with the Confederation of Indian Industry (CII) and Federation of Indian Chambers of Commerce and Industry (FICCI). They organize trainings, workshops, seminars, exhibitions etc. CII's building has acquired Leadership in Energy and Environmental Design (LEED) platinum rating. The Indian Council for Energy Efficiency Business (ICPEEB) was established by private energy service companies (ESCOs) in order to keep the policymakers rightly informed as well as its dissemination. Standards and labeling (S&L) is one of the main activities toward energy efficiency.

- Different types of fiscal incentives are given for encouraging RE projects. The incentives are tax benefits, exemption or reduction of excise and customs duties, allowance of water usage, concessions to industries in backward areas, infrastructural supports, etc.

- Most of India gets 300 sunny days which is equivalent to 5,000 trillion kWh/ year. This is more than the total energy consumption. A family of four members uses 100 lts of hot water per day, which can be met using 2 sq. mt lone collector systems. Such a heater can save 1,500 kWh/ year of electricity along with paying back its cost in just 3–4 years. Uttaranchal, Rajasthan, Karnataka and West Bengal are providing a monthly discount to the people using solar water heaters of ₹ 50 in the electricity bills.

- In industries heating system called solar flat plate collector (FPC) can be conveniently used for drying coffee beans, cereals, spices, tea leaves, mushrooms, papads, fish, vegetables, etc. It also finds application in paper, rubber, leather, textiles and pharmaceutical industries. Solar cooker can save conventional fuels though it cannot replace the other fuel need completely.

- Solar street lighting system (SLS) can be used to illuminate streets and gardens. It comprises a photovoltaic (PV) module, a lead acid battery and a CFL of 11 watt. The solar lighting is crafted to manoeuvre from sunset to sunrise. Such lights routinely illuminates with darkness. An SLS may cost about ₹ 19,000.

- Solar buildings based on solar passive design are able to provide comfort both during and summer. It costs 5–10 per cent more but saves 30–40 per cent of conventional energy.

- MNRE, under the Alternative Fuels for Surface Transportation Programme, has also implemented a scheme for electric vehicles (EVs). It plans in deploying battery operated vehicles (BOV) which has a limitation of a limiting driving range per charge of the battery.

- MNRE has undertaken Biofuel Pilot Demonstration Project in rural areas, such as Menasina, Kalluguga, Sipri, Basghari, etc. to supply energy from non-edible vegetable oilseeds for lighting, farming operations and community-based services. The government targets to use 20 per cent mix of biodiesel with diesel and ethanol with gasoline by 2017.

The time has therefore come to look at alternative, more eco-friendly, clean and green energy sources so that the overall basket of energy mix not only becomes more sustainable but also less polluting.

2.4.2 Future energy sources and hydrogen

Fossil fuels took millions of years to form; hence they are non-renewable and once they release the energy contained in them, they are gone forever. We can conserve fossil fuels by innovatively harnessing energy from other alternative sources like the sun and water. The move is now on to harness the power of hydrogen. At room temperature diatomic hydrogen is a colourless, tasteless and odourless gas. It is lighter than air and inflammable. It occupies roughly 75 per cent of the entire universe and 0.15 per cent of the Earth's crust. It is found to combine with oxygen, nitrogen and other elements on Earth, the most interesting being H_2O. It can exist in liquid form under high pressure and very very low temperatures. Hydrogen should be separated from its combinations in order to make it useful. Currently its use is in manufacture of ammonia, refining of petroleum and methanol production. Hydrogen is the fuel in space ships and fuel cells that supply with heat, potable water and power supply to the space travelers.

Fuel cells are can transform hydrogen into electricity. Hydrogen is the prospective fuel for the future, to meet our increasing needs. The chemical energy in hydrogen can be transformed to produce water and electricity by the reaction between hydrogen and oxygen. Hydrogen can be made from hydrocarbon compounds in 'reforming' process by application of heat. Hence, it can be made from natural gas. Hydrogen can also be obtained from water by electrolysis. Microbes like algae and bacteria, also give out hydrogen under specific circumstances. Hydrogen is high in energy content and does not pollute. NASA is using liquid hydrogen to fuel rockets and propelling space shuttles into path.

Uses of fuel cell: Fuel cells are usually the devices that can utilize the chemical energy of the fuel to convert it into electricity via a chemical reaction. Such devices are endowed with high proficiency to utilize hydrogen for supply of heat and power in houses and as a source of electric energy in automobiles. In a hydrogen fuel cell vehicle, an electrochemical instrument converts aerial hydrogen and oxygen into electric energy, to impel an electrical motor and influence the automobile. They have the potential to replace the internal combustion engines of automobiles. Ideally they can run with pure hydrogen and in the close future they are expected to be used with natural gas, methanol or even gasoline.

This type of fuel reformation will enable us to use the present infrastructure such as gas stations, natural gas, pipelines, etc. In the forthcoming years, hydrogen could also join electricity as a significant energy delivery service that can store, move and distribute energy in a functional form to consumers. It is not impossible to replace 'fossil fuel based economy' by 'hydrogen economy'. The idea into practical application is far away and seems not to materialize in the near future.

Solar power-driven satellites: A significant proposal for the forthcoming years is to position solar power driven satellites into the orbits surrounding Earth, which could gather solar energy from the sun, transform it to electricity and beam it to Earth in the form of microwaves or various other forms of transmission. This power would not emit any types of greenhouse gas, but microwaves might have an undesirable effect on health. Recurrent rocket launches may damage the upper atmosphere. This notion is still far from real for atleast another century.

Case Study 2.5: Solar challenger

It is a solar-powered effective airplane sufficiently robust to support both extended and high trips under normal turbulence. Its efficiency was proven during the cross-channel flight in 1981. The aircraft had a massive straight stabilizer and a sufficient wing expanse to provide accommodations for 16,128 solar cells.

It possessed an in-house computer programme and the propeller was designed aerodynamically. The consequence was a 'suave and compliant' aircraft that can move smoothly up and down steadily. The aircraft was flown with batteries and subsequently by solar energy. The plane flew between Santa Susana, Shafter and EI Mirage airports and then moved to Marana Airpark northwest of Tucson, Arizona.

Summary

- Materials occurring in the nature under diverse environmental conditions are termed as natural resources. Natural resources was classified on the basis of various parameters, the most important being renewable and non-renewable forms of energy that forms the basis of this chapter.

- The Earth's most ubiquitous and potent energy source is the sun reaching the Earth in form of solar radiation. A small fraction of this solar energy that strikes the Earth every minute is ample enough to supply all our energy needs. The wind velocity when harvested with the help of wind turbines generates electricity in 'wind farms'. The tidal energy and waves in motion can be used to drive turbines and generate electricity. Oceans being the largest solar energy collector, it is possible to carry out conversion of solar energy absorbed into electrical energy in OTEC. Geothermal energy is obtained from the stored heat inside Earth's crust.

- Hydropower, a renewable form of energy, can convert the natural water flow of water into electricity and possesses unique operational flexibility to meet the fluctuating demands for electricity. Environmental impacts associated with hydropower have made it a subject of controversy. SHP method introduced requires no reservoir and functions as a runoff river system. Biomass is organic matter from plants and animals containing the stored energy of the sun. Biomass energy is reusable. Biomass may be modified to form biogas. Biogas is a biofuel, and refers to a **mixture of methane and hydrogen** produced by **bacterial decomposition.** Another method of biomass utilization is ethanol production.

- Most important of the non-renewable resources are the fossil fuels; coal, oil and natural gas. Burning fossil fuels is accountable for global environmental issues like increase in greenhouse gases, acidification, pollution of air and water, degradation of land and ground-level ozone accretion.

- Another most important type of non-renewable energy is nuclear energy where matter can be transformed into energy. Nuclear power is usually produced by the process of controlled nuclear fission using Uranium-235 and Plutonium-239 as fuels.

- Energy security is the incessant availability of energy sources for energy consumption at a reasonable price. Since access to cheap energy is necessary for proper functioning of modern economy, it is linked with the national security.

- Conserving fossil fuels can be achieved by innovation and harnessing alternative energy sources like the sun and water. The move is now on to harness the power of hydrogen that comprises about 75 per cent of the entire universe.

Exercise

MCQs

Encircle the right option:

1. Match the following dams to the states they are located:

					Ans:
i.	Tehri	a.	Gujrat	A.	ic, iia, iiid, ivb
ii.	Sardar Sarovar	b.	Maharastra	B.	ia, iid, iiib, ivc
iii.	Loktak	c.	Uttarakhand	C.	id, iic, iiib, iva
iv.	Koyna	d.	Manipur	D.	ic, iia, iiib, ivd

2. Biogas is principally a mixture of:
   A. Methane and Hydrogen   B. Methane and Ethylene
   C. Ethanol and Hydrogen   D. Butane and Oxygen

3. Match the following projects to their uses:

					Ans:
i.	La Rance	a.	Geothermal power station	A.	ic, iia, iiid, ivb
ii.	Haiti	b.	Solar power station	B.	ia, iid, iiib, ivc
iii.	Durgaduani	c.	OTEC	C.	id, iic, iiid, iva
iv.	Landrello	d.	Tidal power station	D.	ic, iia, iiib, ivd

4. Arrange the countries in ascending order of wind energy production:
   A. China, USA, Spain, India   B. India, Spain, USA, China
   C. Spain, India, USA, China   D. USA, China, USA, India

5. Which among these states do not have any wind prospective sites?
   A. Tamil Nadu   B. Kerala   C. Madhya Pradesh   D. Uttar Pradesh

6. Which of these is the main constituent of LPG?
   A. Methane   B. Ethane   C. Ethanol   D. Propane

7. Winds must have a speed above _____ miles/hour to turn the turbines efficiently.
   A. 5   B. 12   C. 20   D. 25

8. The first Small Hydro Power (SHP) in India is:
   A. Sidrapong   B. Galgoi   C. Shimla   D. Manali

9. Which geological period was coal formed?
   A. Carboniferous   B. Triassic   C. Permian   D. Cretaceous

10. Arrange the following types of coal in their ascending order of superiority:

    A. Peat, Lignite, Bituminous, Anthracite   B. Bituminous, Peat, Lignite, Anthracite

    C. Lignite, Peat, Anthracite, Bituminous   D. Peat, Bituminous, Anthracite, Lignite

Fill up the blanks:

1. Resources which cannot be replenished are called _____ resources.

2. _____ is commonly used as a nuclear fuel in a nuclear reactor.

3. _____ is added to LPG for detection of any leakage.

4. SPVs are made of the semiconductor _____.

5. Gasohol is a mixture of gasoline and _____.

State whether the statements are true or false:

1. The least polluting of all the fossil fuel is natural gas. (T/F)

2. Geothermal energy is produced as a result of fission of radioactive substance. (T/F)

3. Big dams are environmentally more sustainable than small dams. (T/F)

4. Tides are produced by the gravitational pull of sun and moon on Earth. (T/F)

5. Clusters of wind turbines are known as wind farms. (T/F)

Short questions:

1. Define natural resource.

2. What is meant by tangible and non-tangible resource?

3. What is a photovoltaic cell?

4. Why don't we get efficient PV modules?

5. How do solar thermal power plants work?

6. What are aerogenerators?

7. State the advantages of wind energy.

8. Which of the Indian states have greater prospect of harnessing wind energy?

9. What are the disadvantages of a tidal barrage?

10. How can we get geothermal energy if water table is not available?

Essay type questions:

1. Compare vertical axis wind turbine with horizontal axis wind turbine.

2. Explain the parts and functions of a wind turbine.

3. Discuss the tidal power station at La Rance in France.

4. How does an OTEC plant operate? What is its advantage?

5. Discuss the pros and cons of conventional hydroelectric project.

Natural Resources – Water, Land, Forest, Food and Mining _____ 3

Learning objectives

- *To identify the various issues associated with natural resource and know the ways to ease such problems.*
- *To realize the immense importance of water and land as resources.*
- *To develop a concern about forest resources.*
- *To know about the present practices of food production and mining and also the judicious ways of using them.*
- *To know about these resources in the Indian perspective.*
- *To be able to take in personal efforts and implement equitable use of resources in natural resource management.*

3.1 Natural Resources and Associated Problems

Growing population: World population is over 7 billion at present. More number of people means more mouths to feed, more crops, more forests to be converted into croplands; more intensive farming, more use of pesticides and fertilizers to boost production and more and more soil deterioration in the long run. A huge amount of water is diverted for irrigation to complement food production and a large number of trees and plants are being cut down for use as fuel and fodder for domestic animals. Human encroachment in forests becomes inevitable. Rapid means of transport and communication depletes natural habitats while rapid urbanization results in severe pollution and health problems.

Rapid pace of industrialization: Production always occurs in anticipation of demand and ahead of it. Industries are working overtime to feed a growing population's needs without paying any heed to the fact being utilized that are used to feed this demand is finite. Wanton and at times even trivial needs are leading to the diversion of scarce resources which is creating a huge amount of stress on the eco systems of the planet Earth. Industrialization and mass production change our world to such an extent that human consumption has increased in tandem with the ability to produce without adequate thought being given to whether production and consumption of such levels is at all required.

Uneven distribution of resources: Earth's resources are distributed unevenly. Fossil fuels, minerals, agricultural resources – all show skewed distribution patterns. While development around the pithead – where the resources are mined – is what should have been the natural result, the fact remains that the history of industrialization and economic development is also the story of how these diverse and scattered resources have been tapped by some nations to fuel their growth and prosperity. A trend reversal, with the nations of the South using the resources at their command to feed their process of development is perhaps playing out currently – a trend that will become more pronounced in the days and years to come.

Over exploitation of natural resources: A new world order, conscious that the pillaging of Earth's resources for long period have ensured that that there is a serious threat of them running out in the foreseeable future, is now talking of sustainable extraction and optimization of resource use. However, the fact remains that most of the resources have been so wantonly over exploited – for example, 'Peak Oil' which is today a chilling reminder – that the world may soon have to cope with terms like rationing and search for alternatives as supply dries up. Over exploitation may cause deforestation, desertification, species extinction, migration, soil erosion, depletion of oil, depletion of ozone, increase in GHG emission, more energy consumption, aquatic pollution and more incidences of disaster.

Measures to conserve natural resources include:

- curbing global warming;
- soil conservation and agricultural management;
- safeguarding general health;
- nurturing sustainable communities;
- management of ecosystem;
- forest conservation;
- revitalizing the world's oceans and seas;
- biological control;
- control of alien species;
- rural and community development; and
- constructing the future with clean energy;
- conservation of endangered wildlife and protection of indigenous places;
- warranting safe and sufficient supply of water;
- assessment of air quality and its improvement;
- environmental and resource management;
- rangelands and grasslands management;
- promoting sustainable development;
- promoting integrated pest management (IPM);
- fertilizer and nutrient management;
- prevention of pollution.

3.2 Water Resources

Water is the most important natural resource and no life can survive without it as for it is a natural resource needed by all living beings not only for survival, but also for growth and development. History reveals that several civilizations have perished due to shortage of water originating from climatic changes.

Here are some important facts about water:

- Earth has about 1.4 billion km^3 of water.
- Out of this, around 35 million km^3, or about 2.5 per cent of the total volume is the freshwater.
- Water in the form of ice in permafrost regions of the mountains, the Arctic and Antarctica are inaccessible to mankind. This amounts to 24 million km^3 or 70 per cent.

- But, around 30 per cent of the world's freshwater is groundwater that comprise about 97 per cent of all the freshwater that is possibly accessible for human usage.
- Lakes and rivers encompass an expected 1,05,000 km³ or around 0.3 per cent of the world's freshwater – this is readily accessible water.
- Approximately 13,000 km³ of water is in the Earth's atmosphere.

The total usable and functional freshwater supply for ecosystems and man is about 2,00,000 km³ of water – that is less than 1 per cent of all freshwater resources which makes this resource abundantly available and scarce at the same time, giving rise to anomalies and strangeness in consumption.

3.2.1 Water use and over utilization

There is a direct correlation between development and water use. At no time in the history of mankind have people consumed the amount of water as they now use. As a matter of fact, water use has been increasing at a rate which is more than twice the population growth rate in the last century.

The world's seven billion plus people are appropriating a huge amount of freshwater accessible in rivers, lakes, streams and underground aquifers.

A cursory glance at the primary reasons of water use reveals all.

World use of freshwater:

- about 70 per cent for irrigation;
- about 22 per cent for industry; and
- about 8 per cent for domestic use.

What is more, this usage pattern is expected to become more pronounced in the years to come, leading to all sorts of demand-supply problems:

- Water extractions are expected to escalate by 50 per cent in developing countries and 18 per cent in developed countries by 2025.
- Presently more than 1.4 billion people reside in river basins where the water usage is more than minimum recharge levels; this leads to drying of rivers and groundwater depletion.
- In 60 per cent of Europe's cities with a population above 1,00,000, use of groundwater exceeds the rate of its replenishment.
- By 2025, 1,800 million people will be facing partial water scarcity and two-third of the world population might be under stark conditions of water stress.

Water for energy production: With fossil fuels under the threat of running dry, nuclear power fraught with dangers and entry barriers making solar power largely unviable, water seems to be the only clean and green option left for an increasingly power hungry mankind to tap into.

Obviously, water requirements for production of energy will rise proportionately with the energy demand. Water is indispensable in power production in mining; transport and treatment of oil, gas and coal; and more importantly in irrigation. It is estimated that water extractions for the purpose of energy generation in 2010 amounted to 583 billion cubic meters (BCM). What is frightening is that the amount withdrawn may not be returned to its source. The projected rise in use of 85 per cent over the time up to 2035 reveals a change towards a water-intensive power production.

The estimated rise in water consumption by 85 per cent, by 2035, echoes a shift towards more water-intensive power production and increasing harvest of biofuels. By now in some regions, water constraints are disturbing the consistency of present operations and they will progressively inflict added costs. In some cases, they could even threaten the viability of projects. The liability of the energy sector to water restrictions is widespread physically and affects power production in parts of China and the United States, India, Canada and Iraq. To manage energy sector's vulnerability to water, better technology and bigger incorporation of energy and water strategies will be required – in short a complete overhaul with a holistic view is needed.

3.2.2 Consequences of over exploitation – water pollution

Not only are people using more water than ever before, people are also polluting the water sources in an alarming rate, ensuring that water, which is also synonymous with life, becomes everything but a life-enhancing force.

- Every day, 2 million tonnes of human wastes are thrown off in water bodies across the world.
- In developing countries, 70 per cent of industrial wastes are disposed without treatment into the water bodies where they contaminate the serviceable water.
- Since 1990, 50 per cent of the global wetlands have been lost.

The developing or less developed countries south of the divide, are worse polluters as resource crunch, lack of awareness and outright callousness ensure that they do not invest in the treatment plants and discharge their wastes in the same sources from which they tap their water. This however is not to say that the developed nations are any better off in polluting their water bodies. The contribution of food sector in water pollutant generation:

- highly developed countries – 40 per cent; and
- developing countries – 54 per cent.

Case Study 3.1: Industrial waste and Narayani river in Nepal

The aquatic life is threatened by the unrestrained industrial pollution at Narayani River and the Chitwan locals face the danger in the neighbouring areas. Two factories, Gorkha Brewery and Bhrikuti Paper factory are primarily accountable for contaminating the river as they discharge their toxic effluents straight into the river. The Bhrikuti Paper mill with 10,000 metric tonnes of paper production capacity per day has augmented the capacity of production to 88,000 metric tonnes after its privatization. Chemicals like rosin, chlorine, caustic soda used in manufacturing paper are released in to the river directly. The court instructed them to treat the effluents before discharging them. Even then, the industrialists have not paid any attention to the escalating problem of river pollution. Rare aquatic animals, like crocodiles and ghariyals, face severe threat besides the ecological menace posed to the humans living nearby. The proportion of ammonia and phosphorus is higher than the normal required for aquatic life while the temperature of the river water ranging between 12 and 20°C is also unsuitable for these rare animals.

3.2.3 Issue of water security

Water security is referred to as 'the reliable availability of an acceptable quantity and quality of water for health, livelihoods and production, coupled with an acceptable level of water-related risks.' A water secure world integrates an intrinsic value of water to a concern for human use along with protection of the environment and reducing the negative impacts of poor water management. Good water governance thus enhances the living standards.

As per Consultative Group on International Agricultural Research (CGIAR), regions like North Africa, South Africa, Central Asia, Middle East, China, Australia, Chile and India are under water stress.

Global initiatives for water security include:

■ The Helsinki Rules on the Uses of the Waters of International Rivers, Helsinki, 1966.

■ Convention on the Law of the Non–navigational Uses of International Watercourses, 1997.

3.2.4 Water security and water resource in india – availability, scarcity and demand

Water is a basic human need, a critical human asset and key to overall socio-economic development. Water concerns are multidimensional. India ranks second in terms of population among all nations with over 1.2 billion people (Census of India, 2011). India is likely to become a water scarce nation by 2050. The National Commission for Integrated Water Resource Development (NCIWRD) has anticipated the total availability of water to be 1,953 BCM of which 432 BCM is groundwater and 1,521 BCM is surface water only. Official records as per Ministry of Water Resources (MoWR) put the total utilizable water at 1,123 BCM against the present-day usage of 634 BCM that reflects a surplus situation (Planning Commission, 2010). In reality, the utilizable water by man comes out to be 654 BCM which is in proximity to the present usage estimate of 634 BCM; thus the condition is indeed alarming (Planning Commission, 2010). Prediction by Water Resources Group states that India is likely to face a huge gap between the present water supply and the anticipated demand. Present supply actually amounts to half of the demand or 754 BCM (Addams, et al., 2009). It is assessed that entire annual usable ground and surface water resources is about 396 km^3 and 690 km^3 in India. The distribution of water resource potential displays that the nation's per capita yearly availability of water in the year 2001 was 1,820 cu. m and is likely to become 1,588 cu. m in 2010.

Surface water: Water resources of a nation comprise rivers, streams and canals, reservoirs, storage tanks, lakes and ponds, other neglected water bodies as well as salt water bodies.

Rivers: The Ganga-Brahmaputra-Meghna basin covers 33 per cent of the area. It's contribution accounts for 60 per cent of the water resources of India. River catchment area that flow west contributes 3 per cent but comprise 11 per cent of India's water resources. Thus, 36 per cent of the area (33 + 3) receives 71 per cent (60+11) of India's water resources. So it seems that 29 per cent (100-71) of the water is available for 64 per cent (100-36) of the area (Verma and Phansalkar, 2007).

The Indian rivers collect a total average annual flow of 1953 km^3 annually. The whole country was divided into 20 river basins. 12 are major basins. They are: Ganga-Brahmaputra-Meghna, Pennar, Indus, Godavari, Sabarmati, Krishna, Cauvery, Mahanadi, Brahmani-Baitarani, Mahi, Narmada and Tapi.

Other remaining medium and small river basins are:

- Subernarekha – combining Subernarekha and other small rivers between Subernarekha and Baitarni;
- east flowing rivers between Mahanadi and Pennar;
- east flowing rivers between Pennar and Kanyakumari;
- area of inland drainage in Rajasthan Desert;
- west flowing rivers of Kutch and Saurashtra including Luni;
- west flowing rivers from Tapi to Tadri;
- west flowing rivers from Tadri to Kanyakumari; and
- minor rivers draining into Myanmar (Burma) and Bangladesh.

Table 3.1: Major Indian river basins (CWC, 2010).

Serial number	River basins	Catchment area (sq. kms)	Average water resources potential (unit: BCM)	Utilizable surface water resources (unit: BCM)
1	a. Ganga	8,61,452	525.02	250.0
	b. Brahmaputra	1,94,413	537.24	24.0
	c. Barak and others	41,723	48.36	–
2	Indus (up to border)	3,21,289	73.31	46.0
3	Godavari	3,12,812	110.54	76.3
4	Krishna	2,58,948	78.12	58.0
5	Mahanadi	1,41,589	66.88	50.0
6	Narmada	98,796	45.64	34.5

Water bodies: The area of rivers and canals has not been acquired; their total length in the country is approximately 2 lakh km.

Rainfall: India receives a long-term average precipitation of 1,160 mm. Considering nations of equivalent dimension, this represents the highest rainfall of the world (Lal, 2001; Kumar, et al., 2005). Rainfall in India is highly variable. Mousinram in Cherrapunji receives the world's highest rainfall, but it also experiences water shortages during the other seasons. The monsoon period usually lasts for about three to four months. As per CWC 2010, the total rainfall varied from approximately 50 cm in the east and west of Rajasthan to about 379.8 cm in the coastal areas of Karnataka in 2008. The annual precipitation together with snowfall is expected to be 4,000 BCM.

Table 3.2: Volume of rainfall in India as per CWC, 2010.

Rainfall in the years	2001	2002	2003	2004	2005	2006	2007	2008	2009
Total (cm)	111	93	123.4	108.6	121.5	116.1	118.1	111.7	95.4
Total volume (in BCM)	3,648	3,200	4,057	3,570	3,996	3,819	3,882	3,674	3,136

Groundwater: As per the last assessment carried out by Central Groundwater Board (CGWB) in 2004, for groundwater resources 433 BCM was the sum total replenishable groundwater resource per year. Over the past 20 years most of the water for irrigation is drawn is from groundwater and a smaller proportion comes from the canals. 369.6 BCM is available for irrigation whereas 71 BCM is available for industry, household and other purposes (CWC, 2009).

Globally, India leads in groundwater consumption with an annual predicted use of 230 km^3 (World Bank, 2010a). Nearly 60 per cent of agriculture and industry requirements and nearly 80 per cent of household requirement is met through groundwater (World Bank, 2010a). Over exploitation resulted in the increase of pumping depths subsequently raising the cost of pumping groundwater. Widespread scarcity drives the farmer to dig deeper and deeper. Another drawback is the contamination of groundwater with increasing concentration of arsenic, fluoride and iron. Such contaminated water adversely affects the health of the people.

As per the estimated demand of water for 2030, the industrial demand is liable to rise four times to 196 BCM (13 per cent). This augments the growth of overall water demand very nearly to 3 per cent per year (Addams, et al., 2009). Currently India possesses more than 20 million modern water withdrawal provisions. One out of four farmer household has tube well, two out of three cultivators have bought irrigation facilities and are supplied by tube well owners.

According to the remote sensing data obtained from the National Sample Survey's (NSS), as a good deal of, say about 75 to 80 per cent of India's land, receives irrigation supply from the groundwater wells (Shah, 2009). Rodell et al. (2009) states that owing to decline in water table, three states namely Punjab, Haryana and Rajasthan have collectively lost about 109 km³ of water between 2002 and 2008. This data was obtained by using NASA's Gravity Recovery and Climate Experiment (GRACE) satellites.

As given by CWC in 2010, with an estimated availability of 1,588 cubic meter (cu.m)/person/ year, India does not feature as a water scarce country; to a certain extent it can be characterized as a nation under 'water stress'. In this regard UN's categorization of a country based on water supply is worth mentioning. If the annual water supply drops below 1,700 cu m per individual, the nation is said to be under '**water stress**'. If the yearly water supply drops below 1,000 cu m per capita, the nation is said to be under '**water scarcity**', whereas falling below 500 cu m labels the country under '**absolute scarcity**'. The water accessibility per person has gone down from 2,309 cu m in 1991 (Sharma and Bharat, 2009) to 1,588 cu m in 2001 (CWC, 2010). In 2025, per person availability of water can further decline to 1,000 cu m with the current predisposition of population growth. It would then become a 'water scarcity' condition.

Table 3.3: Average availability of water per year per person as per CWC, 2010.

Year	Population (million)	Per capita average annual availability (m³/year)
2001	1,029 (2001 census)	1,816
2011	1,210 (2011 census)	1,545
2025	1,394 (Expected)	1,340
2050	1,640 (Expected)	1,140

According to IPCC 2001, 383, salt water intrusion refers to the replacement of the freshwater with the advancing saltwater owing to its higher density. It generally takes place in the coastal and estuarine areas. Soaring population, exceedingly intense human activities, inappropriate resource use and short of suitable management system are some of the reasons for salt water intrusions. Salinity ingress is a significant crisis in the coastal areas of Tamil Nadu and Saurashtra, whereas inland salinity is a main problem in some parts of Rajasthan, Haryana, Punjab and Gujarat.

Water management in India: Water is in the Schedule VII, Entry 17 in List II and Entry 56 in List I of Indian constitution. The earlier Department of Irrigation was renamed as Ministry of Water Resources in 1985, and assigned the nodal role for conservation, management and development of water as a national resource.

A 'River Board Act' has been endorsed for integrated management of interstate rivers. However, no River Board has been established to integrate planning, development and management of water resources of the river basin. This is due to the lack of consensus among co-basin states. Central organizations such as Brahmaputra Board, Narmada Control Authority, Damodar Valley Corporation, Bhakra Beas Management Board, Betwa River Board, Bansagar Board and

Tungabhadra Board have been established for precise purposes to plan, manage and regulate the water resources in specified river basins.

Under the Environmental Protection Act, 1986, the Central Ground Water Authority (CGWA) was set up on 14 July 1997. CGWA is accountable for measuring the groundwater potential of India through hydrological survey, investigation, assessment and monitoring of the groundwater system.

The National Commission for Integrated Water Resources Development Plan (NCIWRD), founded in September 1999, made various recommendations for drinking, irrigation, industrial, flood control, transfer of surplus water to deficit area etc.

There is a 'National Water Resources Council' for laying down National Water Policy and evolving a consensus on water related issues among states with the Honourable Prime Minister as the chairman and all Chief Ministers of the states as its members. Water is the most crucial element and National Water Policy was adopted in September 1987 and revised in 2002 keeping this in mind. A draft of National Water Policy (2012) was formulated to regard water as a scarce resource and also sustainer of life and ecology.

India implemented the world's biggest government sponsored rural water supply programme. Supply of potable water is essentially the liability of the state governments. Yet the central government supports such programme with the help of totally central sponsored Accelerated Rural Water Supply Programme (ARWSP).

The National Drinking Water Mission (NDWM) was established in 1986 and all initiatives were synchronized to be brought under one umbrella of NDWM. This was done with the idea of achieving the objective of International Drinking Water Supply and Sanitation Decade (IDWSSD) by providing cent per cent coverage for the rural India by March 1990.

Organizations under the ministry of water resources are listed below.

Main offices include:

- Central Water Commission (CWC)
- Central Soil and Material Research Station (CSMRS)

Subsidiary offices are:

- Ganga Flood Control Commission (GFCC)
- Farakka Barrage Project (FBP)
- Central Water and Power Research Station (CWPRS)
- Central Groundwater Board (CGWB), New Delhi (It is the national apex organization that carries out and guides scientific development and management of groundwater resources.)
- Bansagar Control Board
- Sardar Sarovar Construction Advisory Committee (SSCAC)
- Upper Yamuna River Board (UYRB)

Public sector undertaking:

- Water and Power Consultancy Services (India) Limited (WAPCOS)
- National Projects Construction Corporation Limited (NPCC)

Independent bodies:

- National Institute of Hydrology (NIH)
- National Water Development Agency (NWDA)

Statutory bodies:

- Narmada Control Authority (NCA)
- Brahmaputra Board (BB)
- Betwa River Board
- Tungabhadra Board

3.2.5 Floods

Floods usually occur in the aftermath of meteorological events. Floods are an overflow of huge amounts of water onto the normal dry land. Floods occur when the overfull water immerses land leading to deluge. It follows an intense and prolonged rainfall spell, remarkably high coastal and estuarine waters due to storm surges, seiches, etc. Floods are often devastating, damaging and deadly. It kills lot of people, damages houses and crops, and cause extensive destruction. On the flip side, they also act as a natural carrier of rich silts that renew and revitalize the flooded lands apart from helping to maintain the ecological balance, the game of life.

Types of floods: Customarily, floods are of the following types – natural, catastrophic and artificial.

A. **Natural floods:** These are caused naturally by the overspill of the banks of rivers, lakes, ponds, oceans, or by heavy rainfall or heavy shower, hurricanes, cyclones, or tsunamis, etc. So they are either riverine floods (caused by rivers) or estuarine floods (caused by a combination of sea tidal surges and heaves and storm-force winds) or coastal floods (caused by tempests, hurricanes and tsunamis).

   The water level rises due to heavy precipitation and rising above the river banks, the water starts spilling, and causes floods. The water overflows into the neighbouring areas adjacent to the rivers, lakes or dams, causing deluge. Floods follow more likely in the regions that get heavy rainfall and heavy snow melting. The rising temperature due to global warming leads to the snow caps to melt faster adding to the cause of floods, whose severity is yet to conquer critical mass.

B. **Catastrophic floods:** These are the floods that are triggered by momentous and unexpected events, for example rupture of dam. Indirectly, floods are also instigated, by *seismic actions*. Coastal areas are highly prone to flooding during tsunamis. Landmass sinking owing to earthquakes diminishes the altitude of landmass and they become inclined to flood near lakes and rivers. Similarly, the elevating of lake and river beds from seismic activities also leads to spilling over of water bodies.

C. **Artificial flood:** Rarely do floods occur uncharacteristically. These are generally because of the human activities such as:

   i. Blasting leads to landslides in the hill slopes which may result in the unintentional obstructing of rivers and streams eventually bringing about water outpourings.

ii.   Erection of provisional dams impedes the river flow and then may cause an overflow.

iii. Failure of hydraulic control structures lead to calamities like the flouting of an embankment or dike causing huge amounts of water to enter in a secure area.

iv. Mismanagement of hydraulic structures or failure to work as per 'operation rule' and may cause unexpected discharge of enormous amounts of surplus water.

Degree of severity of floods depends on magnitude and depth. There can be thus minor or major flooding.

In a minor flooding, accumulation is not due to overbanking and is generally restricted to the flood plain on arbitrary low-lying regions and depressions. Floodwater is typically of shallow depth and there need not be any appreciable flow.

Major flood is caused by the overflowing of rivers and lakes due to severe breakdowns in barrages, embankments, dams and other protecting constructions or by overwhelming discharge of impounded water of dams and by accumulation of excessive runoff. Floodwaters cover a wide connecting area and spread quickly to neighbouring areas of moderately lower height. It is rather profound in maximum parts of the afflicted areas.

Though floods take 12 to 24 hours to build up after intense downpour there is a specific category which culminates in less than six hours or less. This is referred to as flash floods.

Flash floods occur in mountain terrains which have a steep river gradient. The rapid development of flood occurs because of tremendously small concentration phase of the drainage catchment. This means that rainfall showering on a place in the catchment furthest from the river takes very little time to reach the river channel and turn into a portion of stream flow. Thus, the expanse of stream flow increases profusely and, subsequently, raises the water level. When the capacity of the stream flow is exceedingly large, the conduit overspills and the outcome is a flash flood.

Causes of flood: By and large, floods follow more in the low-lying parts or the areas lying underneath the sea level. Rivers streams sluggishly in these regions, the bulk of water upsurges in these areas, the level of water rises and causes floods.

Anthropogenic causes of floods also include deforestation. As result of wanton logging, soil is easily carried away and deposited at the bottom of rivers and seas, which in turn elevates the water level in the water bodies and cause floods.

Floods may also be caused because of poor dams which are incapable of holding the enormous dimensions of water and end up flooding the neighbouring areas.

Consequences of floods: Floods affect both people and societies and also have socio-economic as well as environmental concerns. The effects of flood, both bad and good, vary significantly depending on the place and degree of flooding and the susceptibility and worth of the natural and induced environments they act on.

A. **Social consequences:**

The instantaneous impacts of flooding include loss of human life, destruction of assets, damage to crops, loss of livestock and worsening of health conditions owing to waterborne ailments. Because communication links, thermal power plants, roadways and bridges are broken and disordered, financial activities are greatly hampered, people are left with no other option than to leave their houses which disrupts their normal lives.

Likewise, disturbance to industrial activities lead to disruption of livelihoods. Damage to infrastructure also causes long-lasting effects to clean water supply, treatment of wastewater, transmission of electricity, conveyance, communication and networking, education and health care system. Loss of employments, decrease in buying power and forfeiture of land valuation in the floodplains makes the communities economically weak. Floods can also traumatize victims and their kin for a long period of time.

B. **Economic consequences:**

Flooding in crucial croplands can cause extensive mutilation of crops and loss of livestock. Loss of crops through rainfall, water logging and deferrals in harvesting are additionally exaggerated damaged infrastructure and transportation systems. This in turn raises the food prices due to deficiencies in food supply. On the contrary, flood can assist crop production by revitalizing water resource and by rejuvenating soil fertility through silt deposition.

Floods cause damage to roadways, railway networks and key transportation hubs, such as shipping ports, harbours can have significant effects on local and national economies. Flooding of cities lead to significant damage to private property, such as homes, businesses and offices.

C. **Environment consequences:**

In many normal systems, floods play an imperative role in upholding key ecosystem functions and biodiversity. They connect the river with the land adjoining it, rejuvenate groundwater, fill marshlands, proliferate the connectivity between water habitats and transport both residue and nutrients around the land and into the marine surroundings. For many species, floods activate breeding events, relocation and distribution. Natural systems are quite resilient to these effects except the big floods.

- Floods damage drainage systems in cities, resulting in sewage tumbling out into water bodies leading to contamination and pollution.

- Structures and constructions can be significantly damaged and even demolished. This leads to catastrophic environmental consequences as various toxic substances such as chemicals, paints, pesticides and other contaminants can be discharged into aquatic ecosystems.

- Floods bring out significant amounts of coastal erosion.

- Areas which were highly regulated by human activity tend to suffer more due to flooding. Floods further degrade the already degraded systems. Vegetation removal, increased waterway size, dams, levee bank and catchment clearing all work to degrade natural landscape and increase the attrition and transfer of both residue and nutrients.

- Though cycling of sediments and nutrients is indispensable to a healthy ecosystem, too much sediment and nutrient entering a waterway may have harmful impacts on downstream aquatic quality. Other adverse effects include habitat loss, dispersal of weed varieties, discharge of pollutants, fish production, loss of wetlands and loss of recreation.

- However, floods also have a positive impact on the environment. Floods often spread residue containing helpful nutrients to topsoil that otherwise would might never have arrived.

Case Study 3.2: The River Kosi flood

Kosi is appropriately also known as the 'Sorrow of Bihar. The root cause of frequent floods resides in the unexpected quantity of silt that the Kosi River transports from the Himalayas downstream to the plains of Bihar. The deposition of silt raises the river bed and the enormous force causes the river to strive out for a new path – in this way. It is estimated that the river Kosi may have relocated westwards by an unbelievable 210 km in the past 250 years. In the 1950s, the Indian government embarked on a programme of constructing embankments on the river to control the intermittent shifting of the Kosi's course. This programme is an effort to provide 'permanent rescue from floods'. The purpose of the embankment is to convert unpredictable river actions into somewhat more predictable and therefore more manageable actions. By constructing levees on both the side of a river and trying to curb it to its channel, its substantial silt and sand load is made to settle down within the embanked region itself, thus raising the river bed and the flood water level. The levees too are raised gradually till a limit is reached beyond which it is no longer possible to do so. The people of the adjoining areas are then at the mercy of a wobbly river with an unsafe flood water level, which could at any time overflow over or make a disastrous crack.

As predicted, the ultimate breach was appalling. The Kosi River course moved more than 120 kilometers eastwards in few weeks. In the absence of the banks, such a dramatic shift would have actually taken many years. With time, a progressively greater amount of means were needed to restore the system stability and the subsequent failure was catastrophic. In reality, the flood controlling programme on the Kosi river has turned out to be something else. This was evident in the 2008 Bihar flood which was one of the most devastating deluge in the history of the state.

3.2.6 Droughts

A drought is a prolong period of desiccated weather, when there is lack of rainfall for weeks, months or even years. Several places of the world that anticipate drought each year, generally experience a dry season and a wet season. People tide over the drought condition by water storage and by propagating drought resistant crops.

Three main drought types are common:

A. Hydrological: Many watersheds experience reduced quantity of accessible water. Lack of water in riverine systems and storage systems can impact hydroelectric power companies, agriculturalists, flora and fauna as well as human communities.

B. **Meteorological drought:** A lack of rainfall is the most frequent character of drought and it is this type of drought that is commonly reported in news and the mass media. Many places around the world have their own climatological definition of drought centered round the normal climatic conditions of that area. Normally a rain fed area when gets a reduced amount of rain than normal is considered to be in a drought.

C. **Agricultural drought:** When moisture in the soil becomes scarce or dehydrates, the agro– industry is in deep distress due to drought. Rainfall deficiencies, fluctuations in evaporation and transpiration, depleted groundwater levels can create a state of stress and difficulties for crops production.

Causes of drought: The cause of droughts is simple to understand, but tough to avoid. Based on the location, failure in crop yield, shortage of food supply, high food prices and loss of lives can occur. One of the most terrifying condition of a drought is the time of commencement. A drought often grows gradually much likely due to other severe weather forms or natural disasters.

Droughts result due to drastic reduction of rainfall over a span of time. Such conditions will lead to the regions drying out all over progressively. Droughts are billion dollar meteorological phenomena and are ranked as one of the topmost three threats to the mankind (along with famine and floods). Sometimes it may take decades for a drought to shape up and foreseeing droughts is even more difficult task. Parts of United States experience drought every year. Droughts are entirely natural, but their desolation can be far-reaching and unembellished. The causes of drought can be climate change, fluctuation in ocean temperatures, variations in the jet stream and modifications in the local landscape etc. Selection of plants is another factor. Certain plants with high transpiration rates are often responsible for lowering the water table, for instance, the plantation of Eucalyptus. The Kolar district of Karnataka may be quite advanced in social forestry but the entire district is at the same time very drought prone for such practices.

Impact of drought:

- There are three focal means how droughts influence lives and communities. Firstly, the economic impacts of drought include heavy damages in agricultural production most of which are distributed to consumers in the practice of higher priced commodity.

- Social impacts are reflected in the quality of human life and public safety. Conflicts are common over water shortage, food shortage and lack of fertile land. Additional social impacts may be rejection of cultures, ethnicities, loss of native land, changes in standard of living and increased chance of health risks due to paucity and living conditions.

- Finally, the environmental impacts of drought include biodiversity loss, emigration, degradation of air and water and more soil attrition.

As a mitigation measure mixed cropping can prove to be useful and reduces the chances of crop failures. Social forestry and wasteland development can also be undertaken, but with utmost care to avoid backfire.

Case Study 3.3: Drip irrigation in Israel

Israel is a country prone to water-scarcity. 78 per cent of its arable land is under agriculture, and half of that land is irrigation dependent. So, water management is of great importance.

As with most water-scarce countries, Israel has witnessed how proficient irrigation methods and planned reuse of water can help to safeguard sustainable, consistent crop production. These days, Israel's integrated water resources management system uses the modern water management know-hows, such as pressure irrigation systems, automatic and controlled mechanization and high quality seeds and plants, such as genetically engineered ones that can grow in less water. Israeli water supervisors inspire water preservation and effective usage by all users – particularly agricultural users in the nation. For example, drip irrigation to reduce evapo-transpiration and conservation of water. Drip irrigation was conceived in Israel in the year 1959 as means of using the scarce resource more competently. Since then, Israeli farmers have sophisticated and mechanized the process, accumulating data on temperature, radiation, moisture and soil water content not only to regulate where water is set free, but also the time and quantity needed to meet a plant's necessity for transpiration. Israel accepts treated wastewater as a treasured water reserve. The country's water supervisors have made treatment of wastewater and its reuse a national mission. The farmers are engaged and encouraged to embrace the utilization of treated wastewater by substituting treated wastewater discharge with their consistent freshwater provisions. Presently, Israel uses treated wastewater to meet about 30 per cent of its agricultural water requirement.

3.2.7 Conflicts over water – international and national

With the threat of water shortages around the world, water seems to turn into the fuel of certain disputes in many parts all over the world. 'Water Wars' are obviously the harsh reality in the world's forthcoming years as the misappropriation and overuse of water continues in some countries that share the same water sources. International law has failed in defending the equitable use of common water supplies in many parts of the world aggravating the problem. The rapid population increase has largely affected the quantity of water easily accessible to most people. Water is becoming a very valuable commodity and the fact that freshwater resources are unevenly distributed is leading to increasing tension and even conflicts.

Potential causes: The cutting off or polluting water sources as a potent weapon, has been used by warring nations since time immemorial. The dispute arises over the authority to regulate water which in turn helps in controlling the economy and welfare of the people. Fights on the grounds of water usage comprise military, manufacturing, engineering, farming, household and political usages. These encounters can be further intensified by using water resource as a powerful weapon and as a political objective. Rapid urbanization has exacerbated the demand for water. Unequal distribution of water generates an inequity amongst nations, mostly in developing countries and also results in intra-state disputes.

3.2.7.1 Regions of conflicts:

A. Middle East

The supply of water can be critical in noticeable watersheds like the Jordan River Basin and the Tigris-Euphrates Basin, particularly when they are being shared amidst multiple nations.

Jordan River basin: Portions of Lebanon, Syria, Israel, Jordan and the West Bank of the Jordan River Basin are primarily parched regions. The Jordan River originating in Lebanon has a total flow of 1,200 million cubic meters annually on an average. Jordan River system comprises of the Jordan and Yarmuk Rivers, which start from Syria. The states relying on Jordan River have groundwater aquifers as a source of water supply. In terms of water use, Jordan stands second after Israel. The amount of water use per person daily is the global least. The water release in to river Jordan fell drastically after the construction of dam on Yarmuk River. The Mountain Aquifer beneath the West Bank is a basis of dispute between Israel and Palestine. The conflict is due to the Israel dominion over groundwater which leads to Israel cutting off the Palestine water supplies. Palestinians are charged three times more than the Israelis for water coming from underneath the West Bank.

Tigris-Euphrates basin: The Rivers Tigris and Euphrates originate in Turkey and their watershed supplies a much bigger area than the Jordan River basin. These rivers are important to Syria and Iraq with 85 per cent and 100 per cent of water supply respectively. The Turks and Kurds depend less on these rivers yet they plan to use more of these waters for irrigation; whereas Iraq and Syria profoundly depend on these rivers for water supply. All the countries have raised dams for farming, hydropower production and industry development. Turkey takes the major share and even prevents downstream flow.

The Turks and the Kurds (residing in southeastern parts of Turkey) are not much reliant on the rivers, yet they have strategies for irrigation schemes to elevate their water use on both rivers. The parts Iraq and Syria located near the lower courses of the rivers are mostly dependent on these two rivers for their water resource. Dams installed along the river course by Turkey have barred some amount of the water from flowing to the lower course to these warmer, arid countries. All three countries (mainly Turkey) have built dams across the rivers for purposes of cultivation, hydroelectric power generation and development of industries. Aggressions between Syria and Iraq intensified because of the filling of Lake Assad by Syria, leading to the decline of downstream flow. Iraq began blaming Syria for not releasing adequate water. Among all three countries, the conflict on water resource is equated with their national security.

B. Africa

A large proportion of the African countries are dependent on weather conditions for agriculture to livelihood and fighting over water resource has become a part and parcel of life. The rivers Nile, Volta, Zambezi and Niger are shared by many countries. Conflicts rage on amongst people who have been displaced due to dam construction and water privatization leading to unequal distribution of water supply between the neighbouring countries.

Nile River basin: Nile, the world's longest river, has been the main water source sustaining life in Egypt and Sudan. The Nile and its tributaries along with rivers and lakes across in nine African countries before it opens into the Mediterranean Sea. The Egyptians have used military force to ascertain their control over Nile, as it is their only water source. Countries like Sudan, Ethiopia and Uganda had set up several river projects to boost their annual water extraction, thus, affecting the control of Egypt over Nile. In few instances national governments of Uganda, Sudan and Egypt agreed on a deal to share the waters of the Nile River. Such initiatives can hopefully avoid water shortage and conflict.

C. Asia

The conflict over water sharing stretches from upstream to downstream in the South Asian countries. Water distribution is a raging political issue in the Southeast and Central Asia as they have little understanding of the concept of water sharing and the increasing population further act as a catalyst for such conflicts.

Water shortages in India and China are a socio-economic menace. It is escalating the tension between Nepal, Bangladesh, Pakistan and India. Freshwater resources may seem to be plenty but the distribution is inadequate in the parched regions. Even then clean water available to masses is meager due to massive erosion. The river Ganga, regarded as holy and sacred, is also significant economically and is ground for long contention between India and Bangladesh. A common river system, created by the fusion of the Bhagirathi River and Alakananda River is shared by the nations. But due to increasing demands for industrial, domestic and irrigation use in Kolkata, water conflicts between India and Bangladesh escalated further. Dispute escalated due to the construction of the Farakka barrage in 1962 to increase the navigability of the Kolkata port. To solve the dispute, in 1986, a thirty year long negotiation was signed on the basis of 1985 accord.

Similar is the case of Indus river basin which is a cause of conflict between India and Pakistan. The river and its tributaries straddling across 1,800 miles compose one of the biggest irrigation canal systems in the world. This river basin supplies water to northwestern part of India and Pakistan. Indus Water Agreement was signed in 1960, negotiated by World Bank for over 12 years. Indus, Jhelum and Chenab were allocated to Pakistan while Ravi, Sutlej and Beas to India.

Box 3.1: Major interstate water disputes in India

- **Krishna-Godavari** river water dispute amongst Maharashtra, Karnataka, Andhra Pradesh, Madhya Pradesh and Orissa was resolved by the tribunal passing the verdict in 1976 on equitable allocation of water
- **Yamuna** river water among Delhi, Haryana, Uttar Pradesh
- **Cauvery water dispute between Tamil Nadu and Karnataka**: The watershed is shared between Tamil Nadu and Karnataka. Tamil Nadu is in control over Bhavana and Moyar, tributaries of the Cauvery River. Tamil Nadu located downstream wanted Karnataka to regulate its use in the upstream. The dispute started since 1974 when the 50 year old

agreement between the Madras Presidency and the Mysore State expired on the ground of water supply to Tamil Nadu by Karnataka. Supreme Court reassigned a tribunal in 1991 which ordered Karnataka to release 205 TMC of water to the Mettur Dam of Tamil Nadu monthly from the Cauvery. Karnataka refused to abide stating that a release of more than 100 TMC of water will leave the people of Karnataka in distress. Cauvery basin is known for Samba paddy cultivation in winter and Kurvai paddy in summer, both of which requires huge amount of water. Suitable crop selection, optimizing water use, water distribution control and water tariff were suggested as possible solution to the problem.

- **Ravi-Beas** River waters dispute amongst Haryana, Jammu Kashmir, Rajasthan and Punjab

Though India has the right to impound the waters across the rivers lying within its territory, the treaty demands India to use such waters for non-consumptive purpose without altering its flow and quality. Reservoirs and canals constructed to meet hydropower and irrigation requirements had desiccated regions of the Indus River basin. Such water projects have also resulted in the displacement of inhabitants and to the degradation of the Indus ecosystem.

3.2.7.2 Inter-state river water disputes

Causes: River flows irrespective of the political boundaries. Moreover uneven distribution, variation in rainfall, flood, droughts and regionalization of the national water policy led to water disputes amongst the states of India. The Inter-state River Water Dispute Act (ISRWD), 1956 provides for the establishment of tribunals which give judgment where negotiations have failed.

Case Study 3.4: The selenium curse in Punjab

Hoshiarpur and Nawanshahar, the two districts of Punjab, have been badly hit by a sudden sickness. The residents were suffering from alopecia, swell in the joint, types of distortions and other health ailments. Crop yields and livestock, too, were affected. The root cause of this problem was found to be selenium, a non-metallic element close to sulphur group. The trace mineral is toxic to both health and the environment on account of its excessive existence in that area. In two districts eight villages are known to be affected. In the beginning only 8 hectares (ha) of land was affected. But this rose up to 50 ha. Great amounts of Selenium have been identified in 80 ha of cultivated land in Hoshiarpur. In the village Simbli, 40 ha out of 541 ha of agricultural land is contaminated.

The statistical data of Nazarpur and Panam villages reveal 25 ha out of 103 ha and 15 ha out of 628 ha of contamination respectively.

It is assumed that for the last few years, selenium has been carried through floodwaters from the Shivalik Mountain to these cultivable lands. One of the reasons that seem to have worsened the problem is the harvesting pattern of the region. Some crops such as maize, sorghum and oats are safe, while other crops are known to attract high amounts of selenium. The level of Selenium in the affected area's groundwater have been reported to be as 0.07 parts per million (ppm) while selenium levels in drinking water as per Bureau of Indian Standards is 0.01mg.

For soil, the safe limit is 0.5 ppm. Any

amount above this would result in the plants containing more than 5 ppm of the element. Above 100 ppm, the plants show white patches. Apart from damaging human health, selenium has wrought havoc on the villages' yield and cattle. Plants fail, to ripe properly and the harvest decreased by 25–30 per cent. Milk production, also has seen a fall. The Punjab government's reaction has been highly inadequate. Subsequent to survey in three villages, it was observed that 362 out of 3,958 people displayed symptoms of selenium poisoning. The initiatives taken by the authorities also reflect their callous approach. One such action was the delivery of vitamin B tablets to all the inhabitants. In another initiative, agrarians have been requested to spread one ton of gypsum per hectare every alternate year. Paradoxically, the inhabitants were not consulted in the preparation of the plan. The villagers have clarified that they will only settle for alternate expanses of land and financial reimbursement. The afflicted people should be supplied with fresh land The areas with high selenium should be hedged, utilized for social forestry purposes and these can substantiate as effective long-term actions.

3.2.8 Dams and their impacts

A *dam* is a man-made blockade habitually built across a river to hold water. So, dams primarily serve the purpose of impounding water. Dam stores water which can later be distributed uniformly. Floodgates or levees are used to regulate the water flow in specific areas. Hydel powers are often used along with dams to generate power.

Majority of the dams are constructed for multiple purposes. The dams help to serve an array of household and economic benefits arising out of one investment. Additionally the local benefit lies in the opportunity to employment during the many years of reservoir erecting. As the world's population continues to grow, the requirement for more dam seems inevitable, particularly in the third world countries along with the enormous stretch of desert regions of the world. Though the basic advantage from the dams and reservoir remains water supply, other important uses and benefits include:

- irrigation for agriculture;
- flood control;
- hydropower;
- inland navigation; and
- recreation.

While dams provide significant benefits to the society, their impacts on the surrounding areas, especially the long term impacts on the surrounding ecosystems is only begging to be manifest themselves. They give rise to:

- displacement of indigenous people;
- resettlement and relocation issues;

- deforestation;
- soil erosion;
- loss of other flora and fauna;
- changes in spawning behaviour of fishes;
- reservoirs can be the breeding grounds of vectors;
- exertion of enormous pressure induces reservoir induced seismicity (RIS);
- environmental concerns;
- sedimentation and siltation issues; and
- safety aspects

Dams scenario: The most primitive dam recognized is the Jawa Dam in Jordan situated 100 kilometers northeast of Amman. Constructed in 1957, the Hirakud Dam on River Mahanadi is almost 26 km long and is the world's longest manmade dam. A 55 km Hirakud Reservoir extends beyond the dam. Presently the Three Gorges Dam on River Yangtze in Hubei is the largest and is highly controversial project designed to supply water even during the dry periods. It requires relocation of more than 10 lakh people. As a consequence of such project many historically significant places became vulnerable to flooding.

Till September 2013, as per the National Register of Large Dams in India, there is a list of 5,187 dams of which 4,839 were completed and 348 is under construction. Maharashtra has the maximum number of 1,845 dams in India.

Box 3.2: Few important dams ('Temples of resurgent India')

> Tehri dam is a multipurpose rock and earth fill embankment dam on the River Bhagirathi in Uttarakhand. Bhakra dam on River Sutlej in Himachal Pradesh is the Asia's second highest dam next to Tehri Dam. Its reservoir is known as Gobind Sagar. Nagarjunasagar dam on River Krishna in Andhra Pradesh is one of the earliest multipurpose dams in India.

3.3 Land Resources

The total surface area of the earth is 510,000,000,000,000 sq. meters. About 70 per cent of the total area is occupied by oceans, rocks, ice, etc. 8 per cent forms the tundra, continental shelves, reef, estuaries, lakes, rivers, and so on. The remaining can be used for living and growing food. Earth's land cover has been extensively modified over 50 per cent by humans and this modification is unsustainable. Rising population and the increasing demands forms the most important factor. Changes in albedo and evapo-transpiration due to deforestation and overgrazing resulted in reduced and irregular rainfall. Conversion of land for agriculture and the present practices leads to extinction of the local biota and pollution of air, water and land. Construction of transit system fragments the general habitat. One may lose as much as 15,000 km^2 of land between 2000–30 due to increase in urbanization. All these puts people in the cross border of survival.

3.3.1 Land as a resource

Land is certainly one of the predetermined natural resources. Man has been knowledgeable and equipped enough to familiarize his or her lifestyle in diverse ecosystems, yet, he cannot survive happily for instance on polar ice caps, under the ocean, or in space in the foreseeable future. Mankind does require land for construction of houses, to cultivate crops, to maintain pastures for domestic animals, to develop industries and factories to afford goods and services and to establish townships and cities to sustain industries. Equally important is to preserve and protect the forest, grasslands, marshland, mountains, beaches, etc. and to uphold our precious biological diversity. These landforms are also resource-generating areas on which communities depend. Many traditional farming societies had their traditional means of defending their areas which furnish resources. Take the example of the 'sacred groves' of the Western Ghats. The natives seek the permission of the grove to fell a tree or even extracting resource accompanied by simple rituals. Hence, a sensible use of land requires cautious scheduling, because land becomes a renewable resource only if it is used wisely.

3.3.2 Land degradation, water logging and salination

Land degradation is either a man made or natural method which harmfully affects the land preventing it from functioning efficiently in an ecosystem, such as accepting, storage and reprocessing water, energy and nutrients. Land is degraded in the name of developmental activities which comprise building dams, waterways, rails, vehicles and businesses and by the pollutants they spew. Desertification refers to the process of transformation of average soil in to a desert which happens because of deforestation, soil attrition, grazing and shifting. Livestock leads to more of grazing and thus to soil erosion. The erosion may be checked by rotation of crops, mulching which results in the decrease in the vaporization and increase in the absorption; presence of appropriate passage for carrying water and the seeding of specific crops controls erosion. The planting of trees also checks wearing away of soil. Slash and burn agriculture is relatively common in the ethnic regions especially in the tropical and subtropical parts of Africa and Asia. The trees are felled and set on fire and the crops are grown on the residual ash. This process is also referred to as *jhuming* since it is practiced in the *jhum* forests in the north east India. The roots and stubbles of plants and grasses help to bind the soil. Rigorous irrigation leads to saturated and saline soil, which is unfit for growing crops; land is also transformed into a non-renewable resource when highly noxious industrialized, chemicals and nuclear wastes are disposed on it. Land degradation nullifies out the advantages provided by better quality crop yields and decreased population growth. Land degradation causes are mostly anthropogenic and linked to agriculture.

The most important causes are:

- land clearance and deforestation;
- agricultural quarrying of soil nutrients;
- urban transformation;
- irrigation; and
- pollution and contamination.

The major pressures are:

- enhanced erosion by wind and water;
- deletion of nutrients;
- acidity intensification;
- increased salination;
- alkalization;
- degradation of soil structure; and
- loss of organic substance.

Stark land degradation upsets a substantial proportion of arable lands, reducing the wealth and slowing the pace of economic development of countries. The inter-relationship between a tarnished environment and paucity is straight and close. As the land turns less productive, food safety is compromised and struggle for declining resources upsurges – the seeds of probable conflict are sown. Diversity of species decreases and often vanishes as more land is cleaned and transformed for farming. Thus, a descending eco-social helix is produced with depletion of land and nutrients during management of land based on unsustainability, ultimately leading to perpetual impairment.

3.3.3 Man-induced landslides

A landslide is the downward movement of a mass of rock down the gradient due to the force of gravitation. Classification of landslides can be based on basis of the substances involved (rock, debris, earth, mud) and the nature of drive (fall, collapse, snow slip, slide, flow, spread). Landslides can be triggered when the pressure on the gradient material is in excess of the material's shear strength.

Shallow landslides generally comprise the soil coat only while deep-seated landslides involve rock layer at increased depth. Magnitude and extent of landslide usually ranges from tens of cubic meters to several cubic kilometers for massive landslides. The speed may vary from a few centimeters per year for sluggish landslides to tens of kilometers per hour for fast exceedingly vicious landslides. The phenomenon can be elicited by physical processes such as substantial or protracted rainfall, tremors, volcanoes, quick snow melting, slope undercutting and destabilizing by rivers or wave action and permafrost defrosting. Human activities such as slope diggings, construction of roadways and houses, open-pit mining and mining, deforestation, quick reservoir drawdown, methods of irrigation, vibrations during blasting, water seepage, etc., can also trigger such process.

3.4 Forest Resources

The word 'forest' originated from 'fores' meaning a vast stretch of land shielded by trees. Forests are vital renewable natural resource, ruled by trees and the species composition differing in various parts of the world. Forests are of incredible significance to the humans and closely related with the culture and civilization. They make up the most significant components of nature. They add significantly to the economic development of the country. They are different in different regions and climatic conditions globally – the Amazon rainforests, the North American temperate forests, and northern European boreal forests are some examples.

3.4.1 Structural organization

The species arrangement in a forest is often distinctive to a forest, with forests differing in species richness and species diversity. Forests are also subjected to change due to succession; during which species composition changes. There are some basic structural characteristics that most of the forests share – features that enable one to understand both forests and the wildlife that dwell in them.

Mature forests comprise the following distinct vertical layers:

- Forest floor is often carpeted with decaying leaves, twigs, fallen trees, moss, and other detritus. Fungi, insects, bacteria, and earthworms are among the many decomposers.

- Herb layer is controlled by grasses, ferns, wildflowers, and other ground cover. Vegetation often gets little light due to thick canopies; shade tolerant species are leading ones.

- Shrub layer is characterized by woody vegetation, comparatively near the ground surface. Undergrowth and brambles may develop where light can pass through.

- Understory consists of immature and small trees that is shorter in height but supports a variety of animals.

- Canopy is the layer where the pinnacles of maximum of the forest's trees form a dense layer.

- Emergents are trees whose crests emerge above the canopy.

- These layers provide a montage of habitats and enable organisms to settle into various pockets of habitat of a forest.

3.4.2 Current status – global and indian scenario

Globally, the forest cover is 31 per cent of the total land surface, little more than 4 billion hectares (FAO, 2010b). This is lower than the pre-industrial record of 5.9 billion hectares. According to data obtained from the UN FAO, deforestation reached its peak in the 1990s; on an average the world lost 16 million hectares of forest per year. Simultaneously, forest area also expanded in some parts, either naturally or through planting, bringing down the global net loss to 8.3 million hectares annually. The global net forest loss was reported to be 5.2 million hectares per year in between 2000 and 2010. (FAO, 2010b)

Table 3.4: World forest cover change over 20 years, in million hectares, 1990–2010
from UNFAO, Forest Resources Assessment, 2010

Regions	1990	2000	2010
Africa	749	709	674
Asia	576	570	593
Europe	989	998	1005
North and Central America	708	705	705
Oceania	199	198	191
South America	946	904	864
World	**4,168**	**4,085**	**4,033**

The data on deforestation evidently shows Brazil, Indonesia, Russia, Mexico, Papua New Guinea, Peru, USA, Bolivia, Sudan and Nigeria to have a high annual rate of deforestation. Maplecroft's Deforestation Index in *Climate Change and Environment Risk Atlas*, 2012, shows nine countries in the extreme risk categories. They are Nigeria, Indonesia, North Korea, Bolivia, Papua New Guinea, Democratic Republic of Congo, Nicaragua, Brazil and Cambodia.

Forest degradation results mainly from selective logging, road construction, mining and climate change. Replacing natural forests with monoculture greatly reduces biodiversity and affects the health of the forests. Planted forests currently cover around 264 million hectares, making nearly 7 per cent of total forest cover with an ability of producing 1.2 billion m³ of industrial wood annually. China leads the growth in planted forests. Brazil comprises 13 per cent while Russia's 20 per cent of the world forest. Bolivia and Venezuela, have also felled vast stretches of trees, thus making South America the region with the leading forest loss with about 40 million hectares of forest being lost between 2000 and 2010.

Africa also lost 34 million hectares from 2000 to 2010 mainly due to firewood harvesting and charcoal production. Each year more than 300,000 hectares of forests are cleared in the African nations like Nigeria, Zimbabwe, Tanzania and Democratic Republic of Congo respectively. In the 1990s, Australia switched from a net forest increase to a net forest loss in the following decade. It's continual drought from 2002 to 2010 was devastating for its forests cover: the drought limited forest regrowth along with increasing fire risk. Wildfires supplemented by drought and soaring temperatures burnt huge expanses of forest cover in Australia. The United States supplemented a net increase of 77 lakh hectares of trees between 1990 and 2010, roughly 380,000 hectares per year. US still contribute to deforestation as a major importer of forest products with some $20 billion worth in 2011. Imports of forest products totaled $110 billion in 2011 in Europe. Spain, Italy, France, Norway and Sweden contributed a net 16 million hectares of forested area in between 1990 to 2010.

According to the India State of Forest Report, 2011, by the MoEF, the forest and tree cover of India is 78.29 million ha, which is 23.81 per cent of the country's geographical area. 4,498.73 million cu. m of forests and trees are inside the recorded forest area while 1,548.42 million cu. m are located outside the recorded forests.

Madhya Pradesh presently has the largest forest cover of 77,700 sq. km followed by Arunachal Pradesh with 67, 410 sq. km. Considering the ratio between the forest cover and state area, Mizoram has the highest (90.68 per cent) and Lakshadweep with 84.56 per cent forest cover. The nation aims in achieving forest cover for 33 per cent of country's geographical area by the year 2012.

3.4.3 Deforestation

'Deforestation is the process whereby natural forests are cleared through logging and/or burning, either to use the timber or to replace the area for alternative uses such as housing, agriculture, mining, etc.' Conversion of forests for palm, and soy plantations, building of roads and other infrastructure, forest degradation from wild fires, illegal timbering and logging, harvesting of fuel wood and climate change, are all the causes of deforestation. Estimate says 12–15 million hectares of forest are lost per year.

Causes of deforestation are listed below:

- Increase in the population: Every human being requires facilities such as space, buildings and houses to live. This leads to a large scale cutting of trees.
- Fuels and other resources: Forest is a source of fuel and provides firewood. Wood and trees can be used as a fuel mainly in the rural areas where other resources are unavailable.
- Unsustainable agriculture: Agriculture practices like shifting agriculture, livestock rearing, grazing, crop–plantations, etc. Subsistence farming is accountable for 48 per cent of deforestation whereas commercial agriculture is responsible for 32 per cent of deforestation. (UNFCCC) Slash–and–burn practice entails cutting and clearing a stretch of trees, burning them, cultivating and growing crops till the soil becomes degraded and moving out to a new patch of land.
- Commercial logging and timber harvesting contributes to 14 per cent of deforestation. Making of furniture, boxes, plywood, matchboxes, packing material, chapter, railway sleepers etc.
- Developmental activities like mining industries, human settlement, tourism, infrastructure development (roads, railways dams), hydel power projects.
- Hunting and collecting for food support, cultural uses medicinal plants.
- Trade in food commodities, traditional medicines.
- Accidental activities like trapping, hooking, netting, poisoning.
- Natural hazards like volcanoes, drought, and floods and wildfires.
- Pollution of land and water, global warming.
- Use of forest areas for defense purpose.
- Present–day deforestation may also occur due to corruption of government institutions, faulty policies and implementation.

Effects of deforestation are:

- On removal of forest cover the wildlife is deprived of habitat and becomes defenseless to hunting.
- Deforestation results in 15 per cent of global greenhouse gas emissions. Trees help in carbon offsetting.
- Erratic behaviour of the climate.
- Local climate becomes much dry due to reduction in the loss of water by evaporation, water cycle is disrupted.
- Deforestation speed up rates of soil erosion, by increasing runoff; soil becomes infertile.
- In places with slopes, it results in landslides.
- Disturbance of livelihoods to millions like small-scale agriculture, hunting and gathering, rubber cultivation ultimately leading to social problems.

Case Study 3.5: Sariska Tiger Reserve Case

The Sariska Tiger Reserve stretching about 866 sq. km was acknowledged as a protected area in 1978. It is situated in the Alwar district, in the Thanagazi block of Rajasthan. Before 1947, the forests within the reserve area, a part of the erstwhile Alwar state, were primarily a shooting reserve for the Maharajas. In 1955, this forest area was declared as a state reserve. Further in 1975, to ensure effectual conservation, some neighbouring forest areas adjacent to the reserve were incorporated in the Sariska Tiger Reserve to formally become protected areas.

The protected area is divided into three zones: core zone I, II and III. Nearly 3,000 villagers stay both inside and at the border zone of the tiger reserve. They exert their native rights over use of forest resources which thus interfere with the Sariska Tiger Reserve Authorities, management policies. As a result, the relationship between these people and the authority is bad. The forest staff members usually come from the cities and towns. They humiliate the indigenous people as the lowest level of the social strata. Such outlook and treatment led to the following consequences:

- decline in native people's admittance;
- displacement of local communities; and
- lack of basic social services.

The officials imposed restrictions on the collection of forest products and timber. Though the villagers were allowed to collect firewood, they were not allowed to use forest wood for constructing houses. This was hard for the villagers to acknowledge. However, some sort of permissions was given for cutting of wood from more common and regularly found local trees; that too during a particular time of the year. This gave rise to illegal woodcutting which is a basis of conflict between the adjoining villages sharing common forest resources. When villagers were displaced they were left as paupers, their cultures linking them with biodiversity disappeared.

According to the Wildlife Protection Act, 1972, a reserve, cannot have any permanent human population within its boundary. For Sariska Tiger Reserve, such a rule means dislocation of 17 villages. Virtually all the people are under threat of displacement. Such a plan of relocation seems to echo the objectives of the National Rehabilitation and Relocation Policy, 2007. In reality, such policy aims in reducing displacement and support non-displacing and least displacing alternatives; to guarantee satisfactory and speedy rehabilitation package and also to integrate rehabilitation package into growth, preparation and implementation process. The provisions of the latest policy were highly condemned because it failed to meet the imaginary promises at a realistic level.

Mining leases were allowed in the tiger reserve area, which is a limestone rich area. Mining practices are mercilessly destroying the Aravali mountain ranges, contaminating the environment and endangering the wildlife. Nearly 45 organizations along with Tarun Bhagat Sangh protested under 'Save Aravalis Campaign'. A PIL was filed by TBS in the Supreme Court of India. The court issued an order to discontinue mining. Fresh mining licenses were granted in 2010. This was followed by another petition in the National Green Tribunal. NGT has banned all types of mining and quarrying activities in 84 mines within 1 km radius of the Sariska Tiger Reserve. As per 2006 notification, the 881 sq km area of the core and the buffer zone around Sariska was destined for sustainable land usage. This is meant to afford habitation for the tiger and other wildlife, besides fostering coexistence with mankind.

3.4.4 Timber extraction

Timber logging refers to cutting, skidding, on-site processing, loading of logs onto trucks and transporting to a place outside the forest possibly a sawmill or a lumber yard. Illegal logging of timber refers to the harvest, transportation, purchase or sale of timber with violation of the existing laws. Timber harvesting also involves activities such as building access roads and establishing plantations. Environmental impacts coupled with harvesting trees may lead to undesirable changes to ecosystems, habitat fragmentation, localized fauna loss and alterations in natural flora ecology.

Indian forests produce about 5,000 species of wood, of which about 450 are commercially valuable. Hard woods include important species such as teak, ironwood, mahogany, oak, deodar, silver fir, walnut etc.

3.4.5 Significance of forests

- It is already proven that forest and environment are both interrelated and interactive in such a way that they are inseparable.

- The main product produced by the forest is wood. Wood is the raw material in pulp, paper, board, plywood, and furniture items.

- Forest serves as a source of canes, gum, resins, dyes, tannins lac, fish etc.

- Forests influence temperature, humidity and precipitation. Forests helps in the formation of soil by affecting its composition, conformation, chemical properties and water contents and play an significant role in biogeochemical cycles of water, carbon, nitrogen, sulphur, oxygen, phosphorus, etc.

- Forests help in controlling flood conditions by intervening surface run-off intrusion, vaporization.

- It provides suitable habitats for a number of important organisms – plants, animal and microbes.

- Forests also have aesthetic and tourist values.

- Altering of forestry can ominously affect the environment and economics on which the performance of nature depends.

3.4.6 Forest conservation and management

People can protect the forests for future generation by minimizing the consumption of products made out of paper and wood, use of recycled paper, reclaiming wood, procuring wood lawfully from sustainable plantations, searching for firewood substitutes and shifting towards smaller families.

Forest certification: In order to better manage the forest resources, forest certification is a must as it helps people to first identify and list the resources before concerted efforts are made towards their scientific exploitation and use.

Prevention of Illegal logging: Felling of trees, in the Indian context would mean the right of the sons of the soil and their right to use the resources of the forests for their living. As the forest dwellers, calling on their traditional knowledge are extremely sustainable in their ways, they hardly pose a threat to the ecological balances. However, commercial logging – the use of (read destruction) of the forests for commercial gain is a matter of grave concern though because of the existing legal frameworks they do not pose the kind of threat that they do in the Amazons and in some other forest areas.

Trade reforms: It is demand that creates supplies in its wake and illegal felling of trees is no exception. That is why the need to have trade reforms – wherein the world may live sustainably in tandem with the flourishing forests. While some headway has been made, a lot more remains to be done. The woods indeed are lonely, dark and deep and there is literally 'miles to go'.

Declaration of protected areas: This is another step in the right direction. The mute forests have to be protected as they have no one to talk of their plight, to espouse their cause. To create a protected area is to develop of a global network of ecologically representative, effectively managed and sustainably financed patch. Forests as protected areas, offer better living options to the numerous species which rely on them.

Case Study 3.6: Vedanta mining

On 7 June 2003, a MoU was signed between Vedanta and Orissa Government with an objective to install an alumina refinery. The parent body of Vedanta is Sterlite Industries Ltd. the venture of ₹4,000 crores with an initial capacity of 1 MTPA was upgraded to 6 MTPA. The land required was 721.323 ha for bauxite mining and 723.343 ha of land for refinery. Of this 672.018 ha and 58.943 ha of forest land was required for alumina mining and refinery respectively. For this, the Niyamgiri hills were reserved and allocated to Vedanta.

Niyamgiri involved the livelihood and rights of two primitive tribes – the Kutia Kondhs and the Dongaria Kondhs. This region is also significant from an environmental point of view because of its rich biodiversity and wildlife.

The project was granted preliminary 'in principle' forest and environmental clearance (Stage 1), by the MoEF. There was strong criticism against Vedanta and the state government for infringement of forest and environment legislations and local people's privileges. The head of the expert committee constituted by the MoEF in this matter documented gross defilements of the Forest Rights Act, 2006 both on the part of state government and Vedanta. Even the Forest Advisory Committee (FAC), a legislative committee allied to the MoEF, suggested withdrawing all clearances and authorizations conferred to Vedanta group for quarrying in Niyamgiri on the ground of Forest Rights Act violation as well as violation of forest conservation and management and environment protection laws.

In the above case, the MoEF pulled out forest clearance and authorization for bauxite mining and halted the expansion of refinery. The mandate in the Vedanta case states: 'Upholding the recommendations of FAC, I the Stage 2 forest clearance for Orissa Mining Corporation and Sterlite bauxite mining project on Niyamgiri hills… stand rejected. The primary responsibility of the ministry is to enforce the laws that have been passed in Parliament. For the MoEF, this means enforcing the Forest Conservation Act, 1980, the Environment Protection Act, 1986, the Forest Rights Act, 2006 and other laws. It is in this spirit that the decision has been taken.'

On 1 April 2011, the government of Orissa filed petition in the Supreme Court challenging decision of rejecting environmental approval to the Vedanta Group's bauxite quarrying venture proposed in the Niyamgiri hills that was worth $1.7 billion. The state government lodged a complaint through Orissa Mining Corporation (OMC) against the MoEF's order, on 24 August, 2010, on the ground of denial of project clearance. It was stated that the centre's decision was a violation of the earlier order by the Supreme Court to grant green signal to the project and opposed Centre's order was unlawful and irrational.

3.5 Food Resources

3.5.1 The world food problem and food insecurity

Green Revolution is running out of steam. Rising population, increasing urbanization, the glitches of ecological destruction and resistance to pesticide accompanying agriculture, uncertainties about the influence of GATT on prices of foodstuff and food service, suspicions about the sustainability of intensive farming and systems of irrigation and an deceptive decrease in the level of increase of harvests of the important staple foods, are altogether the essential factors to the food problem. With seven billion plus hungry mouths to feed, the need of the hour is to immediately increase food production manifold and minimize post production losses drastically.

It is projected that the world population will cross 9 billion by 2050 and the demand for agricultural products too will become more than double. The farming systems are already extended to their limit by restrictions on water availability, unpredictable patterns of weather and unstable price, thereby increasing the chance of production shortfalls. Adding up to is the higher frequency of natural disasters as well as other challenges due to climate change, which will even intimidate the resilience of the production systems and small farmers.

Hence, a move towards sustainable development remains no longer an option. One can definitely not ignore the existence of tradeoffs between ecological sustainability and production of food; yet it is imperative to develop a reasonable solution to lessen the pressure that has already strained the foundation of natural resource and local livelihoods. Newer practice of farming may be used combined with various goals. For instance, sizable gain in production, income increase of smallholder and sustainability are needed to offer adaptive buffers. This calls for considerable amount of stakeholders' collaboration along the agricultural value chain including nations, governments, companies, multilateral and civil society organizations, agriculturists, consumers and entrepreneurs.

3.5.1.1 Causes of the world food problem

Table 3.5: Causes of food problems in developed and developing countries

Problems pertaining to the developing world	Problems pertaining to the industrialized world
1. Underdevelopment	1. Too much use of natural resources
2. Too much population growth	2. Pollution
3. Short of economic incentives farmers using inappropriate methods and labouring on land they may lose or can never hope to own	3. Ineffective animal-protein diets
	4. Insufficient research in science and technology
4. Parents ignorant of basic nutrition for their children	5. Too much government bureaucracy
	6. Loss of farmland to competing uses
5. Inadequate government awareness to the rural sector	

3.5.1.2 Problems linking both industrial and developing worlds

- Inappropriate role of multinational corporations.
- Lack of proper planning.
- Politics of food aid and nutrition education.
- Unequal access to resources.

- Inadequate transfer of research and technology.
- Insufficient food support.
- Inappropriate technological research.
- Insufficient emphasis on agricultural development for self–sufficiency.

World food insecurity: According to Food and Agriculture Organization (FAO) in 2012, the number of chronic undernourished people in the world is 870 million out of a total of 7 billion plus people, which means one in every eight people suffer from undernourishment. About 852 million of these are from developing countries comprising 15 per cent of their population and some 16 million people from the developed countries. Food insecurity is quite obviously related to the hunger and malnutrition. World hunger refers to the state of food scarcity whereas malnutrition indicates lack of nutritional requirements necessary for healthy living.

Food insecurity is referred to as 'a condition in which people lack basic food intake to provide them with the energy and nutrients for fully productive lives.'

According to World Food Summit, 1996, in Rome defined food security as 'when all people at all times have access to sufficient, safe, nutritious food to maintain a healthy and active live.'

Food security has three facets:

- Food availability – total food production along with food imports and buffer.
- Food accessibility – food to be within reach of every individual; individuals should be able to afford adequate safe and healthy food.
- Food absorption – depending on potable water, environmental hygiene, heath care and education.

Food security is related to:

- increase in crop production;
- prudent management of natural resources, such as land, water, forest, etc.; and
- ensuring environmental protection

Trends of food insecurity:

- In Asia Pacific, the state of undernourished people decreased around 30 per cent, from 739 to 563 million.
- In Latin America and Caribbean, it went down from 65 million in 1990–92 to 49 million in 2010–12.
- In Africa, hunger grew from 175 million to 239 million.
- Developed regions also witnessed hungry rise from 13 million in 2004–06 to 16 million in 2010–2012.

The new estimates projects that the progress in reducing hunger during the past two decades is better than what presumed; so it might be possible to reach the target of Millennium Development Goals (MDG) by 2015. The International Food Policy Research Institute (IFPRI) has been releasing Global Hunger Index for the eighth year. It presents a multidimensional measure of national, regional and global hunger. It warns 19 countries to have alarming hunger. The UN's World Food Programme (WFP) drew attention towards several global hunger hotspots, such as, Eritrea, Rwanda, Afghanistan, Bangladesh, Chad, Democratic People's Republic of Korea, Somalia, Zimbabwe and Sudan.

3.5.2 Food security in India: Food security mission and National Food Security Bill

As per 2013, India ranks 63 in Global Hunger Index showing a marginal improvement of the score from 22.9 in 2012 to 21.3 in 2013. M. S. Swaminathan Research Foundation and World Food Programme have published an 'Atlas of the Sustainability of Food Security' in 2004 to present a composite index of food security in India as well as its ability to sustain in the future. The food demand is influenced by population growth, income, urbanization, food prices, etc.

The state that has high unexploited resources and stable crop production has higher rank. The first six ranks are occupied by Arunachal Pradesh, MP, Goa, Karnataka, Gujarat and Andhra Pradesh. In Bihar and UP, population stabilization will be crucial to sustainable food security.

A resolution was passed by the National Development Council (NDC) in 2007 to establish Food Security Mission with a policy to increase the rice production by 10 million tons, wheat by 8 million tonnes and pulses by 2 million tonnes by the end of the Eleventh Five Year Plan.

The National Food Security Bill was introduced in 2011 and cleared in both the Lok Sabha and Rajya Sabha by 2013 to address the subject of food security in a comprehensive manner. The Indian Ministry of Agriculture's Commission on Agricultural Costs and Prices (CACP) referred this bill to be the world's biggest experiment on distribution of subsidized food on rights based approach. The important features are:

- Targeted Public Distribution System (TPDS) includes provision for the state government to provide priority households with 5 kg of food grain per month for each person at a subsidized rate. As per National Sample Survey, 5 per cent of the total Indian population are without two square meals a day and are incapable of buying even at BPL rates. Antyodya Anna Yojana was started by the government specifically for these people. The bill is modified to provide household with 35 kg of food grains per month as allotted by the central government to the states.

- Every pregnant and lactating mother is to be provided with free meal during pregnancy and six months after the birth of the child by the local anganwadis together with a maternity benefit of not less than ₹ 6,000.

- Local anganwadis are to provide food to the children between six months to six years as well as malnourished children. Children between 6 to 14 years are to be given one mid-day meal in the school except for school holidays.

- Food security allowance is to be given to people who have not received the benefits as per above mentioned scheme.

- The responsibility of identifying eligible households is left to the states according to their own criteria or they may use Social Economic and Caste Census Data.

- The eldest female member who is of 18 years or above will be the head of the household in absence of whom the eldest male member stands as head.

- There shall be a redressal mechanism at state and district level along with call center and helpline number.

- Eligible beneficiaries to get uniform subsidized prices of ₹ 3, ₹ 2 and ₹ 1 for rice, wheat and coarse grains respectively for the first three years of implementation.

- In case of non-supply, central governments will be providing funds to overcome short supply. There shall be a transparency of records like disclosure, social audits and vigilance commission.

India will spend 20 billion dollars annually on food security as compared to the 400 billion dollars by rich developed nations subsidize their farmers. Agreement on Agriculture (AoA) restricts India from exceeding 'market distorting subsidies' that it gives to its farmers as the farmers livelihood is at stake. After this bill India will have to defend its right to subsidize and have to seek exemptions to uphold its sovereign duties. The only sustainable way to reconcile with the demands is to continuously increase the crop productivity and maintain high stockpiles in good condition; this in turn implies more irrigation investment and creation of efficient supply chains. There is a problem of storage. At present the Food Corporation of India (FCI) and the Central Warehousing Corporation (CWC) have the capacity of storing 87 million tonnes of grain. The new measure will additionally cost ₹ 27,000 crores annually to the exchequer whereas the government puts it as ₹ 21,000 crores by means of subsidy. The question arises whether the government burdened with food, fuel and fertilizer subsidy can withstand an additional programme in a situation where the existing programme, Mahatma Gandhi Rural Employment Guarantee Scheme, has been drilling a hole in the nation's fund? India was been spending ₹ 1.16 lakh crores on various schemes like food subsidy, mid-day meal, Integrated Child Development Scheme, maternity entitlements, etc. this will worsen the fiscal deficit situation and moreover India's trade deficit will hit very badly as this programme will require 70–80 million tonnes of more food annually.

The less dependence on monsoons and improvement in agrarian infrastructure is precarious to curb import distortions during the drought years. To compare, the food grain production reduced from 259.29 million tonnes in 2011–12 to 250 million tonnes in 2012–13 due to pitiable rainfall. The challenge ahead of people is defending their national obligations along with giving credibility to mitigate policy externalities.

If the government is able to deliver this programme rightfully, it could mean an end to malnutrition and poverty. With its implementation this programme which will cover 67 per cent of India's population, the people have a gleaming hope that their struggle for right to food may come to an end. It is also necessary for India to win the bargain at WTO, keeping in mind that the food security bill is a reality and procurement from small farmers will have to be increased from the current 45 million tonnes to 70 million tonnes. This means the minimum support prices will go up and it will not be possible for India to stick to AoA norms.

3.5.3 Changes in land use by agriculture and grazing

Exploding population, degradation due to over exploitation and the random unscientific use of chemical fertilizers have combined to severely affect the productivity of land. Overgrazing without being supplemented with proper nutrients too, has aggravated the problem. This in turn leads to production becoming un-remunerative which in turn leads to agricultural land becoming barren and then pushed into alternative use.

3.5.4 Modern agriculture and its impacts, fertilizer/pesticide problems

The enormous increase in productivity has been the outcome of multiple reasons, such as the use of chemical fertilizers, and insecticides, introduction of agricultural equipment and machineries, production and implementation of high yielding varieties (HYV) such as IR8 in India, and improved scientific information about agricultural management practices. With intensive agriculture, agrarians have turn into skilled agriculturalists being able of develop high yields employing fewer manual labourers, less land. But intensive agriculture is not an unadulterated blessing. The impacts on environment have amplified drastically, including prospective dilapidation of land and water resources essential to both agricultural production and human wellbeing. The problems can better be understood by tracking the advancement of the farming history of India.

Impact of the use of fertilizers on environment: Nutrients are lost from cultivated fields through overflow, drainage, or as add-on to the soil particles that are subjected to erosion. The quantity of nutrient disappeared rest on the type of soil and its organic content, the climatic conditions, the gradient of the land, and depth to groundwater, as well as on the extent and kind of fertilizer and irrigation applied. Most commonly used fertilizers are Diammonium phosphate (DAP), Muriate of Potash (MOP), Urea and Rock phosphate. Such fertilizers provide the key elements, nitrogen, phosphorus and potassium, needed for plant growth. This practice is leading to **eutrophication**. The principal constituents in fertilizers are nitrogen, phosphorus, and potassium (NPK). Nitrogen being highly soluble gets readily lost as nitrates and such leaching from the crop lands increase their concentrations in the groundwater much above the prescribed limits for drinking water.

Increased usage of these chemicals causes salinity and affects the water table. With time the yields per hectare is going down. Soil erosion degrades the soil and nutrient leaching is escalating. As a response, agro-forestry that combines the benefit of forest ecosystem and farming practices can be implemented. Organic nourishments, such as compost, improve the soil fertility by nourishing microbes, decreasing wearing away and keeping soil moist. Grass trimmings, dispersed on the meadow, is a type of fertilizer that delivers phosphorus, nitrogen and potassium and can be obtained for free by trimming the grassland. In the context of pesticide and fertilizer pollution, bio-fertilizer seems to be a good alternative. The concept of bio-fertilizer such as vermicompost is gaining momentum. Bio-fertilizers have the ability to lessen the environmental costs accrued to utilization of fertilizer and pesticides.

Impact of pesticides and herbicides on environment: The tendency of intensive farming has increased the possibility of damage by the insects and other pests and the diseases caused by them. Chemicals like dichlorodiphenyltrichloroethane (DDT), dieldrin, aldrin, heptachlor, endosulphan and herbicides like 2,4,5,T has been out rightly used for controlling pests, diseases and weeds. These chemicals are non-biodegradable or degrade very slowly. Easily they enter the food chains and impair the ecosystem. They become more and more concentrated at higher trophic levels causing **bioaccumulation** and **biomagnifications**. Some pesticides are mutagens and affect the human DNA (Deoxyribonucleic acid). As per reports DDT is lethal even in small doses; it is also carcinogenic to human tissues. Following Second World War, DDT and other organ chlorines were used as pesticides to protect crops and also to control malaria. The harmful effects of DDT was brought to the world attention by Rachel Carson's 'Silent Spring' in 1962. These organic compositions are highly persistent in nature and accrue in tissues of animals, contaminating water, killing aquatic organisms and fishes and lead to drop in several bird populations. Since 1973, DDT was forbidden for agronomic use in the USA. Because of their solubility in water they percolate in underlying groundwater. Because of the rise in the price of fertilizers and pesticides, farmers are opting new ways that are less reliant on chemical based farming. So many farmers have reduced their overall use of chemical compounds and many others are practicing agriculture without using synthetic chemicals. Emphasis has been given to finding out ways and means of restoring quality of environment. One such stepping stone is the beginning of integrated pest management (IPM) that involves a blending of approaches which can lessen adverse environmental impacts.

Both pesticides and herbicides exert two important problems by virtue of its persistence and accumulation in the environment, thereby affecting innumerable plants and animals and also affecting the human health directly or indirectly. In addition to, they contaminate air, water and soil along with exerting detrimental effects on aquatic plants, water and mankind.

A drawback of pesticide is that it not only kills the target pests but also an array of beneficial organisms and the predators of the pests. After the influence of the pesticide fades out, the pest species is likely to recuperate more quickly than its hunters because of alterations in the accessible nutrients source. Additional drawback to the increasing pesticide use is the development of resistance in pest species. The pests that escape and survive carry on breeding, progressively generating a population with enhanced pesticide tolerance to the applied compounds.

3.5.5 Water logging and salinity

Excessive irrigation on poorly drained soils is water logging. This occurs in poorly drained soils where water can't penetrate deeply. Water logging also damages soil structure.

Salinity refers to the existence of solvable salts in the soil and water, comprising both surface water and groundwater. The salts can be of sodium chloride, calcium, magnesium, carbonate, bicarbonate and sulphates. Certain soils and sites are naturally saline, for instance inland salt lakes and lands molded from salty parent substances. This is referred to as natural salinity or primary salinity. On the other hand, secondary salinity is caused due to anthropogenic activities such as clearance of land and excess irrigation. All these activities cause the groundwater to rise to the surface, melting salts and then dumping them in the land. Salinity, though a normal procedure; yet, land use practices, such as clearance and irrigation, have significantly aggravated the problem to severe extent.

3.6 Mineral Resources

Minerals are amongst the valuable non-renewable natural resources. A mineral is naturally formed substance with definite chemical composition and properties in the earth for many years. It has been used as a raw material for household purpose and industries for economic gains. It can be metals, non-metals, rare earths, building materials and gems etc.

3.6.1 Use and exploitation

The minerals to be of use needs to go through the following steps – searching for the existence of ore, assessing the location, size shape quality and economic viability, developing the method of extraction of minerals. Currently, the issues of mine safety and its environmental impacts are given topmost priority.

Table 3.6: Some important minerals and their uses

	Minerals	Uses
1	Aluminium	Automobiles, aircraft, bottling ,canning, electrical industries
2	Barium	Paints, rubber, plastic, glass
3	Beryllium	Nuclear missiles
4	Bismuth	Alloys, ceramics, paints and has replaced lead in many cases
5	Boron	Glass, ceramics, fiber glass, medicines, abrasives
6	Cadmium	Alloys, pigments, plastics, batteries
7	Chromium	Alloy like steel
8	Copper	Electric cables, equipments and utensils, alloys
9	Gypsum	Cement manufacture
10	Manganese	Iron and steel production
11	Phosphate rock	Fertilizers
12	Quartz	Pressure gauges, oscillators, glass, paints and precision instruments
13	Sandstone	Construction
14	Silica	Pharmaceuticals, paper, computer chips, semiconductors, glass
15	Uranium	Nuclear reactors and weapons
16	Zinc	Alloy, medicine, rubber, cosmetics

3.6.2 Mining practices

There are two common mining practices – surface and underground mining. **Surface mining** is becoming more popular nowadays and is used for mining of most minerals. Surface mining starts by removing the vegetation cover, dirt and bedrocks to reach the deposits. It can be in form of open pit mining, strip mining, quarrying etc. **Underground mining** involves digging tunnels into the earth's crust to reach the ore. The ore and waste rocks are brought to the surface for processing and disposal respectively. It can be drift mining, slope mining or shaft mining. Mining of rare earths is usually done by **in-situ leaching** method.

3.6.3 Mineral resources in India

India is one of the most naturally gifted nations in the world in terms of mineral wealth. India is abode to abundant minerals which is key to the country's development. India constitutes one-quarter of the world's known mineral reserves. Out of 87 minerals that India produces, 4 are fuels, 10 metals, 47 non-metals, 3 atomic and 23 minor minerals.

Table 3.7: Important minerals and their location in India

Nos.	Minerals	Place
1	Aluminium	Kerala
2	Asbestos	Karnataka, Rajasthan
3	Bauxite	Jharkhand (Ranchi, Palamau), Maharashtra(Belgaum, Thane), Chhattisgarh (Mandya, Bilaspur)
4	Cement	MP (Katni, Jabbalpur), Rajasthan (Lakheri), Andhra Pradesh (Guntur)
5	Coal	West Bengal (Raniganj), Jharkhand (Jharia, Bokaro)
6	Diamond	MP (Panna)
7	Gold	Karnataka (Kolar)
8	Gypsum	Rajasthan (Bikaner, Jodhpur), Tamil Nadu, Gujarat, Himachal Pradesh
9	Iron ore	Jharkhand (Singbhum), Orissa (Keonjhar, Mayurbhanj), Chattisgarh
10	Petroleum	Assam (Digboi, Patharia)
11	Zinc	Rajasthan (Udaipur)

3.6.4 Environmental and social impacts of mining

A. Impact on water resources

Possibly the most important impact of a mining is its effects on quality of water and its availability. Often mining activities lead to the contamination of groundwater which becomes not only un-potable but also becomes hazardous to the marine and the other species dependent on it for their existence.

i. Acid mine drainage and contaminant leaching

Metals such as gold, lead, copper, zinc, silver, iron and molybdenum frequently occur in rock as sulfides. After extraction the sulphides on exposure to water and air form sulfuric acid. The neighbouring harmful metals often dissolve in this acidic water form acid mine drainage which might runoff into rivers or streams or seep into the groundwater. Mined materials after extraction and exposure to oxygen and water can form acids if profuse iron sulfide minerals are present. This acid will dissolve metals and other contaminants to form an acidic solution extremely hazardous to the entire ecosystem. Arsenic, selenium, can leach even in absence of acidic conditions. Increased cyanide and nitrogen levels are also found in the waters at mine sites. If mine debris is acid-generating, the impacts can be severe with a pH value of 4 or even lower – akin to battery acid. Aquatic flora and fauna cannot survive in such water bodies.

Acid mine drainage (AMD) also dissolutes copper, aluminum, cadmium, lead and mercury, from the adjacent rock. Of all these, iron specially, may coat the base of the stream with an red/orange/yellow coloured slime called 'yellow boy'.

ii. **Erosion of soils and mine wastes into surface waters**

Since a large area of land is disturbed during mining the possibility of soil and sediment erosion is enormous. Erosion can be a major concern at hard rock mining sites. Erosion may cause significant loading of sediments to nearby water bodies, this may lead to lowering of pH, metal loading to the surface waters and contamination of groundwater to an extent that there can be loss of habitat and vegetation.

iii. **Impact of tailing impoundments**

Tailings are high-volume waste often comprising arsenic, lead, cadmium, chromium, nickel, and cyanide etc. all of which are highly toxic. The effects of wet tailings, waste rocks, mounding dump on quality of water can be ruthless. Impacts include groundwater contamination if the bottommost level is not lined with an impermeable material.

iv **Impact of mine dewatering**

If water table is higher than the underground mining, mine water is formed which must be pumped out of the mine. This is no problem when the mining operations are on. Once over, mine water pumping also stops; it gets accumulated in the rock crevices, tunnels and open pits thereby its unending exposure to the environment.

B. **Impact of mining projects on air quality**

Airborne emissions are evident during each step of mining. Mining operations generate large amounts of small dimension particles that simply disperse in the air. The biggest sources of atmospheric pollution in mining operations include discharges from mobile sources due to the use of transport and gaseous emissions from fuel combustion during explosions and processing of mineral such as heavy metals (mercury, arsenic), carbon monoxide, sulfur dioxide and nitrogen oxides. It also includes fugitive emissions produced during activities such as constructions, blasting, tailing piles etc. Noise pollution associated with mining may be of various types from the use of various types of equipments.

C. **Impact of mining projects on wildlife**

i. **Habitat loss**

Survival of species is dependent on soil condition, neighbouring weather, altitude, and other characteristics of the neighbouring locale. Mining affects wildlife directly and indirectly. The impacts arise primarily from upsetting, removing and redistributing the soil. Impacts are referred to as short-term if they are restricted to the mine site; or may be long-term. Mobile wildlife usually leaves the place. But sedentary animals are brutally affected.

If water bodies are drained with mine wastes, fish, aquatic invertebrates and amphibians too are affected. Predators fall short of food supply due to disappearance of habitat.

ii. **Habitat fragmentation**

If large stretches of land are divided up into smaller and smaller patches, the migration routes are cut off, dispersal of species from one patch to another becomes unfeasible. Isolation may lead to inbreeding and finally the extinction of the species.

D. Impact of mining projects on soil quality

Since mining contaminates soil over a large extent adjoining farming activities are particularly affected by exposing the surrounding landscape to erosion of exposed soils and sediment loading to surface waters and drainage system. In addition, spills and leaks can contaminate soil.

E. Impact of mining projects on social values

The social impacts of mining are controversial. Mining creates wealth and brings affluence and prosperity. Mining projects may open up opportunities for jobs, roads, schools, and increase the load of goods and services in distant areas but the benefits and costs are unequally shared. Mining projects can direct to social stress and violent conflict. Mining communities often feel helpless when their relationship with their authorities are poor, or when ecological impacts of mining affect the survival and employment of local people. Mineral projects must guarantee the basic human rights of the individual and communities such as right to utilize land; the right to safe, environment and livelihood; the right to be free from threats and aggression; and the right to be reasonably compensated for dislocation, occupation and locale loss.

i. Human displacement and resettlement

Sometimes the entire community may be uprooted and required to migrate elsewhere. In doing so, communities not only lose their land, but also their livelihoods. Such forced dislocation is disastrous for native communities who have very strong cultural and spiritual bonds to their ancestral place.

ii. Impact of migration

Another type of problem is the inflow of people in and around mining sites; invasion of newcomers can have a deep impact on the original population, leading to disputes over land and the other benefits.

iii. Lost access to clean water

Impacts on water quality and mass are among the most debatable aspects of mining projects with increase in human loads in a given area.

iv. Impact on livelihoods

Habitually mining degrades soil and water quality, wildlife and forest resources, which are precarious to the sustenance of native people. Without contamination being controlled, these costs will be transferred to other economic activities like farming, aquaculture etc. The situation worsens if mining occurs in the areas inhabited by marginalized or excluded tribes. In theory it does ensure all types of basic rights to the indigenous people. However, exploitation and a wanton disregard of the law of the land is often the rule as opposed to being the exception.

v. Impact on public health

Toxic substances and trashes in atmosphere, lithosphere and hydrosphere exert severe, deleterious impacts on public health like anthracosis, silicosis, coal miner's disease and increase of death rate.

vi. Impact on cultural and aesthetic resources

Mining ventures may affect sanctified lands, ancient structures, and natural benchmarks. Potential impacts include:

- total demolition of the resource through superficial disruption or diggings;
- dilapidation due to physical or hydrological changes, or from soil attrition, sedimentation;
- unlawful removal of relics or sabotage as a consequence of more access to areas previously unreachable; and
- pictorial impressions due to clearance of foliage, large quarries, dirt and dust as well as the existence of large-scale machineries, and automobiles.

F. Climate change considerations

Large-scale mining ventures have the probability to alter overall carbon equilibrium and are effective with impressions that have far reaching outcomes.

Case Study 3.7: Mining in Goa

Mining operation in Goa is concentrated in four *talukas* – Bicholim, Salcete, Sanguem and Quepem. Numbers of mines are ever increasing. Till 2002–03 around 400 mining leases had been granted covering 30,325 ha of land. This comprises about 8 per cent of the total geographical area of the state. Since 2007, 120 mining proposal have been submitted to MoEF for EC and almost half of the projects were recommended by the MoEF for clearance. The rest were not rejected but are pending. The projects that caught special attention are the Mahavir Bauxite Mine and Devapan Dongar Iron and Manganese mine. For the former, the lease area fell within CRZ and only iron ore mining was given permission in Devapan Dongar mine. Since 2007 the mining proposals submitted for clearance covers a total area of 9,404 ha. If this area is added to the existing mining area the total area becomes 10.5 per cent area of the state. On an average 1 tonne of iron extraction generates 2.5 to 3 tonnes of waste. The quality of EIA's are very poor and most of them are copy paste of each other. If all the projects get clearance 55 tonnes of waste would be generated annually. If the projects were given clearance, it will significantly deplete the forest cover and engulf a lot of agricultural areas. The region experiences heavy rainfall and this will cause overflow of mining waste and thus causing extensive damage to the neighbouring cropland and water bodies. The environmental impacts include deforestation, land degradation, groundwater pollution, surface water pollution, damage to sea beaches and dust.

A battle is ranging in the Supreme Court over illegal mining. It is stated in the Shah

Commission's report the unrestricted and unregulated mining in Goa resulted in a loss of ₹ 350 billion. Iron ore mining stands next to tourism in terms of economic contribution. Goa is the largest exporter of iron ore in India. The closure of US $4 billion Goan iron ore affects one in every three Goans. The state has lost 35 per cent of its revenue since the ban. Nearly 93,000 workers have lost their jobs. Thousands of trucks and barges lie idle since the ban. More than a decade of ruthless and senseless mining has brutalized the natural environment. Environmental issues are right in its way and even there are also pressing human needs. People have taken loans for various purposes centered around mining, but, now they are unable to repay it back. The banks are facing non-performing assets as there is no business. The appraisal for these projects should be done scrupulously and mining could then start with sufficient checks and balances.

Case Study 3.8: Uranium mining project in Nalgonda

Activists have protested the plan of Uranium Corporation of India Limited's (UCIL's) to start mining and processing at Nalgonda in Andhra Pradesh but have earned favour from a sect of localities. The company is facing similar protests in Domiasiat in Meghalaya. This ₹ 500 crore project, is only 6 km away from the tiger Sanctuary. The proposed mining area is a part of Yellapuram Reserve Forest in Rajeev Gandhi Tiger Sanctuary. It is located just 4 km above the Nagarjuna Sagar Reservoir. The processing plant was to be sited at Mallapuram village 4 km away from Akampally Reservoir. This is meant to supply wate to both Hyderabad and Secunderabad along with 600 villages. The reservoir supplies both drinking and irrigation water. UCIL plans to mine 1,250 tonnes of Uranium per day for the next 20 years.

Both the uranium isotopes and their decay products are toxic uranium miners show five times more incidence of cancer. The half-life period of Uranium is approximately 80,000 years and is extremely hazardous. It affects the brain, kidneys, liver etc. the dust damages the lung causing silicosis and tissue scarring.

There were reports of congenital defects, mental retardation, crippling, polydactyl and syndactyly from the radioactive mining areas such as Jaduguda in Jharkhand. The project would affect more than 1 lakh of people. People will be exposed to radon gas that will be released from the mines and this can spread over 150 acres. Above this uranium mining will add to their woes. It will also affect the wildlife of the area. The area is already suffering from fluoride contamination in water. The ore dumping seeps in to the soil and ground water. There is a possibility of contaminating the Nagarjuna Sagar Reservoir and the downstream of Krishna River. The leachates from the processing plant are likely to contaminate both the soil and groundwater of that locality. The tailings could be blown away by the wind and further rain water could wash it away to distant water systems.

UCIL defends by saying that the granite bedrock in the mining area will be able to block the seepage into Nagarjuna Sagar Reservoir and there is no possibility of contamination in Akampally as the processing plant and uranium tailing ponds would be at a lesser height. Moreover the tailing ponds would be lined with impermeable sheets to block the seepage. UCIL claim to be functioning at zero discharge concepts. UCIL also claims of earmarking ₹ 80.93 crores for the purpose of afforestation in order to compensate for the loss of forest cover that might happen.

3.7 Role of an Individual in Conservation of Natural Resources

Global warming and climate change are issues that concern mankind and considering the huge impact that they have on the ecosystem, man cannot shrug it off as a problem that will be resolved by governments and communities alone. The time has come for every individual to become conscious and to lead a life that is not only just and equitable, but also sustainable. If one were to go to the crux of the problem, one would see that the entire problem emanates from man's greed and his desire for instant gratification. While a simple sustainable lifestyle is enough to meet his needs and ensure that he maintain his normal lifestyle, mankind's greed and desire to exploit nature to meet his pleasures and the urge to satiate his perceived needs on an immediate basis is what has brought the world to this state. If humans do not mend their ways immediately, if they do not embrace sustainability and continue on the path of conspicuous consumption and mindless exploitation, they will not only fail to reverse the trends that have been set on motion, but will also aggravate the problem and put mankind in serious jeopardy.

3.8 Equitable Use of Resources for Sustainable Lifestyles

One of the biggest problems that face mankind stems from the fact that the Earth's resources are scattered unevenly. It is also a historical truth that these resources, in spite of their locations, have been traditionally accessed and utilized by certain groups of people. Nations that have exploited them for their economic growth and development often at the expense of the people/nations to whom these resources belonged.

The tea that the British brought to India came from China, paid for by opium and cotton from India which was forced to destroy its own weaving ecosystem to make way for the mill produced cloth from England. Similarly, coal, iron ore, gold , silver, agricultural commodities among others were mined and produced in one part of the world to fuel and feed the development pangs of people in other parts. This exploitation and the ability to subjugate, pillage and develop also instilled in the victors as sense of never ending déjà vu, which in simple words ensured that they neither bothered about over exploitation, nor about the equitable usage.

The result is for all to see. Earth's finite resources have already been stretched to such a limit that today man is faced with serious threats – peak oil is already a thing of the past and the prospect of feeding seven billion plus hungry mouths and providing them with safe drinking water and sanitation facilities is a task that borders on the impossibility. The desire for a bigger, faster, higher lifestyle has already seriously compromised the ecosystem – the air that people breathe is toxic, the food they eat, the water they drink is polluted and the world that they live in is dangerously heating up with consequences that are too frightening even to imagine.

And in this backdrop, the **North-South divide** – gap between the rich and the poor is becoming wider by the hour. While the rich continue to consume with wanton disregard of the planet (often frittering away valuable resources for frivolous pursuits), the poor are finding it increasingly difficult to eke out a living in an environment that is becoming increasingly hostile.

The answer is not merely sustainable development as is being made out by the advanced west. The answer is in a holistic approach, in a completely reformed thought process in which those who have sinned against nature accept the fact and consciously become responsible Earthlings. The **nexus** approach refers to systemic thinking and a pursuit of integrated and inclusive solutions to guide people in decision-making regarding resource use, development and progress along with minimizing externalities. In fact, **environmental nexus approach** is having a growing recognition as the best approach towards sustainability given the complex linkages and feedbacks involved. The problems that mankind now face are not demarcated by or restricted within man made geographical boundaries – they are common problems that affect all of us. Let's say, climate change is certainly to increase pressure on the resources and thereby add vulnerability to the inhabitants and ecosystems in marginal regions. A nexus approach to assist climate mitigation world involve Reducing Emission from Deforestation and Forest Degradation (REDD+) or Carbon Capture and Storage (CCS), becoming more water smart, less energy intensive, etc.

Figure 3.1: Environmental nexus approach to show linkage between environment, poverty and sustainable development

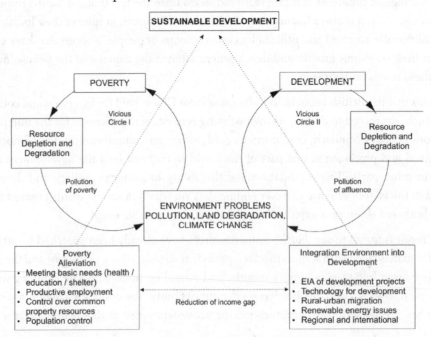

The response too has to be a uniform one, one in which mankind as one (and to the last man standing) provide the response. And this response will necessarily begin when there is a just and equitable system of resource use. Air, water, sunlight and other natural resources belong to everyone. The resources like fossil fuels and minerals too should be approached with such a vision so that they may be harnessed for the common goal, as opposed to being tools in the hands of transnationals with the motive of profiteering.

Summary

- Natural resources are associated with problems of growing population, rapid industrialization, uneven distribution of resource, over exploitation of natural resources, etc.

- Water is another most important natural resource required for survival and also for growth and development. Less than 1 per cent of all freshwater resources is the resource available for use. There is a direct correlation between development and water use. Overuse of water is also polluting the water sources in an alarming rate.

- Water security is referred to as the reliable availability of an acceptable quantity and quality of water for health, livelihoods and production, coupled with an acceptable level of water-related risks. Globally, India is the largest consumer of groundwater with an estimated usage of 230 km^3 per year.

- In the present day food insecurity is of pivotal importance and related to the hunger and malnutrition. The enormous increase in crop productivity was the outcome of multiple reasons, such as the use of chemical fertilizers, and insecticides, introduction of agricultural equipment and machineries, production and implementation of HYVs. Exploding population, over exploitation and the random unscientific practice severely affects the productivity of land. This leads to eutrophication, water logging and salinity.

- Minerals being one of the most precious non-renewable natural resources; mining practices affects on quality of water and its availability.

- One of the biggest problems that face mankind is the unequal distribution of resources. It is also a historical truth that these resources have been traditionally been accessed and utilized by certain groups of people. The resources like fossil fuels and minerals too should be approached with such a vision so that they may be harnessed for the common goal, as opposed to being tools in the hands of trans-nationals looking aged by the motive of profiteering.

Exercise

MCQs

Encircle the right option:

1. Match the following: Ans:

    i.   Krishna–Godavari   a.   Tamil Nadu and Karnataka   A. ic, iid, iiib, iva

    ii.   Yamuna river   b.   Haryana, Jammu Kashmir,

                        Rajasthan and Punjab   B. ia, iid, iiib, ivc

    iii. Ravi–Beas   c.   Maharashtra, Karnataka,

                        AP, MP and Orissa   C. id, iic, iiib, iva

    iv. Cauvery water   d.   Delhi, Haryana, Uttar Pradesh   D. ic, iia, iiib, ivd

2. The amount of water in solid form is approximately:

    A.   70 per cent   B.   50 per cent

    C.   90 per cent   D.   20 per cent

3. The amount of freshwater is approximately:

    A.   5 per cent   B.   50 per cent

    C.   2.5 per cent   D.   0.25 per cent

4. Most of the freshwater is used in:

    A.   Industry   B.   Domestic

    C.   Irrigation   D.   Mining

5. Which of these is the main reason for Narayani River Pollution?

    A.   Brewery   B.   Paper factory

    C.   Thermal plant   D.   A and B

6. Match the following river basin to the countries of dispute: Ans:

    i.   Jordan River Basin   a.   Turkey, Iraq and Syria   A. ic, iia, iiid, ivb

    ii. Tigris–Euphrates Basin   b.   Uganda, Sudan and Egypt   B. ia, iid, iiib, ivc

    iii. Indus River Basin   c.   Jordan and Israel   C. ic, iia, iiie, ivb

    iv. Nile River Basin   d.   India and Pakistan   D. ic, iia, iiid, ivb

                        e.   India and Bangladesh

7. Arrange the river basins in ascending order in terms of area:

    A.   Ganga, Indus, Godavari, Krishna   B.   Ganga, Godavari, Krishna, Indus

    C.   Indus, Ganga, Godavari, Krishna   D.   Indus, Ganga, Cauvery, Godavari

8. Which among these is the largest river basin in India?

    A.   Indus   B.   Ganga

    C.   Godavari   D.   Cauvery

9. Which of these can be described as the condition of India's water resource?

    A. Water abundance   B. Water stressed
    C. Water scarcity   D. Absolute scarcity

10. The subject of water is in _____ list of Indian constitution:

    A. List I   B. List II
    C. List III   D. List I and II

Fill in the blanks.

1. _____ coal has the maximum carbon content.

2. _____ is the country to have the highest per capita use of water.

3. Plant roots are deprived of _____ is in water logged soil.

4. A material that can be transformed into valuable and useful product is called _____.

5. Prospecting, exploration, development and exploitation are the steps involved in _____.

State whether the statements are true or false.

1. Amount of groundwater is less than freshwater. (T/F)

2. Organophosphates are more toxic pesticides as compared to organochlorines. (T/F)

3. Desertification is only caused by overgrazing. (T/F)

4. SYL canal dispute is between Punjab and Haryana. (T/F)

5. Wood pulp is used for making chapter. (T/F)

Short questions.

1. What is a natural resource?

2. What is the present scenario of global water resource?

3. What are the global initiatives towards water security?

4. What are the main sources of water in India?

5. Name the major river basins in India.

6. Which is the largest river basin in India?

7. What is salt water intrusion?

8. Briefly describe the unpredictability of rain in India.

9. How does UN categorize water stress, water scarce and absolute scare regions?

10. What are the main uses of ground water?

Essay type questions.

1. What are the issues related to the exploitation of natural resource?

2. Suggest a few measures for conservation of natural resource.

3. What are the impacts of overexploitation of water?

4. With the help of a suitable example explain the impacts of industrial water pollution on environment.

5. Discuss the role of the government of India towards water security.

Ecosystems 4

Learning objectives

- *To define and build up the concept of ecology and ecosystem.*
- *To be acquainted with abiotic and biotic components of the ecosystem.*
- *To know about their structural arrangement like food chain, food webs and ecological pyramids.*
- *To find out how these structures are related to functions like energy flow and nutrient recycling.*
- *To know about the process of transition in ecosystem (ecological succession).*
- *To classify ecosystem.*
- *To know the characteristic features and distribution of biota in various ecosystems and the reason behind such distribution.*
- *To develop a comprehensive understanding about the various species interactions.*
- *To enumerate the ecosystem services that people receive.*

4.1 Definition and Concept of Ecology

The word ecology has emanated to the forefront of human consciousness. It is the branch of science that incorporates the basic conceptions of civilization. The word was introduced by Hans Reiter, a German scientist, by combining two words 'Oikos' (house) and 'logos' (to study). Thus, it deals with the study of animal and plant inter-relationship and also their relation to the environment. Ernst Haeckel was the first to define ecology as the systematic study of organisms in their natural home. It was Odum (1963) who defined ecology as 'the study of structure and function of nature'. It is the study of any or all inter-relationships between organisms in relation to their surrounding or environment. Ecology can be studied as plant and animal ecology. Often the tern 'bioecology' is used when equal importance is given to the study of flora and fauna.

4.1.1 Ecology as a discipline

Ecology was regarded as an offshoot of biology as it deals with all forms of life and explores the riddles at different hierarchies of living systems. But presently, it is a multidisciplinary science that requires input from the various streams like physics, chemistry, biology, geography, geology, economics, international relations, policies, legislation and uses mathematical, statistical and computational tools to explain naturally observed phenomena with relevance to their varied consequences.

4.1.2 Basic ecological principles

There are four fundamental doctrines of ecology.

- It encompasses a web of inter-relations of its constituents.

- These interconnected networks form an arrangement that involves both abiotic (non-living) and biotic (living) components.

- The system of networks exhibits energy flow and flow of nutrients.

- Solar energy is the prime regulator in the nutrient and energy flow.

From these main beliefs arise the fundamental ecological ideas and concepts like population, community, habitat, territory and ecosystem. Thus, through ecology one strives for understanding and ways to explain the various life processes, adaptation, species distribution and abundance, the various stages in the development of communities and the material and energy flow through ecosystem.

An ecosystem maintains a biological equilibrium or balance of nature between its various components referred to as **homeostasis.** It is the dynamic equilibrium among the biotic members of an ecosystem with changing environmental conditions of water, air, wind, soil and nutrients with time. This ecological balance is maintained by a number of factors such as carrying capacity, ability to recycle waste, etc.; the balance is maintained through intricate and interlocking sequences. One component of the ecosystem keeps a check on the other component through feedback systems and this may be either positive or negative. For example, increase in plant population lead to increase the population of herbivores. Such a control is known as positive feedback. In another case when the increased population of insectivorous animal, predates on herbivorous insects, it is called negative feedback mechanism.

4.1.3 Scope of ecology

Ecology plays significant role in conservation and management of land and its resources like minerals, forests etc.; livestock, water and pisciculture; urbanization and town planning; population control; risk assessment and disaster management.

Ecology can also be studied as:

- **Autecology**: when individual organisms with their environment are studied, for example, study of population.

- **Synecology**: when groups of organisms in relation to environment are studied, for example, study of community.

4.1.4 Sub-divisions of ecology

There are three levels of integration in ecology –

- individual;

- population: assemblage of interbreeding organisms belonging to same species inhabiting a given area, for example, *Homo sapiens*; and
- community: Group of populations interacting in a given habitat for example, forest floor.

Figure 4.1: Levels of organization

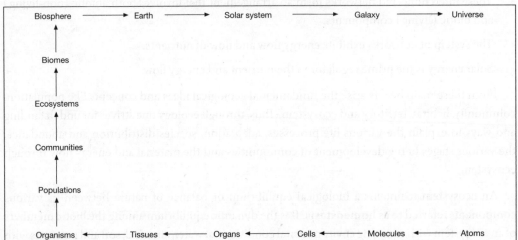

Box 4.1: Ecological footprint

It may be defined as the impact of the human activities measured in terms of biologically productive land and water that is needed for the production of goods and services and to assimilate the generated waste. It is a measure of natural capital in comparison to Earth's ecological capacity to rejuvenate. The human ecological footprint was estimated to be 1.5 earth, for the year 2007. Ecological footprint can also be measured for individual countries.

4.2 Concepts of an Ecosystem

4.2.1 Definitions

Ecosystems were initially defined as '*units of the earth's surface, i.e., the whole system including the organisms and the physical factors that form the environment.*' (Tansley, 1935)

'*Ecosystem is a complex in which habitat, plants and animals are considered as one interesting unit, the materials and energy of one passing in and out of the others.*' (Woodbury, 1954)

Later on, ecosystems were grouped by their structure and function with much stress on integration and interactions. The ecologists were classified into two groups – those who were engaged with quantifying an ecosystem's input and output relationships (flows of matter and energy; Evans, 1956) and those that were concerned with particular populations (Levin, 1976).

'*Populations do react to environmental stimulus and so ecosystems can also be defined by biota and by the environment.*'(Chapin et al., 1997)

An ecosystem is the basic functional unit in ecology, as it comprises both organisms and their abiotic environment. Organisms cannot exist without their environment. Ecosystem thus symbolizes the highest level of ecological integration and is based on energy. A forest, a lake, a mangrove, paddy field and even laboratory culture can represent ecosystems. Thus, an ecosystem is a specific unit of all the organisms inhabiting a specified space that interacts with the physical element of the environment to produce discrete trophic structure, biodiversity and nutrient cycling. **The term ecosystem was first proposed by the British ecologist A. G. Tansley.**

4.2.2 Basic/Elementary processes in an ecosystem

- **Nutrient cycle** comprises the transfer of inorganic substances between living beings and the environment. The plants are able to prepare complex organic materials from simple raw materials. This organic matter is finally released as raw material after their death and is reverted back to the environment.

- Continual energy inflow is another fundamental requisite of the ecosystem. Solar energy is trapped by the green plants by photosynthesis. Sun is the ultimate source of energy. Herbivores procure their nutrition and energy from the plants. This energy intake passes on to other organisms. In this way the energy gets transferred from one organism to another. This is known as **energy flow in ecosystem**.

4.3 Components of Ecosystem

4.3.1 Abiotic components

The term abiotic refers to non-living substances like air, water, land, elements and compounds. These are innate but become a part of biotic world once they enter the body of living organisms. The oxygen required in for metabolism comes from the air people breathe. The hydrocarbons that form our body come from the hydrogen in water and carbon dioxide in air. Most of the necessary elements are found as ores such as calcium in limestone, iron in magnetite etc. The most important aspect of these substances is the entry into living organism and their release from the living organism, i.e. their recycling.

Based on Odum's classification the abiotic components are grouped into three:

- Inorganic: carbon, hydrogen, oxygen, nitrogen, sulphur, phosphorus etc.
- Organic: carbohydrates, proteins, fats, vitamins, coenzymes etc.
- Climatic factors: light, rainfall, temperature, soil types etc.

4.3.2 Biotic components

The diverse groups of living organisms comprise the biotic components. These organisms interact to allow the unidirectional flow of energy that was fixed by the autotrophs and also the nutrient cycling. Bulk of the energy trapped by the producers is dissipated in the environment as heat energy.

■ **Producers**: They are autotrophs ('self-feed'); they manufacture both the food and energy they require. Examples include all photoautotrophs (planktons, algae, cyanobacteria and all green plants) and chemoautotrophs (*Sulfolobus, Thiobacillus, Thiothrix*). Chemoautotrophs can use inorganic energy sources, such as hydrogen sulphide, elemental sulphur, iron (ferrous), molecular hydrogen, ammonia etc; most of them are either bacteria or archaea that live in the hostile environments such as hydrothermal vent, hot springs, volcanic fumaroles and geysers.

■ **Consumers:** These are heterotrophs that obtain food and energy by feeding on other creatures. Examples – herbivores, carnivores and parasites. Animals are unable to manufacture their own food and are hence dependent on plants and/or other animals. Consumers can be of three types:

Figure 4.2: Flow of energy and nutrient recycling between the biotic components of ecosystem

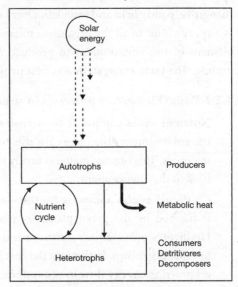

a. **Herbivores**: They are the animals that eat only plants (also known as **primary consumers**).

b. **Carnivores**: The animals that consume other animals are carnivores.

   i.   Carnivores that consume herbivores are called secondary consumers.

   ii.   Carnivores that consume other carnivores are referred to as tertiary consumers.

   iii.   Omnivores: They are animals who nourish on both plants and animals. Example – man, crow.

   iv.   Parasites: These are organisms that derives their food and shelter at the expense of living hosts. Example – tape worm, round worm, plasmodium.

   v.   Scavengers: They are animals that eat on dead animals. Example – vultures; however, the biggest assemblage of scavengers are the insects.

c. **Decomposers/Detritivores:** Detritus refers to the decaying organic material and can range from the dead meat or carrion to bacteria, including organic leftover from vegetation and animals. Heterotrophs that acquire energy from deceased or decomposed organic remains and wastes of other organisms are known as decomposers. They assist in returning the nutrient back to the producers or autotrophs. Detritivores essentially refer to small insects, earthworms, bacteria, fungi etc. They get their energy requirement by devouring dead plants and animals as well as animal waste. Organisms such as bacteria and fungi transform these wastes materials and dead remains into nutrients that can be reused again by the plants and animals.

It is to be noted that scavengers and decomposers are also heterotrophic organisms, but instead of devouring living flora and fauna they devour on detritus. In reality, detritivores and decomposers nourish at each trophic level.

Box 4.2: Schematic representation of the ecosystem components

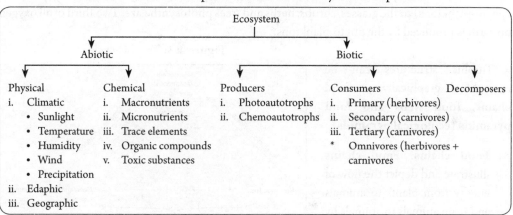

4.4 Structures and Functions of an Ecosystem

4.4.1 Structure implicates

- Composition and arrangement of biotic community: this comprises species in the community, their abundance, their biomass, life history, distribution in space, etc.
- Amount and allocation of non-living substances, such as different nutrients, water, etc.
- Range or gradient of the factors of survival, such as temperature, sunlight, pH, etc.

Functions of an ecosystem include:

- Rates of **energy flow,** i.e. the productivity and rates of respiration in the community.
- Rates of materials flow or **nutrient cycles,** such as carbon, nitrogen, phosphorus etc.
- **Biological or ecological regulation** that comprises both control of organisms by environment and vice versa.

Hence in any ecosystem, their structure and function are inseparably and intricately linked with each other and thus to be considered together.

Trophic structure: Trophic literally means 'to feed'. The trophic structure comprises the 'feeding levels' or tiers and the feeding relationships between the ecosystem components. It states the pattern of organization in which the organism exploits the food sources and helps in transfer of energy within an ecological system. Every ecosystem rests on autotrophs that make organic food matters for themselves and other members of the community. Some bacteria are capable of utilizing chemical energy to make food (chemoautotrophs). Organisms possessing chlorophyll are able to use solar energy through photosynthesis. They use water from the soil and carbon dioxide from the air for photosynthesis to make organic materials like glucose and releases oxygen as a byproduct. Macromolecules like starch, proteins and vitamins are assembled from the simple molecules along with the minerals (sulphur, nitrogen, phosphorus) present in the soil.

Animals consume these plant products and breathe in atmospheric oxygen. They oxidize these food substances with the help of oxygen and release carbon dioxide and water, a mechanism known as cellular respiration or metabolism. All green plants starting from microscopic plant (phyto-planktons) to the grasses, shrubs, herbs and trees photosynthesize. Two third of all oxygen on earth is produced by the phyto-planktons.

Trophic structures can be exemplified graphically as **food chains, food webs and food pyramids (ecological pyramids)**.

Figure 4.3: Trophic cycle

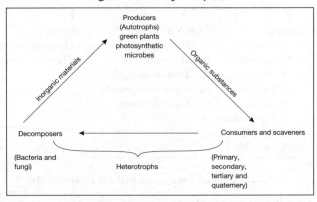

A. **Food chains**: Food chains illustrate and depict the flow of energy from plants to animals and from animals to animals by the route of eating and being eaten relationship. Green plants are the **producers** since they can manufacture their own food. Plants undergo photosynthesis to make carbohydrates. **Consumers** devour on vegetation and other creatures. Each such nutrient step in a food chain is referred to as **trophic level.**

Table 4.1: The trophic levels

Trophic level	Type of organism	Energy source	Examples
1st	Primary producer	Sun	Trees, shrubs, algae
2nd	Primary consumer	Primary producers	Grasshoppers, deer (herbivores)
3rd	Secondary consumer	Primary consumers	Frogs, lizards, crabs (carnivores)
4th	Tertiary consumer	Secondary consumers	Eagles, tigers, sea otters (top carnivores)

A food chain is a linear arrangement of who is eating whom in any ecosystem. Otherwise, a food chain is a **straight line sequence** that comprises links in a food web commencing with a trophic species or organisms that consumes no other species in the network and concludes with a species that is again eaten by no other species or organisms in the web. Food chain seems to be basic abstractions of the food webs operating in reality, but complex in their dynamics and consequences.

Usually, food chains are restricted within four or five trophic levels. For example, a food chain consisting of a plant, a frog, a snake and finally peacock consists of four levels, whereas, a food chain comprising grass, a grasshopper, a mouse, a snake and finally eagle consists of five levels. Thus, a food chain can vary in lengths.

Food chains can be of two types –

i. **Grazing food chain:** This type of food chains always begin with autotrophs followed by grazers.

ii. **Detritus food chain:** This type of food chains begin with dead matters followed by detritivores.

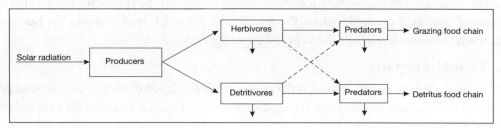

Figure 4.4: Inter-relation between grazing and detritus food chain

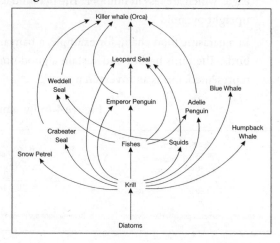

B. Food web

It was Charles Elton who presented the notion of food web what he referred to as food cycle. Food web refers to anastomosing or interwoven food chains with numerous producers, consumers and decomposers operating simultaneously. It is the real depiction and illustration of the feeding relationships amongst species in a community. It is also a way of displaying how food energy flows between various organisms of a community as a consequence of feeding relationships. In a food web the species are interconnected by means of arrows called links.

Figure 4.5a: Food web in a grassland ecosystem

Figure 4.5b: Food web in Antarctic ecosystem

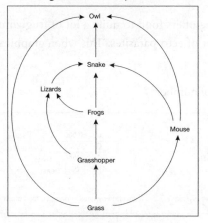

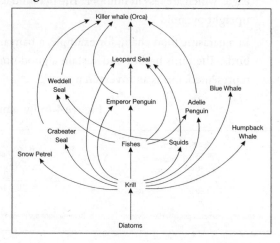

A food chain shows only a partial picture of the food web consisting of a simple, linear sequence of species (e.g., plant, herbivore and a predator) coupled by feeding relations. A food chain exhibits linear **monophagous** pathway and differs from a food web which is actually a **complex polyphagous network** of feeding links that are grouped into trophic species. The objective of a food web is to portray a more **inclusive image** of the feeding inter-relationships. Simply it can be referred to as a system of numerous interconnected food chains operating within the community. All the species in the similar position in a food chain represent a nutrient or trophic level within the food web. For example, all of the green plants in the food web make up the first trophic level or 'primary producer', all herbivores consist of the second trophic level or 'primary consumer' and carnivores that consume the herbivores make up the third trophic level or 'secondary consumer'.

C. Ecological pyramid

Trophic structure and trophic function of an ecosystem may also be graphically represented by means of 'ecological pyramids' where the producer or autotroph level occupies the base and successive consumer levels form the steps and create the apex.

i. Pyramid of numbers

'Pyramid of numbers may be defined as the graphical representation of number of organisms per unit area in the various trophic levels arranged stepwise with producers forming the base and top carnivores the tip'.

The shape of the pyramid of numbers may vary from one ecosystem to another ecosystem.

- In **grassland** and **aquatic ecosystems**, producers are present in massive numbers per unit space. They sustain a reduced number of herbivores, which in turn sustain fewer carnivorous animals. Hence, the manufacturers are the maximum in number but usually of smaller body size while, top carnivores are lesser in number and larger body size. Therefore, the number pyramid in aquatic and grassland ecosystem is always **upright and erect.**

- In a forest ecosystem, large sized trees bear enormous number of insects which in turn are eaten by frogs, lizards etc.; these are again consumed by carnivorous birds like hawks and eagle, which are less in number. The pyramid assumes a spindle shape and is known as **semi-upright** pyramid.

- In a parasitic food chain, for example, a banyan tree offers food to quite a lot of frugivorous birds. The birds harbor and sustain a good number of ecto-parasites. This when graphically represented, forms an **inverted** pyramid.

Figure 4.6: Pyramids of number

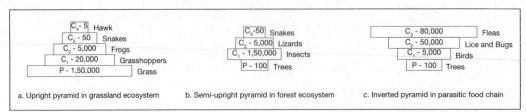

a. Upright pyramid in grassland ecosystem   b. Semi-upright pyramid in forest ecosystem   c. Inverted pyramid in parasitic food chain

ii Pyramid of biomass

The overall quantity of living or organic substance in an ecosystem at any moment is termed biomass. Pyramid of biomass is the graphical illustration of the organisms biomass present per unit space in the different trophic levels, with producers occupying the base and top carnivores making up the top spot.

In the biomass pyramids a trophic level seems to have more energy than it does in reality. For instance, the exoskeletal structures like beak, feathers and skeletons, though constitute the biomass are not consumed by the next higher trophic level. These portions of the body would be still measured and shown even though they do not add to the process of energy flow.

- In a **land ecosystem**, the maximum biomass befalls in producers followed by a sequential and progressive decrease in biomass as one moves from lower to higher trophic levels. Thus, in a land ecosystem, the pyramid of biomass is **upright**.

- In an **aquatic ecosystem** the pyramid of biomass is either **inverted or semi-upright.** Here the biomass of each nutrient level depends upon the biotic potential and lifespan of the member in the ecosystem.

Figure 4.7: Pyramids of biomass

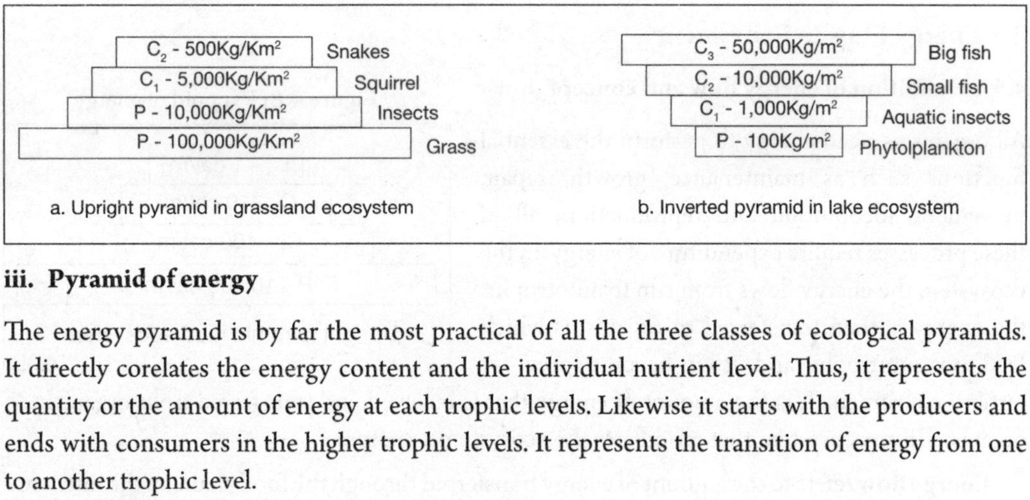

C_2 - 500Kg/Km²	Snakes
C_1 - 5,000Kg/Km²	Squirrel
P - 10,000Kg/Km²	Insects
P - 100,000Kg/Km²	Grass

a. Upright pyramid in grassland ecosystem

C_3 - 50,000Kg/m²	Big fish
C_2 - 10,000Kg/m²	Small fish
C_1 - 1,000Kg/m²	Aquatic insects
P - 100Kg/m²	Phytolplankton

b. Inverted pyramid in lake ecosystem

iii. Pyramid of energy

The energy pyramid is by far the most practical of all the three classes of ecological pyramids. It directly corelates the energy content and the individual nutrient level. Thus, it represents the quantity or the amount of energy at each trophic levels. Likewise it starts with the producers and ends with consumers in the higher trophic levels. It represents the transition of energy from one to another trophic level.

In an healthy ecosystem, the energy pyramid shall always appear like 'standard ecological pyramid'. For the ecosystem to be self sustaining, lower trophic levels should have more amount of energy than the higher trophic levels. This help the lower level organisms to maintain a stable population, which would be fed by the organisms at upper trophic levels, thus aiding in transfer of energy up the pyramid.

During the transfer of energy to the subsequent trophic level, merely around 10 per cent of it is utilized to assemble bodymass at the next trophic level and become stored energy. Remaining 90 per cent is lost in metabolic activities. This is referred to as **Lindemann's data or 10 per cent law.**

The **advantages** of the pyramid of energy are as follows:

- It represents an account of the rate of production over a period of time as each trophic level represents energy per unit area or volume per unit time. A simple example of unit can be – cal/m²/year.
- The energy content of two species bearing same mass or weight may be different; in such case biomass may be a misleading parameter whereas energy is truly comparable.
- In energy pyramid, the relative energy flow can be compared and so also different ecosystems can also be compared using energy pyramids.
- Energy pyramids can never be inverted.
- The input of sunlight can be taken into account.

The **disadvantages** of the pyramid of energy are:

- Complete combustion of a sample is required in order to obtain the energy value for an organism of given mass.
- Difficulty exists in assigning organisms to a specific trophic level as there is a problem in assigning the decomposers and detritivores to a particular trophic level.

The best way of presenting of the feeding relationships of a community is to do with the help of energy pyramids.

4.5 Energy Flow in Ecosystem

4.5.1 Definition of energy flow and concept

All organisms need energy to perform the essential functions such as maintenance, growth, repair, movement, locomotion and reproduction; all of these processes require expenditure of energy. In the ecosystem the energy flows from sun to autotrophs, the energy is then transferred to organisms which feed on autotrophs and itself become prey for tertiary consumers. It is the amount of energy that is received and transferred from organism to organism in an ecosystem that modulates the ecosystem structure.

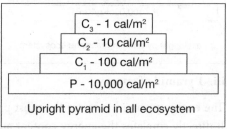

Figure 4.8: Pyramid of energy

C₃ - 1 cal/m²
C₂ - 10 cal/m²
C₁ - 100 cal/m²
P - 10,000 cal/m²
Upright pyramid in all ecosystem

Energy flow refers to the amount of energy transferred through the food chain up the nutrient/trophic level. Energy flow is alternatively known as calorific flow since energy input or the energy entering the ecosystem is measured in joules or calories.

Sun is the biggest source of energy. Energy left unutilized in the ecosystem is ultimately lost as heat. When one organism feeds on another organism, both energy and nutrients are transferred through the food chain. The surplus or the leftover energy in an ecosystem is then devoured by the decomposers. While nutrients are recycled through an ecosystem, energy is simply lost over time.

Energy flow in any ecosystem commence with the autotrophs that would trap the energy from sun. Herbivores feed on these autotrophs to transform the energy within the plant into their usable form. The carnivores then feed on these herbivores to obtain their energy, lastly, other top carnivores prey on the lower level carnivores. In each such step, energy from one trophic level is transferred to the next higher trophic level and every time some amount of energy is lost in the form of heat into the environment. This happens because each one utilizes part of the energy they obtain from the other one for their sustenance. The top consumer receives the least amount of energy.

Earth is constantly hit by the radiation coming out of the sun about 93 million miles far-off which provides energy not only to the air, water and land, but also to the objects capable of absorbing energy; in simple words, radiant energy is transformed to heat energy. The winds and water currents are caused by unequal heating; i.e. thermal energy is transformed into kinetic energy. Water evaporates into air due to warming thus setting on the hydrological cycle. Potential energy is formed with the vaporization of water which when starts flowing from kinetic energy. Nevertheless, the most important radiant energy mediated course of action with respect to the living beings is photosynthesis.

4.5.2 Solar constant

The amount of radiant energy emanating out of the sun and striking the earth's atmosphere is called solar constant and has a value of 1.361 kW/m^2 or 1.952 calories per minute per square centimeter, (approximately, 2 cal/min/cm^2) on the upper atmospheric region of the Earth. The value may vary as a consequence of Earth's elliptical path. The energy left after reflection by the Earth's surface is known as the net radiation. An astronomical unit is used for doing the calculations of solar constant. The mean distance between Earth and Sun is referred to as the Astronomical unit (AU); 1 (AU) is almost equal to 149,604,970 km.

. The temperature of the earth is mainly controlled by evaporation and convection of air. The air movements enable the heat energy to be given out into the space, which, otherwise would make earth unbearable due to overheating and life would smother. Such interaction is also very useful to maintain the polar ice caps. On the other hand, a **decrease in the solar constant by 2–5 per cent would be sufficient to lead to a second ice age.** The uncontrolled burning of fossil fuels can also increase earth's temperature. The polar ice caps also facilitate in reflecting back a portion of Earth's radiation which is very crucial in the maintaining of present day climatic conditions.

Energy budget: About 340 watts/m^2 of solar light is incident on the Earth (only one-fourth of the total solar irradiance). Earth moves the heat energy from the surface and lower atmosphere back to space. Such incoming and outgoing flow of energy is known as **Earth's energy budget**. To have a stable temperature over considerable periods of time, the incoming energy should be equal to the outgoing energy or the energy balance at the top should balance at the top of the atmosphere. This state of balance is referred to as **radiative equilibrium**.

Of the energy that falls on the earth, nearly **29 per cent** of the solar radiation arriving at the upper part of the atmosphere is reradiated or **reflected back** to space by the clouds (20 per cent), air particles (5 per cent) and the ground surface like ice or snow (4 per cent). Remaining **71 per cent** of the total incoming solar radiation is thus **absorbed** by the Earth system. Of this, water vapour, dust and ozone in the **atmosphere absorb 23 per cent** and the **surface absorbs 48 per cent.**

On absorption of energy by matter, the particles (atoms and molecules) that comprise the matter gets excited and sets in rapid movement. This increased movement also increases the material's temperature. The temperature does not rise indefinitely since, apart from absorbing solar energy they also radiate infrared radiation. According to **Stefan–Boltzmann Law,** the energy radiated by a black body per unit surface area is directly proportional to the fourth power of the thermodynamic temperature of the black body.

$j = \sigma T^4$ where j is the energy radiated, T the thermodynamic temperature and σ is the Stefan–Boltzmann constant.

So, if Earth's temperature rises, it emits an increasing amount of heat to the space. This is known as **radiative cooling.**

The atmosphere and the surface together absorb 71 per cent of the solar light. The **atmospheric absorption** is **23 per cent** while **surface absorption** is **48 per cent. To balance they must radiate almost equal amount of energy to keep earth's temperature stable.** In this case the contribution of the atmosphere and the surface is **asymmetric.** Satellite observation reveals that the **atmosphere radiates 59 per cent infrared energy** and the **surface radiates only 12 per cent.**

Most solar heating takes place at the surface. The question arises about how and where the reshuffling of energy takes place between the surface and the atmosphere. Let's look at the phenomenon at the **surface** first.

Of the 71 per cent, 23 per cent of solar radiation absorbed by the atmosphere is radiated back. Therefore, 48 per cent of energy remains that is absorbed by the surface. To strike a balance, the surface must radiate this 48 per cent back; energy primarily leaves the surface by evaporation, convection and thermal infrared energy.

25 per cent of the incoming radiation leaves the surface through **evaporation** of water; nearly **5 per cent** of the incoming radiation leaves the surface through **convection** of air; a net amount of **17– 18 per cent** of the incoming radiation leaves the surface as thermal **infrared** energy radiated by the ground particles thus amounting to 25 per cent + 5 per cent + 17 to 18 per cent = 47 to 48 per cent.

Coming to the **atmosphere, evaporation and convection transfer 25 per cent and 5 per cent** of solar radiation from the surface to the atmosphere. The **atmosphere** (clouds, aerosols, water vapour, ozone) **absorb 23 per cent** of the sunlight. Thus, a total (25 per cent+ 5 per cent+ 23 per cent = 53 per cent) is transferred to the atmosphere. As per the satellite measurement 59 per cent of the energy is radiated back by the atmosphere. Therefore, the remaining fraction of 5–6 per cent (56 per cent – 53 per cent) comes from the earth's surface.

Major proportion of the atmospheric gases like oxygen, hydrogen and nitrogen are transparent to both incoming solar radiation and outgoing infrared radiation. But the molecules like water vapour, CO_2, CH_4 are opaque to infrared energy. Of the net 17 per cent infrared heat, **12 per cent directly escapes to the space** and the rest 5–6 per cent is transferred to the GHG molecules in the atmosphere which absorbs that energy. This raises its temperature. The GHGs radiate infrared heat in all directions. Some of this energy comes down, gets in contact with the Earth's surface and is absorbed. The temperature of the surface gets warmer. This additional heating of the surface by the atmosphere is the natural green house effect that raises the Earth's temperature to an average of nearly 15 ^{0}C.

Figure 4.9: Earth's energy budget allowance

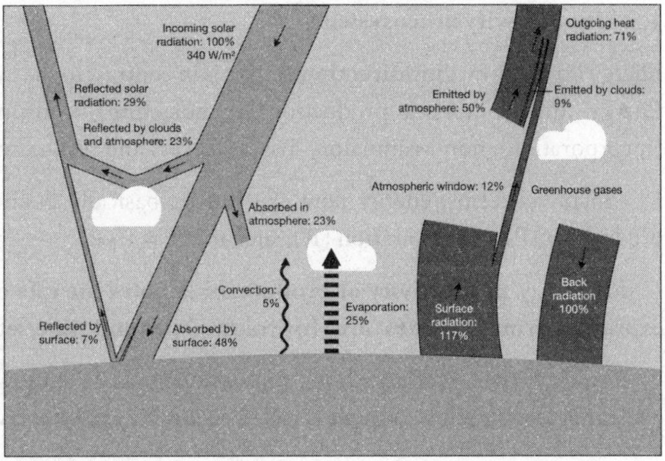

4.5.3 Laws of thermodynamics as related to ecology

The thermodynamics laws are elemental doctrines and theories to all the chemical processes of this world. They are not only imperative in chemical sciences, physical sciences but also in many biological phenomena. The laws state how energy is transformed from one form to another; this is applicable to ecology as energy transfer and is all that drives metabolism.

Zero Law of Thermodynamics: Of all the three laws, this seems to be the most comprehensible. It merely states that if there are three objects X, Y and Z and the temperature of object X and Y are equal, and also temperature of object Z and Y are equal, then the temperature of object Z is also equal to the temperature of object X.

The First Law of Thermodynamics: It states that energy cannot be created nor be destroyed, but can be converted from one form to another form. Therefore, in a system the overall influx of energy should be equal to the overall outflow of energy.

Second Law of Thermodynamics: The randomness or 'entropy' of the universe is constantly rising. The energy flow from one trophic level to the next cannot be 100 per cent efficient. Energy transfer is always followed by a dissipation of energy into other forms. Majority of ecological applications is interpreted by this thermodynamic law. The second law also propounds a dire consequence that there will be a **'heat death of the universe'** taking place as all the energy of the universe gets uniformly distributed, though this will not happen for at least 10,100 years.

Third Law of Thermodynamics: According to this law, the entropy of the system decreases as the temperature of a system approaches absolute zero (0°K). Correctly, any system at absolute zero is said to have zero entropy. This is theoretically impossible because absolute zero cannot be reached experimentally. This explains the reason why substances turn into gases at high temperatures as entropy increases and freeze at low temperatures as entropy decreases. Decomposition or breakdown also proceeds at a higher rate at elevated temperatures for the same reason.

4.5.4 Productivity in ecosystems

Energy flow is always **unidirectional.** This is in contrast to the cyclic flow of material in ecosystems. Energy flow comprises production (manufacture), consumption (utilization), assimilation (incorporation), non-assimilatory losses (faeces) and respiration (maintenance costs).'

'In broad terms, energy flow (E) can be basically defined as the sum total of metabolic production (P) and respiration (R), such that E = P+R'.

In ecology, **productivity or production denotes the rate of generation of biomass, usually expressed in mass per unit area (or volume) per unit time,** let's say grams/square meters/day.

Productivity at producer level, such as plants is called **primary productivity**, while that at consumer levels such as animals is called **secondary productivity**.

Primary productivity: Primary productivity is the manufacture of new organic materials from inorganic molecules such as H_2O and CO_2. It is achieved by the process of **photosynthesis** which utilizes sunlight to manufacture organic molecules such as sugars. A small fraction of primary production is also achieved through **chemosynthesis.** Organisms responsible for primary production include green plants, phyto-planktons, blue green algae, sulphur bacteria etc.

Theoretically one is able to determine the energy uptake by the autotrophs by calculating the total sugar produced since the **entire radiant energy trapped by the producers is converted into sugars.** This amount is referred to as **Gross Primary Production (GPP)** as it happens in the autotrophs or producer level. But more practical and relevant is to assess the **Net Primary Productivity (NPP).**

In ecological system **NPP refers to the rate at which the dry mass is built up by the producers, generally expressed as kilograms/m²/year** or the energy added in unit time (kilojoules/m²/year). It is this reserve of energy that will serve as the prospective food for the consumers up the trophic levels in ecosystem.

'**NPP can be measured as the difference between the rate of photosynthesis by plants and their rate of respiration. Glucose manufactured out of photosynthesis has two main outcomes:**

- Some of it provides energy for anabolism such as growth, maintenance, repair and reproduction besides a certain amount of respiratory energy being lost as heat.

- The residual or surplus amount of glucose is stored in and around cells and signifies the dry mass (NPP = GPP – R).

Factors limiting net primary production:

Terrestrial factors include:

- sunlight;
- moisture and water; and
- heat and temperature;
- availability of nutrients.

Aquatic factors include:

- sunlight;
- availability of nutrients.
- heat and temperature; and

Global warming has been immensely affecting the Net Primary Production of aquatic ecosystems, such as decline of the coral reef due to coral reef bleaching. Coral reef supports one-fourth of the marine biota.

Secondary productivity (SP): 'It is the total amount of chemical energy assimilated by consumer organisms'.

Secondary production is the production of biomass in the subsequent trophic levels by the consumer organisms in ecosystem. This is the rate at which primary organic matter is made into animal tissue per unit area in a given time. This includes consumption of primary producers by herbivorous or carnivores. Custodians of secondary production include protozoans, fungi and animals.

Figure 4.10: A simplified energy flow model in ecosystem with three trophic levels in a linear food chain

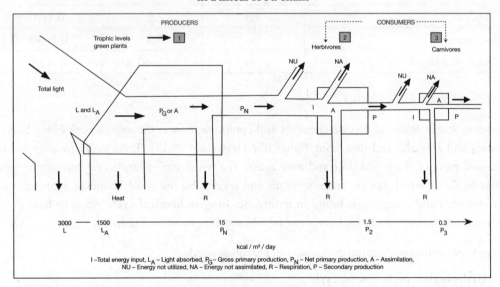

I –Total energy input, L_A – Light absorbed, P_G – Gross primary production, P_N – Net primary production, A – Assimilation, NU – Energy not utilized, NA – Energy not assimilated, R – Respiration, P – Secondary production

All of the biomass that is being consumed is not utilized by the animals. Undigested matter comes out as faeces. Gross production in animals is simply the biomass or the assimilated energy minus faecal matter.

Similar to the producers, the assimilated energy of the consumers is utilized in the cellular processes by way of respiration while the surplus is available for new biomass. The amount is known as Net Secondary Production.

So, Net secondary productivity (NSP) = Foodstuff consumed – faecal matters – respiratory energy.

NSP = GSP- R

Table 4.2: Productivity and biomass in different ecosystems

Vegetation type	Sunshine Kcal/m²/yr	Biomass t/ha	Productivity t/ha/yr
Tropical rainforests	3.0	250–400	10–50
Estuaries	-	0.1–40	2–40
Temperate deciduous forests	2.1	70–250	3.5–10
Taiga coniferous forests	1.8	25–70	4–20
Upwelling zones	-	0.2	4–10
Savannah grasslands	3.0	5–35	3.5–5
Continental shelves	-	0.01	2–6
Prairie grasslands	2.4	1–18	1–4
Deserts	2.7	0–1	0–2.5
Tundra	1.3	5	0.1–4
Estuaries	-	2–40	0.1–40
Cultivated lands	-	1–40	4–120

4.6 Biogeochemical Cycles

The constituent atoms of diverse elements and compounds form the unit of both living beings (plants and animals) and non-living things like (water, air, rocks). These units are recycled in different parts of the world over and over again. The word 'bio' means living organisms; 'geo' refers to Earth, which can be inclusive of air and water; and the word 'chemical' represents all the elements and compounds found in nature. So, **biogeochemical cycle** refers to how these chemicals move between living beings and non-living things and back to living.

Three types of biogeochemical cycles are usually categorized:

- **Hydrological cycle** – water cycle

- **Gaseous cycle** – The recycling of those elements, that spends most of their lifetime in a gaseous form in the air, known as gaseous cycle. Elements like carbon, nitrogen and oxygen fall in this category. Carbon generally occurs in the atmosphere as carbon dioxide (CO_2), nitrogen as nitrogen gas (N_2) and oxygen as oxygen gas (O_2).

- **Sedimentary cycle** –These elements mostly occur in a solid form. For example, phosphorus and sulphur.

4.6.1 Water cycle

Water makes up 80 to 90 per cent of all living organisms. Water acts as the most important buffer in all living forms and the environment, preventing heat shocks. Human body comprises 70 per cent of water. Water is plentiful, but more than 97 per cent of this is salty water and the remaining is freshwater. 2 per cent of this freshwater is trapped in ice caps and glaciers; 0.31 per cent is reserved in deep groundwater reserves. Less than 0.01 per cent is existing in the rivers and lakes. Water is unique in its properties like latent heat, density etc, and an excellent solvent. These properties are attributable to its molecular composition and arrangement.

Water is cycled through the mechanisms of evaporation from any exposed surfaces, respiration, transpiration from plants parts (mainly the leaves) into the atmosphere, and precipitation in the form of rain, snow, sleet etc back to Earth. Huge amounts of water disappear from the exposed surfaces of the Earth's oceans and seas by evaporation. Approximately one thousand gigatons of water evaporate from the ocean surface per day. Water molecule has an average residence time of about eight days.

It is predicted that 41,000 Km³ of water comes back to the sea from the land, harmonizing the transfer of water from sea to land.

Water is thought to be the bottomless sink but acute water problem becomes evident due to pollution like acid rain, eutrophication, dam construction and irrigation and consumption patterns. The uneven distribution of water gives rise to severe conflicts and water war like situation. The problem needs to be addressed and handled carefully.

Figure 4.11: Water cycle

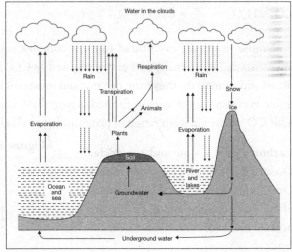

4.6.2 Carbon cycle

Carbon forms the structural element of all living organisms; it exists as coal and limestone deposits in the lithosphere, carbon dioxide in the air and in water. Atmospheric lifetime of carbon is approximately nine years. Air contain 750 billion tonnes of carbon chiefly in the form of CO_2; living plants and animals contain 560 billion tonnes, buried organic matter in the soil contains about 1,400 billion tonnes, ocean and seas contain 38,000 billion tonnes as dissolved CO_2 and about 11,000 billion tonnes are locked up in methane. With development and industrialization, burning of fossil fuels increasing, huge amounts of CO_2 (22 billion tonnes/year) are released. Deforestation adds a further amount of 1.6–2.7 billion tonnes.

The main stores of carbon are the sedimentary rocks, fossil fuels (coal, petroleum and natural gas), oceans and biosphere.

Carbon goes primarily through three cycles with different time constraints:

- a long-term cycle linking the sediments and the lithospheric depths;
- a cycle involving the atmosphere and the lithosphere; and
- a cycle relating the air and the oceans.

The cycles 2 and 3 are faster and subjected to anthropogenic interference.

Carbon cycle 1 – between air, oceans and sediments: This cycle occurs between air, oceans, and sediments and entails a sluggish dissolution of carbon in the air and rock carbons by weathering into the oceans. Oceans contain huge deposits of carbon as calcium carbonate deposits. The carbon in the sediments dissolves slowly and some of the sediments are reverted into the air in the course of volcanic eruption. This cycle is extremely long and takes over hundreds of millions of years.

Carbon cycle 2 – between air and land: This cycle between the atmosphere and biosphere may range from few days to decades. Carbon dioxide serves is the basic food ingredient to the living beings and thus the biosphere plays an instrumental role in this cycling. Plants fix atmospheric carbon through photosynthesis. The chemical may be represented as:

$$CO_2 + H_2O \xrightarrow[\text{chlorophyll}]{\text{sunlight}} CH_2O + O_2$$

Simple sugars are assimilated to form complex substances like starch and cellulose. Plants are consumed by animals; part of it is respired as CO_2 and part turned into animal tissue. When plants and animals die they form detritus and most of it is decomposed into inorganic forms. Over millions of years, fossil fuels are formed as partially decomposed matter. Combustion of fossil fuels adds CO_2 to the air. Respiration of plants and animals also releases CO_2 in the air.

Carbon cycle 3 – air and sea cycle: Oceans are vast deposits of carbon. Over thousands of year the carbonates of the rocks dissolve and makes up the oceanic biomass.

The summary of the three cycles together is shown in Figure 4.12.

4.6.3 Nitrogen cycle

Nitrogen is the most abundant element in the air and makes 79 per cent by volume of air. Nitrogen forms the building blocks of protein monomers

Figure 4.12: Three carbon cycles integrated

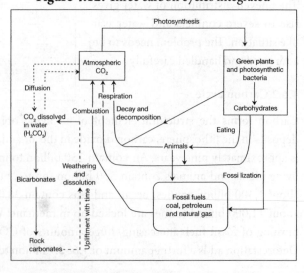

– the amino acid and also forms the constituent of bases in DNA and Ribonucleic acid (RNA) – adenine, guanine, cytosine and thymine.

The nitrogen cycle is subjugated by the N_2 in the air. NO_2 is the second most familiar form. N_2O, normally referred to as laughing gas, is a greenhouse gas (GHG). Nitrogen gas is inert and so it is able to offset the high reactivity of oxygen, the other major atmospheric component. Nitrogen can form a range of compounds owing to its variable valency. For instance, it combines with oxygen in variable proportion to form N_2O, NO, NO_2, or N_2O_5. Collectively, these oxides (except for N_2O_5) are represented as NOx.

Nitrogen is an essential element of all living beings. Amino acids, the monomer, are the building blocks of proteins. They contain nitrogen in the form of the 'amino' group (NH_2). The four nitrogenous bases of DNA and RNA [Adenine (A), Cytosine (C), Guanine (G), and Thymine (T)/ Uracil (U)] consist of either single or double rings of carbon and nitrogen atoms, with various side chains. Nitric oxide (NO) is also a neurotransmitter.

Nitrogen is inert in the air. All living organisms want considerable amounts of nitrogen for sustenance. Neither plants nor animals are able to fix atmospheric nitrogen. Nitrogen from the air is fixed by three processes.

- by microbes;
- by lightning; and
- by Haber's process.

Microbial fixation can be symbiotic or asymbiotic. Symbiotic fixation is between the leguminous plants like peas, beans, alfalfa, clover and the symbiotic bacteria like *Rhizobium*. *Rhizobium* traps atmospheric nitrogen, converts it into nitrates and gives it to the plants. The plants give finished food to the bacteria. Aymbiotic fixation is mediated by blue green algae like *Nostoc* and *Anabaena*.

Lightning momentarily raises the temperature causing the atomic nitrogen and oxygen to react together form nitrogen monoxide and further into nitrogen dioxides. The oxide dissolves in water to form nitric acid that precipitates down into the soil and water along with rain. This enriches the nitrogen content in the soil as nitrites.

In industry, nitrogen and hydrogen is combined under controlled temperature and pressure and in presence of catalyst to form ammonia. The process is known as Haber's process.

Plants take up soluble nitrates from the soil, along with other minerals, and utilize them to build up plant tissues. Animals in turn get their nitrogen requirement by nourishing

Figure 4.13: Nitrogen-fixing bacteria on legume roots

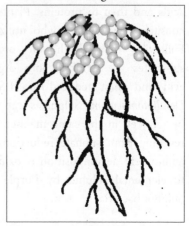

on plants. Plant and animal wastes and their dead remains are transformed into ammonia by the process of ammonification with the help of *Clostridium*. The use of fertilizers like ammonium nitrate is to compensate for the nitrogen deficit in the soil. Ammonia undergoes the process of nitrification in two steps aided by soil bacteria. In the first step, *Nitrosomonas* converts ammonia into nitrite and subsequently in the second step, *Nitrobacter* converts nitrite into nitrates. Nitrates are then be easily taken up by the plants with the help of plant roots. Denitrifying bacteria like *Bacillus* convert the nitrates back into elemental nitrogen making it unavailable once again.

4.6.4 Oxygen cycle

Oxygen is a basic element of life. It constitutes about 21 per cent of the atmosphere. Oxygen is omnipresent. It also occurs in combined state in Earth's crust and mantle and in water. Oxygen is highly reactive. Dissolved oxygen in water supports aquatic life. Photosynthesis releases oxygen in the atmosphere whereas respiration, burning or combustion and decomposition utilizes oxygen.

Figure 4.14: The nitrogen cycle

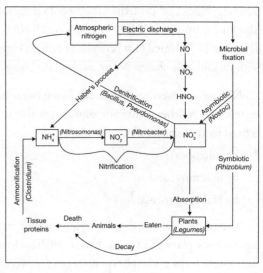

4.6.5 Sulphur cycle

Sulphur, a yellow coloured solid chiefly occurs as sulphates in the rocks and as elemental sulphur. Sulphur is one of the essential elements in living organisms. It occurs in three amino acids like methionine, cystine and cysteine. Sulphur in elemental form cannot be utilized by the plants and animals. Sulphur is oxidized to sulphates by bacteria like *Thiobacillus thioxidans*. Sulphates assimilated by plants are incorporated into amino acids and then to proteins. Plants are consumed by animals who utilize it. Following their death, the organic substrates are oxidized into organic acid and then reduced to H_2S by *Desulfotomaculum*. Sulphates can also be reduced in the same manner. H_2S resulting from suphate reduction and amino acid decomposition is oxidized to elemental sulphur by Purple and Sulphur bacteria.

Figure 4.15: The oxygen cycle

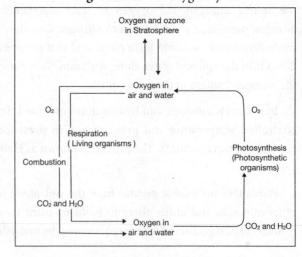

4.6.6 Phosphorus cycle

Weathering of phosphate rock supplies phosphates to the soil. The dissolved phosphorus is absorbed by the plants and animals do obtain phosphorus from the plants. Phosphorus is an essential ingredient to nucleic acids, Adenosine Di Phosphate (ADP), Adenosine Tri phosphate (ATP), Nicotinamide Adenine Dinucleotide Phosphate (NADP) and phospholipids.

Plants and animals during their lifetime excrete and die. Decomposers breakdown the organic phosphates and return phosphorus to the soil. A part of it is leached to the oceans through the rivers and streams and deposited in the shallow sea. A part of these deposits are utilized by marine animals and after their death they contribute to rocks. Weathering may increase the availability of phosphorus to the biotic community. Much of the shallow deposits are lost to the relatively deep sea deposits.

The biogeochemical cycles are illustrations of the **Law of Conservation of Mass**; matter is neither created nor destroyed but simply transforms from one form to another.

Figure 4.16: The Sulphur cycle

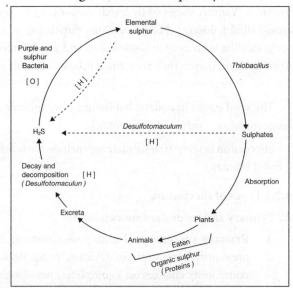

Figure 4.17: The phosphorus cycle

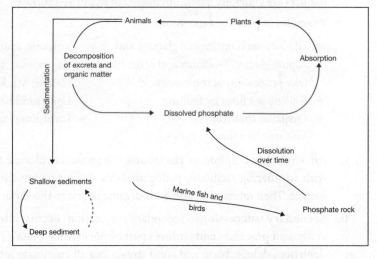

4.7 Ecological Succession

4.7.1 Definition and concept

This refers to the sequence of changes that the biotic community goes through as it matures towards a stable condition known as climax. These changes are systematic, progressive and more or less predictable. It involves stepwise replacement of one community by another community over time. The term succession was first put in use by Henry David Thoreau.

Successions that initialize in aquatic habitats like lake, ponds, marshes etc. are referred to as hydrarch. Various stages of hydrarch progress make up hydroseres. The different stages in salty areas called haloseres. For successions initializing in arid conditions like sand dunes and bare rocks etc., the term xerarch is often used and the different stages are known as xeroseres. Xeroseres developing on barren rock are termed lithoseres; while xeroseres on sand are called psammoseres. Succession beginning on inter–habitat between water and sand is termed mesarch.

The seral stages in aquatic habitat may be: submerged stage – floating stage – reed swamp – sedge meadow – shrub – tree (climax)

Succession in terrestrial habitat may include the following seres: lichen – moss – herbs – shrubs – forest (climax)

4.7.2 Types of succession

A. Primary and secondary succession

i **Primary succession**: Primary succession is a 'change in vegetation which occurs on previously barren terrain' (Barnes, *et al.*, 1998). Primary succession is the sequence of community changes on a completely new-fangled habitat that have never been colonized previously. So, if succession progresses from a nude area, i.e., area uninhabited by organisms or an area unchanged by living beings, is known as primary succession. Such habitats are generally newly uncovered or deposited surfaces.

Examples of primary succession:

- Glaciations (continental glaciers and Alpine glaciers) – Plants and other life take possession of the bare rock.
- Raising sea floor by faulting.
- Quarried rock fronts.
- Volcanic lava flows – Hawaii.
- Volcanic eruptions – Iceland volcanic eruptions, Mt. Pinatubo, Mt. St. Helens, Mt. Kilimanjaro
- Lofty sand banks and sand dunes.
- Landslides.

After volcanic eruption or glaciations, the pioneers colonize the barren land, this along with weathering facilitates pedogenesis or soil formation through interaction with the surface. Their interaction also adds organic debris to the surface.

ii. **Secondary succession**: 'Secondary succession occurs after a disturbance disrupts ecosystem processes and removes part of the existing biota' (Barnes, *et al.*, 1998). Forest depletion, deluge, blaze and wind stream can all culminate into secondary succession. It takes place on formerly occupied, disturbed and damaged habitat. It proceeds from a state where other organisms are still present such as seeds, left over stumps and root system and the effects are obvious. Vegetation recovery followed by forest fire is an example of secondary succession. Since the soil is already developed, vegetation was already present, vegetation transformation occurs quickly and rapidly.

Examples of secondary succession:

- tree felling, clearance of woodland; and
- forest fires.

B. Autotrophic and heterotrophic succession

The succession can be illustrated as **autotrophic and heterotrophic succession.** The autotrophic succession is marked by early and sustained domination of autotrophs. It begins in mainly inorganic setting. The heterotrophic succession on the other hand is marked by an early dominance of heterotrophic organisms and this usually starts in primarily organic environment. The last stage of succession is known as the **climax community.**

C. Autogenic and allogenic succession

The stages in succession is caused by both endogenous (internal) or exogenous (external) factors depending upon whether such changes are brought about by the actions of the foliage by themselves or by outside factors. Transitions caused by endogenous factors or internal factors are termed autogenic, whereas those caused by exogenous factors or external factors are referred to as allogenic. Primary succession is the typical case of autogenic change, in that the plant is partly the reason for the development of soils. On the other hand, allogenic succession is determined by intermittent disturbances.

4.7.3 General process of ecological succession (mechanism of succession)

The process of primary succession moves all the way through a number of orderly steps, which pursue one another.

- **Nudation** is the progress of an exposed surface without any living forms. It may be due to topographical factors (wearing away of soil, landslides, earthquake etc.) or climatic factors (glaciations, hailstorm, fires etc.) or even biotic factors (manmade activities, epidemics etc.). Plants who will be able to arrive and colonize will survive. They must be able to thrive and withstand the conditions at the new place in order to survive. They are known as pioneers. For example moss and lichens are lithophytes that colonize on the bare rocks.

- **Invasion** refers to the successful founding of a species in a bare or exposed area, which occurs in three steps.

  a. Migration: transport of seed or spores in such a nude area through agents like wind, water etc.

  b. Ecesis or 'establishment' means adjustment of the migrated species with the existing conditions of the region. The pioneers after arrival alter the environment. The old ones are replaced by the new ones, the roots penetrate deeper and deeper breaking the particles, increasing the water retention capacity and making enough space for others to move in.

  c. Aggregation: After establishment, the living organisms multiply and grow in number by reproduction.

- **Competition and co-action:** Intra-specific and inter-specific competition for food and space begins as the plants increase in number. Larger and tall plants grow and overshadow the smaller ones. The species enters into various types of interactions.

▪ **Reaction**: The living organisms modify the environment and new seral stage replaces the existing community. The process is repeated.

▪ **Stabilization**: When the community becomes stabilized with the climatic conditions of a given area, it is said to be the final or terminal community often known by the name of climax. This community is in equilibrium with the prevailing conditions of a given area. The climax dominates for a long time except for catastrophic condition. Succession may be arrested by several factors and thus the progress into the climax community may be inhibited. Sometimes this arrested development leads to subclimax like condition. If the arresting phase prolongs, a completely different type of vegetation might develop.

4.8 General Classification of Ecosystems

Ecosystems can be classified into two major groups: terrestrial ecosystems and aquatic ecosystems.

Box 4.3: Scheme of ecosystem classification

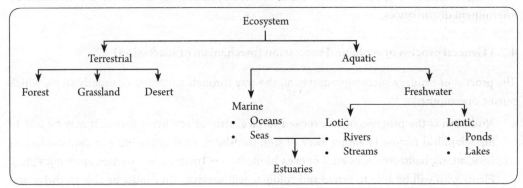

4.8.1 Terrestrial ecosystems

Terrestrial ecosystems comprise the relationships between the abiotic and biotic components on the landmasses such as continents and islands. It covers 28 per cent of the Earth's surface. They are different from aquatic ecosystems because aquatic habitat is the prime requisite of aquatic ecosystems. The dominant species here are the pteridophytes, angiosperms, gymnosperms, burrowing organisms like annelids, insects, reptiles, birds and mammals.

Classification of terrestrial ecosystems: Terrestrial ecosystems can be classified as –

Grassland ecosystem (Temperate and Savannah): It predominantly consists of various types of grasses. Temperate grasslands are known as the pampas in South America, the prairies in North America, Murray–Darling Basin in Australia, the Steppes in Central Asia and the Veldt in Africa.

Summers are warm and moist. Rainfall is not enough to support the growth of forest. The humus content is high. The climatic condition is cooler to a great extent than the Savannahs with an average summer temperature of 18°C and winter temperature of 10°C.

The primary producers are the grass. Fauna includes hares, deer, Saiga antelopes, foxes, gophers, sheep, goat, cow, wild dogs and bison occur in the temperate grasslands. They act as natural pastures for the grazers.

Savannahs' are intermediate between grassland and forests. The grasslands are interspersed with trees. Rainfall varies from 100 to 150 cm. Fire serves as a limited factor. *Panicum, Imperata*, etc., are the common species of grasses.

In India *Deyeuxia* and *Arundinella* can be found in the Himalayas, *Cymbopogon* in the Western Ghats, *Dichanthium* in the Malwa Plateau and *Saccharum* in Sunderbans and Cauvery delta.

Forest ecosystems: Forests cover roughly 30 per cent of land and 9.4 per cent of all the planet Earth. It refers to the section of land that is thickly covered with trees may be defined as a forest. They are also known as woods, weald or woodlands. A forest is a unit of living organisms on land that exhibits unique environment and many small undefined micro environmental conditions within that covers broad to very small areas. The environmental 'common denominator' of forest community is the tree and obeys all the natural cycles of energy, water and nutrients flow. There is substantial amount of detritus and huge number of decomposers. Fire is another limiting factor in the forest. Many plants develop adaptations to re-grow even after fire events. For example the presence of thick bark, dormant buds at the base of the trunk and epicormic buds under the bark. If the leaf canopy is removed they are stimulated to spring up. The tree cover generally reaches at least 2 m height at maturity.

Temperate and tropical forests: Tropical Forests occur close to the equator. The distinctiveness lies in the presence of only rainy and dry season lack of winters. Tropical forests may be tropical rainforests and tropical deciduous forests.

Temperate forests occur in between polar and tropical regions in both Northern and Southern Hemisphere. They exhibit modest and reasonable climatic conditions. Temperate forests can be grouped as deciduous, coniferous and broadleaved evergreen forests. The kinds of trees form the basis of classification. It is the abode to many faunal species, like rabbits, mountain lions, giant pandas, bears, kookaburras etc.

Forests ecosystems are dynamic and constantly changing community. Forests are treasured for social, environmental, cultural, and economic factors. Forests provide wood and non–timber products and services; plays key role in combating climate change, contributes to the economy and provides superb opportunities for amusement and tourism.

Importance of forests:

- Globally over 1.6 billion people are reliant on the forests directly for their livelihoods – foodstuff, attire, shelter, and conventional medicine.
- Industries dependent on forest gives employment to over 60 million people all over the world.
- They serve as essential source of raw materials like timber and non-timber products. Timber products consist of wood, pulp and paper and other wood-based products. Approximately, 3 billion people all over the world are dependent on fuel wood for heating and cooking. Forests

offer medicines, saps, spices, rubber, and oils are also a significant part of the forestry industry. Berries, nuts, seeds, mushrooms etc are edible.

- Tress removes carbon dioxide through photosynthesis and gives out plentiful of oxygen thus purifying air. Forests trap and store huge amounts of carbon dioxide –which assist in mitigating climate change. Tropical forests provide carbon-offsetting service, which may value up to $140 billion annually.
- Trees offer shade during the daytime and protects from the wind streams in the winter.
- Tree roots play a vital role in soil binding thus preventing weathering of soil by wind or rain.
- Forests also play a significant role in scavenging pesticides, fertilizers and other pollutants from local water bodies.
- Forests form the habitat for about 90 per cent of the world's terrestrial organisms.
- Forests are aesthetically beautiful.
- In many religions, trees are considered to be sacred and important parts of local tradition and mythology.

Deserts ecosystem: Deserts are usually hot during the day and cold at night. Deserts are generally found in the heart of the continents, generally between 25° to 40° north and south of equator. Deserts have extreme temperatures. Deserts may range from hot and dry deserts to cold deserts. Plants and animal life vary from desert to desert. During the day the temperature may be as high as 50°C, whereas night temperatures may fall to below 0°C. Rainfall is less than 250 mm per year and can be unpredictable. Animals and plants that reside in deserts have adapted to live in these adverse conditions. Major plants are the shrubs that are hardy and resistant to droughts such as sagebrush and cactus. The principal desert animals include desert snakes and desert lizards, foxes, dama gazelles etc.

Three factors cater to the formation of deserts:

- existence of **high pressure** resulting in cloud-free conditions;
- **cold ocean currents;** and
- **mountain ranges** to create rain shadows areas.

   A few of the world's deserts are – Sahara Desert, Sonoran Desert, Mojave desert, Atacama Desert, Thar Desert, Kalahari Desert, Gobi desert, Siberia (cold desert) etc.

4.8.2 Aquatic ecosystem

The interactions between biotic and abiotic components in oceans, seas, lakes, streams, creeks, marshes, pools and ponds form the aquatic ecosystem. The primary habitat in this case is water. Aquatic organisms can be classified as:

- **Planktons**, that flows with the water current – phytoplanktons (sargassam, diatoms, dinoflagellates) and zooplanktons (copepods, shrimps, jelly fish).
- **Nektons** or strong swimmers – squids, lobsters, crabs, fishes.
- **Neustons** or organisms that rest or swim on the water surface – Collembola, Gerris, Flying fish.

- **Benthos** or bottom dwellers –sea stars, sea cucumbers, sea anemones, brittle stars, clams and oysters.
- **Periphytons** that are attached to the submerged surfaces – crustaceans.

An aquatic ecosystem can be marine or freshwater ecosystems.

Marine ecosystems: Marine ecosystems cover approximately 71 per cent of Earth's surface. Diverge habitats ranging from coral reefs to estuaries, open pelagic sea to the deep benthic and abyssal region, and constitute this largest aquatic ecosystem in the planet. Major examples of marine ecosystems are:

- **Oceans and seas**: Characterized by the presence of salt water, this ecosystem is additionally classified into important oceans and smaller seas. They are huge reservoirs of water and support more than 2.5 lakh of marine species. The five major oceans are Pacific Ocean, Indian Ocean, Arctic Ocean, Atlantic Ocean and Southern Ocean. Notable seas are Red Sea, Caspian Sea, Black Sea etc.
- **Intertidal zone**: The area which remains under the water at high tide conditions and transforms into terrestrial habitat at low tide is referred to as the intertidal zone. Rocky cliffs and sandy beaches all fall under intertidal zones.
- **Estuaries**: Areas lying between riverine and marine environments, those which are vulnerable to tides and inflow of both freshwater and saline water. Estuaries are extremely rich and diverse as it is an ecotone. Estuaries exhibit constant mixing of water which stirs up the silt and increase the availability of nutrients. Due to the inflow of both fresh and marine water, estuaries have elevated levels of nutrients and exhibit a high biodiversity. Such a phenomenon is called edge effect. Organisms show a wide range of tolerance to temperature and salinity. They are often rich in migratory fishes such as salmon, eels, etc. Estuaries are alternatively known as inlets, lagoons, harbors. Examples are Chesapeake Bay, San Francisco Bay, Thane creek, Vasai creek, Vellar, etc.
- **Coral reefs**: Referred to as the 'rainforests of the sea'. They are mounds found in marine waters as a consequence of accretion of calcium carbonate deposited by oceanic organisms like corals and shellfish. Coral reefs represent the most speckled marine ecosystems on this earth, but comprize less than 1per cent of the world's ocean. Nonetheless, reefs support around 25 per cent of marine animals including varieties of fishes, sponges and mollusks. E.g. – Great Barrier Reef.

Common species found in marine ecosystems include:

- Marine mammals such as seals, dolphins, whales and manatees.
- Different species of fish including halibut, sea horse, mackerel, sardine, flounder, salmon, dogfish, sea bass, etc.
- Organisms such as brown algae, diatoms, corals, mollusk, echinoderms, etc.

Presently, marine ecosystems are susceptible to environmental issues such as climate change, pollution and overfishing and tourism.

Freshwater ecosystems: Freshwater ecosystems cover only 0.8 per cent of the Earth's surface. The water is non–saline. About 41 per cent of all the fishes are found in freshwater. Freshwater ecosystems faces threat because of the speedy extermination rates of several nonchordates and chordates, mainly because of pollution, overfishing, and other harmful activities.

Freshwater ecosystems can be of the following types:

Lotic ecosystems – Lotic systems refer to the swift flowing waters that move unidirectionally. Rivers and streams are the best examples, which shelter numerous species of insects, crustaceans, snails, slugs and fishes. Crayfish, crabs, clams and limpets, crocodiles and fishes are very common in streams and rivers. Lotic mammals include beavers, otters and river dolphins.

Lentic ecosystems: Lentic ecosystems refer to immobile, still or stagnant waters such as lakes and ponds. Ponds are usually of shallow depth and holds key to the village activities. Ponds and lakes supports a wide variety of organisms such as algae, rooted plants, floating – plants, protozoans, crabs, shrimps, crayfish, clams, frogs, salamanders, alligators and water snakes. Presently, ponds and lakes are subjected to pollution as a consequence of human activities such as washing clothes and utensils, bathing, cattle bathing and for drinking. Lake are much more deeper than the ponds and usually have a shallow littoral zone, an open water zone called limnetic zone with effective light penetration and a deep water or profundal zone without light.

On the basis of temperature lakes can be stratified as the epilimnion (warm surface water) and hypolimnion (cold non-circulating bottom water) with thermocline in between.

Lakes can be classified as:

- Oligotrophic lakes with poor nutrients and clear water – Lake Vostok, Lake Superior.
- Eutrophic lakes with excess nutrients – Lake Erie.
- Dystrophic lakes with low pH and high humic acid – Lake Matheson, Humic Lake.
- Endemic or ancient deep lakes – Lake Baikal, Lake Tanganyika.
- Desert salt lakes with very high salt concentrations and high evaporations – Great Salt Lake, Utah and Sambhar.
- Volcanic lakes receiving water from volcanic eruptions – Lake Toba, Lake Chagan.
- Meromictic lakes with permanent stratification – Round Lake.
- Artificial lakes constructed by man – Nagarjuna Sagar, Govind Sagar.

Case Study 4.1: Coral reefs, coral bleaching and Agatti Conservation Reserve, Lakshadweep, India

Coral reefs survive best between the temperatures 25°–29°C. The reefs stretch over an area of 280,000 sq. km and support about 2 million other marine species. It is the shelter for more than 25 per cent of fishes and a nursery bed for the juveniles. Corals serve as important input to the food chain and help in nutrient recycling. As per the World Meteorological Organization (WMO), tropical coral reefs provide global goods and services worth more than US $30 billion per year. Coral reefs assist the environment by:

They protect the shores from storms and wave actions.

- They have a prospect for food and medicines

- They are attractive tourist spots resulting in economic benefits,.

As per the 'Status of Coral Reefs around the world' in 2004, 20 per cent of the reefs have been destroyed with no prospects of recovery and 24 per cent of the reefs are under the risk of collapse.

The corals reefs faces threat from inland pollution, climate change and temperate rise, over fishing by bombing and cyanide poisoning, dredging practices, coastal development, industrial and tourism transports, mining of coral reef rocks, ocean acidification and lack of good governance.

Warmer temperatures result in expulsion of the algae (zooxanthellae) from the tissues of the corals turning the corals completely white, a phenomenon known as coral bleaching. The coral are not dead with such bleaching event but become highly stressed and more vulnerable to mortality.

When a coral bleaches, it is not dead. Corals can survive a bleaching event, but they are under more stress and are subject to mortality. Such weak reef cannot act as a buffer to the shorelines and becomes visually less appealing both to the divers and tourists. There may be an accelerated change in the species composition and reef community. Many susceptible species may die as a result of bleaching leading to loss of coral islands.

The Lakshadweep Islands are a cluster of islands about 200–300 km from the coast of Kerala in the Arabian Sea. The islands are very rich in biodiversity with thousands and thousands of marine species including 150 coral species. It has 11 inhabited and 14 uninhabited islands. The area suffers severe human threats as a consequence of human population, pollution, climate change, erosion, tourism pressure, beach armouring, fishing and fishing vessels, coral collection, fishing methods for tuna, etc. After monitoring the island for few years, changes in lagoon ecosystem, reef integrity and giant clam population were observed. In 2008, the Agatti Conservation Reserve was established in collaboration with BNHS, Lead International and funding from Darwin Initiative with an objective of conserving the giant clams, improving the local peoples livelihood through sustainable use of natural resources, creating opportunities for eco–tourism and reducing the direct human impacts on the reef. A local group called the Sandy Beach Cultural and Ecotourism Society has developed a glass bottom boat business to cater eco-tourism and a programme for turtle nest protection.

Wetlands: Wetlands include swamps, bog lands and marshes, where the water is usually of shallow and low depth. They are regions between land and water. As per Ramsar convention, wetlands can be defined as

'areas of marsh, fen, peat land or water, whether natural or artificial, permanent or temporary, with water that is static or flowing, fresh, brackish or salt, including areas of marine water the depth of which at low tide does not exceed six meters'.

Wetlands can be marine, estuarine, lacustrine, riverine or palustrine. Wetlands are excessively diverse and provide shelter to numerous animals and plant species such as black spruce, sundri, water lilies, mangrove, goran and tamarack. A range of reptiles, amphibians, birds and mammals are found in wetlands. Notable wetlands are Sundarban, Rashikbeel, Bhitorkanika, Vembanad, etc.

Case Study 4.2: The Sunderban mangroves

Mangrove forests comprise less than 1 per cent of all the forest areas all over the world. Sunderbans are one of the carbon rich tropical forests with a very high carbon sequestration potential. They are a transition from marine to freshwater and terrestrial systems. The Sunderban ecoregion lies in the delta formed by the confluence of the Rivers Ganges, Brahmaputra and Meghna, extending across the southern Bangladesh and West Bengal of India. The region experiences heavy rainfall of over 350 cm during monsoon and temperatures as high as 48°C. The unstratified and dense forest is characterized by *Heritiera fomes, Avicennia sp., Sonneratia apetala, Rhizophora, Nypa, Xylocarpus*, etc. Numerous fishes and lower animals spend their juvenile periods amongst the pneumatophores. The extensive mangrove food pyramid has a huge reserve of phytoplanktons in water and detritus in the forest floor. The tigers, at the pinnacle of the pyramid, are active swimmers. They predate on Chital, Munjtjac deer, wild boar and even Rhesus monkey. Often the people, moving into the thick forests for collecting honey, fish or wood fall prey to these tigers.

The plan of setting up of a 1,320 MW coal fired power plant at Rampal, located 14 kilometers from the Sunderbans raises many questions on the viability of these ecologically critical area. Some 4.75 million tonnes of coal would be burnt annually. The EIA report admits the daily air emissions of about 142 tonnes of SO_2 and 85 tonnes of NO_2. About 5 lakh tones of sludge would be produced. Power plant construction would involve site clearance, land filling, sand lifting, etc which will surely affect the aquatic habitats and diversity. The bherry cultures have already damaged the ecologically valuable juveniles. Such practice will lead to the collapse of the ecosystem. The air emissions will affect the kewra tree leaves on which the deer thrives. The impact on deer population will in turn spill over to the tiger population.

Environmentalists are worried over the threats from pollution, ignorance, hunting and poaching, illegal trade in wild life, lack of monitoring, all of which will jeopardize the future of Sunderbans. Such mangroves serve as major barrier to climate change and protect the coastal area from tsunamis and cyclones. What Sunderbans needs at present, is the conservation of habitat, policy and advocacy, information dissemination, capacity building, adaptation to climate change, lessening man wildlife conflict where humans can live in harmony with nature.

Table 4.3: Characteristic features of the various ecosystems

	Forest	Grassland	Desert
Characteristics	Dominated by trees. For example, Tropical rainforests of the Amazon basin, the temperate forests of eastern North America, and the Boreal forests of northern Europe. Mature forests often exhibit distinct vertical layers – forest floor, herb layer, shrub layer, understory, canopy, emergents.	Dominated by grass. Grasslands once occupied up to 25 to 40 per cent of the earth's land surface, but many of these grasslands have been used up for crop production. Grasslands can be classified as temperate or tropical. Temperate grasslands have cold winters and warm to hot summers; have deep, fertile soils. Tropical grasslands are warm throughout the year, having marked wet and dry seasons with annual rainfall amounting of 50 to 130 cms; have more number of woody shrubs and trees than temperate grasslands.	Deserts are dry places with sandy soil, too hot or too cold, availability of water is an important factor that determines the distribution of organisms; on an average having less than 25 cm of precipitation per year; strong tendency to lose water by evaporation and the potential for water loss exceeds the annual rainfall. All this result in scanty vegetation.
Abiotic components **Inorganic**	Calcium, Phosphorus, Sulphur, Potassium, Magnesium, Nitrogen, Oxygen, Carbon dioxide	Sulphur, Calcium, Potassium, Magnesium, Iron, Nitrogen, Phosphorus,	Minerals are comparatively less.
Organic	Chlorophyll, Carbohydrates Proteins, Fats, Humus	Chlorophyll, Carbohydrates, Proteins, Fats	Chlorophyll, Carbohydrates, Proteins, Fats
Climatic	Light, Temperature, Humidity, Rainfall	Light, Temperature fluctuations, low to moderate rainfall	Sandy soil, dry, porous, very low rainfall
Biotic components **Producers**	Maple, walnut, junipers, Pine, Spruce, Teak, Sal, Rosewood, palms, orchids etc	Cynodon, Digitaria, Setaria, Aristida	Acacia, Calotropis, Opuntia, Euphorbia, Zizypus

	Forest	Grassland	Desert
Consumers	Beetles, ants, spiders, deer, elephants, Nilgai	Cow, Buffalo, Goats, rabbits, hare	Kangaroo rats, cattles, camels
	Frogs, fox, snakes, Wolves	Snakes, Lizards, jackals, foxes	Sphryna, Lizards
	Lion, tiger, leopard		Vultures
Decomposers	*Clostridium, Bacillus, Nitrobacter,* *Fusarium, Agaricus, Physarum*	*Mucor, Rhizopus, Aspergillus*	Thermophilic bacteria
Examples	Tropical evergreen, tropical deciduous, tropical rainforest,	Prairies, steppes, Pampas, Veld	Thar, Gobi, Atacama, Sahara, Sonoran desert

	Freshwater	Estuary	Marine
Characteristics	Freshwater represents very small proportion of the earth and includes lakes, ponds, rivers, streams etc. Salinity is less than 1per cent.	It is the ecotone between freshwater and sea water, Mixing of two opposite currents; the unidirectional flow of river water mixes with the oscillating tidal currents of the sea.	It covers two–thirds of the surface of the Earth and is the largest of Earth's aquatic ecosystems. They include oceans, salt marsh, seas, coral reef etc.; they are marked by high salinity.
Abiotic components	Calcium, Phosphorus, Sulphur, Potassium, Magnesium, Nitrogen, Oxygen, Carbon dioxide	Sodium, Chlorine, Calcium, Phosphorus, Sulphur, Potassium, Magnesium, Nitrogen, Oxygen, Carbon dioxide	Sodium, Chlorine, Calcium, Phosphorus, Sulphur, Potassium, Magnesium, Nitrogen, Oxygen, Carbon dioxide
Abiotic components	Chlorophyll, Carbohydrates Proteins, Fats Turbidity of water and light penetration, pH, temperature, pressure, hardness in water	Chlorophyll, Carbohydrates Proteins, Fats Turbidity of water and light penetration, pH, temperature, variable salinity	Chlorophyll, Carbohydrates Proteins, Fats Turbidity of water and light penetration, pH, temperature, pressure,
Biotic components	Blue green algae, *Euglena, Chlamydomonas, Oscillatoria, Chara, Chlorella, Spirogyra*	Sea grasses and sea weeds	Dinoflagellates, Diatoms, Red algae (*Chondrus*) , Brown algae (*Sargassum*)

	Freshwater	Estuary	Marine
	Lemna, Wolffia, Eichhornia, Hydrilla, Vallisnaria, Sagittaria	Sea grasses and sea weeds	
	Protozoans, Daphnia, Cyclops, insects	Oysters, *Neries, Arenicola*	
	Small fish, frogs		
	Large fish, aquatic birds	Shoe crab	Crustaceans, Molluscs
		Bacteria and fungi	Herring, Shad, Mackerel
			Shark, Whale
			Marine barophilic bacteria
Examples	Baikal, Chilka lakes, ponds, Ganga, Tapti	Hudson Bay, Chesapeake Bay, Vellar	Indian Ocean, Pacific, Atlantic, Caspian, Red Sea

4.9 Species Interactions in Ecosystems

The fundamental life processes such as growth and development, nutrition and reproduction be governed by the interactions between individuals of same species (intraspecific) or between different species (interspecific). Certain interactions or associations are beneficial to each other and some are harmful or some may be neutral. The various types of possible interactions/associations can be categorized as:

A. Neutralism: This is the most familiar type of interspecific interaction (between individuals of different species) where neither population affects each other. These inter-relations are held to be indirect or incidental.

B. Beneficial association/interactions:

i. Proto-cooperation:

These types of interactions are mutually beneficial but non-obligatory between two species. For example, birds taking away pests from the bodies of bovine creatures etc. In absence of such interactions however, the birds have the choice of finding substitute nutrients sources and so do the bovine animal which is not dependant on the birds for survival.

The Crocodile bird (*Pluvianus aegyptius*) frequently goes inside the oral cavity of the crocodile and nourish on the leeches which are parasites of the crocodile. In this way, the bird procures its food and the crocodile is relieved of the sanguinivorous parasites.

ii. Mutualism (Symbiosis):

It is mutually beneficial, absolutely obligatory association between two species. For example, the association between bees and flowers; the honey bee gets food in the form of nectar, flowers in turn gets its pollen transported to distant areas by the bee. Without the flower the bee was not able to collect pollen and the plant would not have its pollination that will affect its survival ability. Benefit in terms of nutrients exchange is often termed as 'syntrophism'.

Lichen (association of algae with fungus) is the association in which algal part gets protection and simple nutrients provided to it by the fungal hyphae. The fungus obtains and uses CO_2 given out by the algae during the process of photosynthesis.

Microorganisms may also form symbiotic relationships with plants. An example is the nitrogen fixing bacteria, Rhizobium-legume association; the plant is benefited by getting readily available nitrate released by the bacterial partner, whereas Rhizobium is getting protection and finished food from the plant. A similar type of interaction exists between *Anabaena and Azolla. This* association is of enormous significance in paddy fields, where nitrogen is habitually a limiting nutrient. Symbiosis between actinomycetes (*Frankia*) with the roots of *Casurina and Alnus (non-*legumes) is widespread in temperate forest.

Another example of symbiosis subsists between the fungus and the roots of higher plants called *Mycorrhiza.* The fungus gets essential organic nutrients and protection from the plants. The plants absorb phosphorus, nitrogen and other inorganic minerals made readily available by the fungi. *Zoochlorella* subsists on the outer tissues of the Poriferan sponges. The alga manufactures food by photosynthesis and gives out O_2. In return the host offers matrix that serves as shelter to the algae and nitrogenous wastes as raw material.

iii. Commensalism:

In this type of interaction, one organism/partner is benefited and the other partner is neither benefitted nor harmed. For instance, many fungi are capable of degrading cellulose to the monosaccharide, glucose, which is then used by many bacteria. Remora gets attached to the body of the shark and is taken to new feeding sites. In due course, Remora is also able to feed on the food pieces falling from the shark's prey. Epiphytes clinging on trees receive mechanical support but, barely affect the trees. Red billed ox-pecker (*Buphagus erythrorhynchus*) frequently feed on the ecto-parasites such as lice, ticks and mites etc on the skin of rhinoceros. Rhinos are hardly affected by such act. Hermit crab (*Eupagurus prideauxi*) is found to live inside the empty shell of gastropods and allows sea – anemone (*Adamsia pallicata*) to fix and anchor on its shell. The sea-anemone is responsible for disguise and camouflage (protective colouration) and defends the crab from its enemies, while the crab aids in the speedy transit of the sea anemone and helps the anemone to find new feeding grounds.

iv. Scavenging:

This is a direct food linked interspecific interaction in which the scavenger or saprobiont consumes the deceased bodies of other animals, which have died naturally or is killed by some other creatures.

Scavengers sanitize the surrounding and the prevailing food is eventually disposed of so that the bulk of nutrients move in the nutrient recycling process. Animals such as vultures, foxes, hyenas, etc. are regular scavengers. Dogs, crows and ants are infrequently observed to act as scavengers.

C. Negative (harmful) associations/interactions:

i. Antagonism/Ammensalism:

The interaction involves the environment where one species is inhibited by another species. The inhibition may be direct or indirect; very common in case of antibiotic production. Antagonism may be of three types, i.e. antibiosis, competition and exploitation.

In the case of antibiosis, the metabolites or antibiotics synthesized or made by one organism inhibit or impede the growth and survival of another organism. *Bacillus* secreting an antifungal agent blocks the growth of numerous soil fungi.

ii. Competition:

Active competition may exist among the organisms either for the available nutrients or for the available space. The food and space may act as a limiting factor for one species and may result in favouring one species over another. Hence, competition is usually defined as 'the injurious effect of one organism on another because of the removal of some resource of the environment'. Take for instance, tiger and leopard competing with each other for preying upon a deer.

iii. Parasitism:

A hetero–specific association where one organism lives inside or on the body of another for the purpose of food and shelter. The parasite is reliant on the host and establishes metabolic relationship with the host. Consequently, in this host–parasite relationship, one organism (parasite) is benefited while other organism (host) is harmfully or adversely affected, even though not inevitably killed. Parasite can exist as ectoparasite or endoparasite. E.g. leech, *Ascaris, Fasciola*.

iv. Predation:

In this type of association/exploitation one organism (predator) kills and feeds on the other organism (prey). Generally, the predator is stronger and stout as compared to the prey, for example, tiger predating on an antelope.

Table 4.4: Summary of species interaction

Interaction	Effect on species A	Effect on species B	Nature of interaction
A. **Neutralism**	0	0	None is benefitted, none is harmed
B. **Positive**			
Scavenging	+	0	Beneficial to scavenger, who feeds on flesh or dead organisms
Commensalism	+	0	One is beneficial, other is unaffected
Proto– cooperation	+	+	Beneficial to both, but not obligatory
Mutualism or symbiosis	+	+	Beneficial to both, obligatory

Interaction	Effect on species A	Effect on species B	Nature of interaction
X. **Negetive**			
Ammensalism	+	–	One is beneficial, other is harmed
Competition	–	–	Negative
Predation	+	–	One is beneficial, other is killed
Parasitism	+	–	One is beneficial, other is harmed

4.10 Ecosystem Services

Ecosystem function is the ability of the ecosystem to provide goods and services through natural processes to fulfill human requirements, either directly or through indirect means. Ecosystem functions are understood to be a subset of ecosystem structures. Natural processes are the outcome of complex relations between living beings and the physio-chemical constituents of ecosystems through the universal forces of matter and energy.

There are four main groups of ecosystem functions:

A. Regulatory functions: refers to the capacity of the natural and seminatural ecosystems to regulate life support systems like photosynthesis, respiration and ecological processes like nutrient cycles, energy flow, evaporation, precipitation etc. This in turn helps to provide fresh air, water and land for sustenance.

B. Habitat functions: Natural ecosystems provide refuge and substrate to the wildlife. It contributes to the conservation of biological diversity at all levels and to the evolutionary process.

C. Production functions: The vital process like photosynthesis enables the autotrophs to trap solar energy and convert it into food. These are then consumed and utilized by the secondary organisms for their survival and a variety of other functions.

D. Information functions: Human evolution took place in the wild and nature serves as the reference point for the multidimensional development of a person. The beauty, serenity and diversity in nature provides excellent platform for spiritual improvement, intellectual advancement, refreshment and amusement, etc.

The ecosystem function progresses toward a complete and inclusive view of the natural goods and services which is shown in Table 4.5.

Table 4.5: Ecosystem services according to de Groot *et al.,* 2002

1. Regulation functions and related ecosystem services	1. Regulation of gaseous components 2. Regulation of climatic pattern 3. Prevention of any disturbances 4. Regulation of water cycle 5. Supply of water 6. Retention of soil 7. Formation of soil 8. Recycling of nutrients 9. Treatment of waste 10. Pollination 11. Biological control
2. Habitat functions and related ecosystem services	1. Act as refugium i.e. , a region unaltered by changes at has habitat for relict flora and fauna 2. Fostering and rearing
3. Production functions and related ecosystem goods and services	1. Supplies food 2. Supplies with raw material 3. Provides with genetic resources 4. Excellent source of medicines 5. Source of orchids and ornamental resources
4. Information functions and related ecosystem goods services	1. Aesthetic information 2. Eco-tourism and recreation 3. Cultural and artistic inspiration 4. Spiritual development and historic information 5. Dissemination of scientific and educational information

Summary

- Ecology is the branch of science that incorporates the basic conceptions of civilization. The word was introduced by combining two words 'Oikos' (house) and 'logos' (to study). Ecology plays significant role in conservation and management of resources, urbanization and town planning, population control, risk assessment and disaster management.

- An ecosystem is the basic functional unit in ecology, as it comprises both organisms and their abiotic environment. Elementary processes in an ecosystem include the nutrient cycle and continual energy inflow. Solar energy is trapped by the green plants by photosynthesis.

▪ Components of ecosystem comprise the abiotic factors that refer to non-living substances like air, water, land, elements and compounds. The diverse groups of living organisms comprise the biotic components that interact to allow the unidirectional flow of energy. Bulk of the energy trapped by the producers is dissipated in the environment as heat energy. Heterotrophs obtain food and energy from the autotrophs. The left over energy in a dead organism is devoured by decomposers.

▪ The flow of energy can be depicted by food chains from producers to consumers by eating and being eaten relationship. On the other hand, food web depicts anastomosing food chains with numerous producers, consumers and decomposers operating simultaneously. Trophic structure and trophic function of an ecosystem may also be graphically represented by means of 'ecological pyramids'

▪ The average amount of energy from the sun reaching the Earth's atmosphere is measurable as solar constant. The thermodynamics laws state how energy is transformed from one form to another and is applicable to ecology as energy transfer drives metabolism. Energy flow is always unidirectional, which is in contrast to the cyclic flow of material in ecosystems as represented by the biogeochemical cycles.

▪ Ecosystem reflects series of transitions known as ecological succession as it matures towards a stable climax condition.

▪ Ecosystems are broadly categorized into terrestrial and aquatic ecosystems. Terrestrial ecosystem refers to the interaction amongst the biotic and nonliving objects on land masses like islands and continents. Aquatic ecosystems commonly refer to interaction of living organisms and non-living entities in oceans, seas, lakes, streams, marshes and ponds. The primary habitat in this case is water.

▪ The fundamental life processes such as growth and development, nutrition and reproduction be governed by the intraspecific or inter-specific interactions. Certain interactions are beneficial to each other and some are harmful or some may be neutral.

▪ Ecosystem, thus, provides goods and services through natural processes to fulfill human requirements, either directly or through indirect means.

Exercise

MCQs

Encircle the right option:

1. The term ecosystem was coined by:
   A. Haeckel   B. Tansley   C. Wilson   D. Fourier

2. The chapter of interaction between living organisms and environment is called:
   A. zoology   B. ontology   C. ecology   D. physiology

3. Ecosystem comprise of:
   A. biotic components   B. abiotic components
   C. biotic and abiotic components   D. none

4. Nutrient level in an ecosystem is called:

   A. food chain   B. food web   C. trophic level   D. none

5. The study of groups of organisms is known as:

   A. Synecology   B. Limnology   C. Autecology   D. Pedology

6. The dissipation of energy from one level to the next in an ecosystem follows _____ law of thermodynamics:

   A. Zeroth   B. First   C. Second   D. Third

7. Pyramid of numbers in a parasitic food chain is:

   A. Upright   B. Inverted   C. Semi upright   D. all at times

8. The amount of solar radiation utilized by the producers is:

   A. 2 per cent   B. 10 per cent   C. 8 per cent   D. 50 per cent

9. 'The 10 per cent law' is by:

   A. Maslow   B. Grinnel   C. Lindemann   D. Malthus

10. Competition relationship is denoted by:

   A. + +   B. + 0   C. – +   D. – –

Fill in the blanks.

1. The chapter of interaction between living organisms and environment is _____.
2. The number pyramid in parasitic food chain is _____.
3. _____ is the main source of energy in ecosystem.
4. Lakes with algal blooms are referred to as _____.
5. GPP is highest in _____ forest.

State whether the statements are true or false.

1. Food chain exhibits eating and being eaten relationship in ecosystem. (T/F)
2. Both energy flow and nutrient flow are cyclical in nature. (T/F)
3. Normally ecosystem is in the state of equilibrium known as homeostasis. (T/F)
4. NPP is always less than GPP. (T/F)
5. Detritivores are also known as saprotrophs. (T/F)

Short questions.

1. Define ecology.
2. State the principles of ecology.
3. Compare autecology and synecology with proper examples.
4. What is ecological footprint?
5. Define ecosystem.

6.   State the difference between photoautotroph and chemoautotroph.
7.   What are omnivores?
8.   How are scavengers different from decomposers?
9.   What is trophic structure?
10. Define food chain.

Essay type questions.

1.   Describe the various components of ecosystem and give suitable examples.
2.   What are ecological pyramids? Describe the various types of ecological pyramids with suitable examples.
3.   Energy flow follows the law of thermodynamics. Justify.
4.   Explain the water cycle.
5.   Explain the Carbon cycle.

Biodiversity and Its Conservation ⎯⎯⎯⎯ 5

Learning objectives

- *To understand the various aspects and levels of biodiversity.*
- *To know about the various biogeographical regions and enormous wealth of India along with global perspective.*
- *To realize and develop a value and appreciation for the biodiversity services we receive.*
- *To draw attention towards the threats that biodiversity confronts.*
- *To know about scientific biodiversity management strategies.*

5.1 Introduction and Concept

The primary aims of the Convention of Biodiversity (CBD) are:

- conserving biodiversity including protection and management;
- sustainable and viable employment of the constituents of biodiversity; and
- rational and justifiable benefit distribution that arise out of the utilization of genetic resources.

India being a signatory party to the CBD also pursues the same objectives.

According to **Convention of Biodiversity, CBD**, 1992 'Biological diversity' means the variability among living organisms from all sources including, inter alia, terrestrial, marine and other aquatic ecosystems and the ecological complexes of which they are part: this includes diversity within species, between species and of ecosystems'.

The term *'biological diversity'* was invented by Thomas Lovejoy in 1980 while the word *'biodiversity'* was invented by Walter G. Rosen in 1985; however, it appeared for the first time in 1988, in an article by EO Wilson. Diversity can be defined as the total number of diverse objects and their relative frequency in which they occur on earth. In other words, **biological diversity refers to the variety (multiplicity) and variability (changeability) among the sum total of all organisms along with the environmental domains in which they usually occur.**

According to **World Resources Institute (WRI), World Conservation Union (WCU), and United Nations Environment Programme (UNEP)**, in 'Global Biodiversity Strategy,' 1992: *'Biodiversity is the totality of genes, species, and ecosystems in a region...'*. Biodiversity can be represented by three hierarchical or tiered categories, i.e., in terms of genes, in terms of species, and in terms of ecosystems that refer to and describe relatively different aspects of living organizations that scientist's quantify in diverse ways.

In biodiversity, the organization of living beings is considered at various levels, starting from the complete ecosystems right up to the chemical assemblies that constitutes the hereditary units at molecular level. Hence, biodiversity includes various ecosystems, various species, various genes and their comparative abundance. Or to rephrase, total number and diversity of species, ecosystems and the genetic variability they comprise (Office of Technology Assessment, 1987).

Simply defining, biological diversity is the variety of life and its processes; and it includes the variety of living organisms, the genetic differences among them, and the communities and ecosystems in which they occur. **Biodiversity encompasses ecosystem or community diversity, species diversity and genetic diversity.**

5.1.1 Genetic diversity

Genetic diversity refers to the variation of genes that occur within the species. This comprises discrete populations of the same species (for example, varieties of indigenous paddy *Oryza barthii, Oryza sativa indica, Oryza sativa japonica*) or genetic variation within populations (which is found to be high amidst Indian rhinos and low amongst the cheetahs).

Indian lion (*Panthera leo persica)* populations are inherently different from that of African populations (*Panthera leo senegalensis, Panthera leo bleyenberghi)* in terms of genetic constituents. Among two populations of same species, adjustments and adaptations to local circumstances often culminate into genetic variations.

5.1.2 Species diversity

Species diversity is the unevenness and diverse types of organisms that exist in a given space. A given area of the Himalayan region may harbour more variety of species than in Western Ghats of a similar size. Such diversity is measurable in several ways. Often, 'species richness' or the number of species in an area is employed as a measure. To be precise, 'taxonomic diversity' is a more perfect measure. It reflects the relationship of one species in relation to each other. Take for example, an island with five species of birds and two species of mammals possess greater taxonomical diversity than another island with ten species of birds and no mammals.

5.1.3 Ecosystem diversity

Ecosystem diversity embraces the different types of abode occurring within a given region or the different areas found within a landscape. The array of habitats and the environmental impulsions that comprise the Sunderbans is a significant example. This Mangrove ecosystem comprises forests, wetlands, rivers, estuaries, fresh and salt water. **Ecosystem diversity** is often difficult and tough to determine than both species diversity and genetic diversity, given the fact that community boundaries, confederations and correlations of species and ecosystems are mysterious.

5.2 Measuring Biodiversity over Spatial Scales

- **Alpha (α) diversity:** number of species found in a small, homogeneous area. It is usually expressed by the number of species (i.e., species richness) in that ecosystem.
- **Beta (β) diversity:** a comparison of diversity between ecosystems, usually measured as the amount of species change between the ecosystems. The greater the beta diversity, the greater will be the distinctiveness of the two communities.
- **Gamma (γ) diversity:** a measure of the overall diversity within a larger geographical scale.
- **Delta (δ) diversity:** diversity of higher rank (macro-regional) complexes of ecosystems, for example, global biogeographical regions, presented below;
- **Epsilon (ε) diversity:** variety of life environments (marine, land based);
- **Omega (ω) diversity:** phylo-genetic diversity/diversity of the global taxonomical hierarchy.

Table 5.1: Distribution of species to the five known kingdoms

Types of organisms	Approximate number recorded	Defined, labeled and categorized
Animals	7.77 million species	953,434
Plants	298,000 species	215,644
Fungi	611,000 species	43,271
Protozoa	36,400 species	8,118
Brown algae, diatoms and water moulds	27,500 species	13,033

Box 5.1: Biogeographical realms of the world

According to **Alfred Wallace**, the entire Earth is divisible into the following realms:-

1. **Palearctic** – comprises the whole of Europe, Northern Africa up to Sahara desert and Asia with the exceptions of India, Pakistan, and southeastern Asia along with Middle East.
2. **Nearctic** – whole of Canada, USA, and Mexico up to the tropics.
3. **Neotropical** – rest of Mexico from the tropics up to South America and the Antilles.
4. **Ethiopian or Afro–tropical** – whole of Madagascar, Africa lying south of Sahara, southern region of Arabian Peninsula.
5. **Indo-Malayan or Oriental** – Pakistan, Indian, Southeastern Asia, Philippines, part of Indonesia lying west of Wallace's line.
6. **Australian** – whole of Australia, New Guinea, Tasmania, portion of Indonesia lying to the west of Wallace's line.
7. **Oceanian** – Isolated small islands.
8. **Antarctic** – Antarctica.

Box 5.2: World's Major biomes (Source: WWF)

A biome is an ecological unit. According to WWF, a biome is *'large unit of land or water containing a geographically distinct assemblage of species, natural communities, and environmental conditions'*. These eco-regions do not have sharp distinct boundaries but comprises an area important for ecological and evolutionary processes. The WWF classifies the **terrestrial system** into 14 biomes. They are:

1. Tundra
2. Boreal/Taiga forests
3. Temperate broad leaved and mixed forests
4. Temperate coniferous forests
5. Montane grasslands and shrublands
6. Mediterranean forests, woodlands and the scrub
7. Flooded grasslands and Savannahs
8. Temperate grasslands, savannahs and shrublands
9. Tropical and subtropical grasslands, savannahs and shrublands
10. Tropical and subtropical moist broad leaved forests
11. Tropical and subtropical dry broad leaved forests
12. Tropical and subtropical coniferous forests
13. Deserts
14. Mangroves

WWF classifies the **freshwater system** into 7 biomes:	WWF further classifies the **marine water** into five biomes:
1. Large river ecosystem	1. Temperate upwelling
2. Small river ecosystems	2. Tropical upwelling
3. Large river headwater ecosystems	3. Tropical coral
4. Large lake ecosystems	4. Polar eco-regions
5. Small lake ecosystems	5. Temperate shelf
6. Large river delta ecosystems	
7. Xeric basin ecosystems	

'Distinct species' seems to be a simpler documented feature of biodiversity. Not many species are plenteous. Others have small populations that may even confront threats and risk to extinction. Conservation of biodiversity thus comprises continuing and maintaining the species in nature in numbers along with their spatial distribution which in turn will guarantee a high prospect of uninterrupted survival.

'**Species association**' within and between the species are an imperative to biodiversity. 'Associations' are commonly referred to the organism communities generally recognized and accepted as distinct stands, areas, or places. The biotic parts of ecosystems are the communities. The variation of species in any ecological unit is a function of its organizational and operational uniqueness and the multiplicity of its ecological processes and the physical environment.

To conclude, at large geographic scales, starting from *watersheds* to the total *biosphere* – comprises varieties in ecosystem, their plan and organization, their associations across their arrangements and relationships across regions. It is from these enormously big, regional landscapes that people must obtain sustainable yields of resources along with maintaining and propagating numerous samples of organically dissimilar ecosystems.

The echelons of the biodiversity constituents and operations are artificial having a discrete human relevance. Nevertheless it makes for a conception that is substantially wide-ranging and dynamic which must be dealt in the context of the total range of requirements and aspirations of mankind.

Globally a total of 1,750,000 species has been reported and named, the total number inclusive of the undescribed ones may be 10 times more (*World Atlas of Biodiversity*, 2002, Goombridge and Jenkins).

According to recent UNEP–WCMC (World Conservation Monitoring Centre) (23 August 2011 from UNEP Information Centre), the most recent sum total of species on Earth is around 8,700,000, of which about 6.5 million species are terrestrial and 2.2 million (about 25 per cent of the total) are marine.

The 8.74 million species, when assorted and distributed to the five known kingdoms, can be represented in Table 5.1.

5.3 Bio-geographic Classification of India

The biogeographical classification of India used four levels of planning.

- Bio-geographic zone – comprises units with similar ecology, representation of biome, community and species.
- Biotic province – particular communities within the zone separated by barriers and exhibit a gradual transition in environment.
- Land region– various landforms within the province.
- Biomes – an ecological unit.

Bio-geographic classification of India was accomplished by Rodgers and Panwar (1988), describing 10 bio-geographic zones in India with 25 bio-geographic provinces. Further revision in 2002 using GIS techniques by Rodgers, Panwar and Mathur classified India into 10 zones and 26 provinces based on factors such as altitude, moisture, topography, rainfall, etc. Bio-geographic zones were used as a basis for planning wildlife protected areas in India. Bio-geography deals with the geographical distribution of plants and animals. Communities of plants and animals in different geographical areas of the world vary widely from each other. There are two divisions of biogeography: (a) phyto-geography; and (b) zoo-geography. Phyto-geography deals with origin, distribution and environmental inter-relationships of plants; whereas zoo-geography deals with the migration and distribution of animals.

The 10 bio-geographic zones are described as follows:-

A. Trans-Himalayan region

- The Himalayan ranges, covering around 25,00,000 sq. km, a vast stretch of the cold, desert, lie to the north of Himalayan range, encircling the entire Tibetian Plateau, Karakoram, Ladakh, and Zanskar range. K2, Hidden park, Broad Peak are the significant peaks while Siachen, Biafo, Hispar are some of the important glaciers of this region.

- The region exhibits scarce vegetation and is the richest habitat of wild sheep and goats.

- Tibetan Gazelle, Tibetan antelope, snow leopard *(Panthera uncia)*, blue sheep, and migratory black–necked crane *(Grus nigricollis)*, bear, wild deer, wild yak, ibex, marbled cat, Himalayan Pit Viper and wild ass occur in these parts of Tibet Himalayas.

- Cashmere wool or pashmina is the main economy for the Ladakh inhabitants.

B. Himalayas

- This region stretches across J&K, HP, Uttaranchal, Sikkim, WB, Arunachal Pradesh, Mizoram, Nagaland, Manipur, Assam, Tripura and Manipur. Thus, it stretches from NW region of Kashmir right up to the east of 'NE FA' and occupies 6.4 per cent of Indian geographical area.

- It represents the world's youngest and highest mountain chains.

- There is no vegetation above the snowline. The vegetation is mostly tropical rainforests in the Eastern Himalayas and thick subtropical and alpine forests in the Central and Western Himalayas.

- Rhododendrons attain tree height on the eastern part. The widespread growth of evergreen tall trees along with grasses adds intensity to these thick forests. Predominant trees found in the Himalayan region are oak, pine, deodar, chestnut and other conifers.

- Several interesting animals including ibex, shrew, tapir and panda are found here. Sambar and muntjac/barking deer mainly occur in the subtropical foothills; serow, goral and the Himalayan Tahr inhabit the temperate and subalpine regions; whereas snow leopard and brown bear dominate the alpine region.

The Himalayan region comprises three vegetation zones governed by latitudinal, altitudinal and precipitation regime. They are:

- Sub-mountain or lower region – a tropical and subtropical region having a height between 5,000 ft to 6,000 ft. *Dalbergia sissoo* and *Acacia catechu* are the important trees here.

- Temperate zone stretches from 5,500 ft to 12,000 ft. *Pinus roxburghii* is predominant.

- Alpine zone – upto 1,200 ft, restricts the growth of trees, hence also referred to as timber line.

The lofty altitude, steep rise and the rich temperate vegetation have together contributed a inimitable persona to the Himalayan region.

C. Semi-arid areas

- It forms the transitional area between the desert on one hand and dense forests of the Western Ghats on the other hand. This zone comprises the states of MP, Chhattisgarh, Gujarat and the port of Odisha and is adjacent to the North-western desert.

- The vegetation is discontinuous often with areas of exposed soil and is characterized by soil-water deficit throughout the year. The forest has grown into thorny, mixed deciduous and Sal type being influenced by precipitation. Thorny scrubs, grasses and a small number of bamboo trees are also present in some parts of this region along with xerophtic herbs and ephemerals. Vegetations are *Tectona grandis, Dalbergia*, scattered Acacia, Calotropis, Salvadora, Tamarix, Euphorbia, etc.

- Birds, jackals, leopards, eagles, snakes, foxes and buffaloes are found in this region. Asiatic lion is the endemic species of this province.

D. Western Ghats

- Geographically, this region stretches from the southernmost tip of the Indian peninsula (8°N) to the north about 1,600 km upto the mouth of Tapti River (21°N). Thus, it occupies the parts between Gujarat in the north to the Cape Comorin in the south inclusive of the Western Ghats mountains and Malabar Coast.

- The average height of the mountains here lies amid 900 to 1,500 m from mean sea level. The southwest monsoon winds intercepts to produce a rain shadow area to their east.

- Globally significant, one among the 34 biodiversity hotspots declared, it is characterized by soaring levels of endemism at both higher and lower taxonomical levels.

- The region is well-known for harbouring Nilgiri langur, lion tailed macaque, Nilgiri Tahr, Malabar grey hornbill; 14 endemic species of caecilians occur in this region out of the 15 recorded species.

- The region receives profound rainfall. It exhibits moist tropical evergreen forests, subtropical or temperate evergreen forests, mixed deciduous forest and mangrove forests.

The varied landscape and mottled climatic conditions results in a diverse compilation of habitats that bear only one of its kinds of flora and fauna.

- With many native people dwelling in these forests, the cultural diversity supplements the biological diversity as well.

- It also shares many floral species of Sri Lanka.

- Cultivation of cash crops such as areca nuts and pepper is followed by rice cultivation. The high elevations are very thinly populated by the indigenous people.

- Myristica swamps are formed below a height of 100 meters with slow flowing streams and poorly drained valley floors.

- The large pockets of the primary forests have been replaced by expanding traditional agricultural practice and increase in artificial rubber, tea, coffee and forest tree plantations.

E. Northwest Desert regions

- Kutch, Thar and Ladakh including parts of Delhi comprises the desert zone.

- Summers are very hot and dry while winters are cold; precipitation often less than 700 mm.

- Xerophytic plants like *Acacia nilotica*, *Tecomella undulata*, Babul, Kikar, etc. and wild palm grows in areas with modest rainfall. Rann of Kutch serves as the breeding place for the flamingoes.

- Great Indian Bustard (endangered), camels, wild asses, desert foxes, chinkara, black buck, gazelle, nilgai and snakes are found here.

F. Deccan Plateau

- It covers Deccan plateau (south) Central plateau, East Plateau, Chhota Nagpur and Central Highland; precipitation approximately 100 mm.

- Being situated beyond the Western Ghats, it is mostly semi-arid. It falls in the rain shadow area of the Ghats. It is the largest unit of the Indian Peninsular Plateau.

- Northern, central and southern part of the plateau is characterized by tropical dry deciduous forests.

- The highlands being covered by wide varieties of forests produce a huge variety of forest products.

- Tigers, sloth bear, wild boar, gaur, sambar and chital dominate the zone along with small groups of wild buffaloes, elephants and barasingha.

G. Gangetic Plain

- It extends up from the Himalayan foothills in the north and covers of Uttar Pradesh, Bihar and West Bengal. This forms the biggest unit of the Great Plains and Ganga is the most important river. The Great Plains cover about 72.4 million hectares area with the Ganga and the Brahmaputra establishing the chief drainage system.

- The alluvial sediments vary considerably in thickness. The physical features vary significantly. It may be arid and semi-arid areas of the Rajasthan Plains or the moist, damp regions of the Delta and Assam valley in the east.

- The region holds the highest population density and is highly dependent on agriculture based economy in a number of of these areas.

- The important trees are teak, sal, shisham, mahua, khair etc.

- The animals include elephants, black buck, gazelle, Indian rhinoceros, Bengal florican, crocodile, freshwater turtle and a thick waterfowl population.

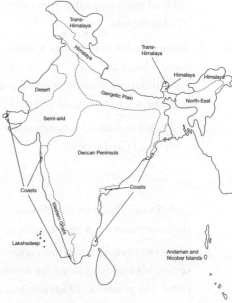

Figure 5.1: Biogeographical zones of India

H. Northeast India

- Within the country, northeastern India is in fact the richest region in terms of vegetation with rainfall as much as 1,000 cm.

- Mostly it is evergreen and semi-evergreen rain forests.

- It supports diverse types of orchids, bamboos, ferns and other plants. The wild varieties of banana, mango, citrus fruits, pepper and other cultivable flora is generally found.

- Dense elephant population is found here.

I. Islands

- Islands of Lakshadweep and Andaman and Nicobar comprise the islands. The two clusters of islands one in the Arabian Sea and the other in the Bay of Bengal respectively exhibits a wide variety of coastal vegetation of mangrove plants.

- Lakshadweep and Andaman and Nicobar are both distinct in terms of their origin as well as in topographical features.

- The Arabian Sea islands (Laccadive and Minicoy, etc.) represent the ruined remains of the old land mass and subsequent coral formations. The Bay of Bengal islands stretch about 220 kilometers.

- Though Lakshadweep has a width of 58 km, yet it exhibits evergreen forests. Some of these islands are bordered by coral reefs. 200 endemic higher plants are found. Five types of forests, namely, tropical evergreen forests, tropical semi-evergreen forests, moist deciduous forests, littoral and mangrove forests are exhibited.

- The fauna includes Andaman water monitor, turtles, wild boar, Narcondam hornbill and a stretch of coral reef.

J. Coasts

- India has an elaborate coastline of about 7,516.5 km. The Indian coastlines are speckled in their distinctiveness and structures.

- The western coast is narrower with the exception of Gulf of Cambay and Gulf of Kutch but somewhat extensive along the southern part of Sahyadri. Coconut trees grow all along the coast. The presences of back waters are distinctive of this coastal region.

- The plains on the eastern coast are considerably wide owing to the deposits from the east flowing rivers. Rivers like Mahanadi, Godavari, Krishna and Cauvery have extensive deltaic systems. The estuarine tracts are marked by distinct mangrove vegetation.

- The soil covering the coastal plains are fertile and suitable for growing various crops. Rice is the most important grown in these areas. Rice is the main crop cultivated in these areas.

- Dugong, dolphins, crocodiles and freshwater and marine turtles and tortoises and birds are dominant. The region has the highest tiger population.

5.4 Value of Biodiversity

Throughout the ages ecosystems are sufficient to adapt and cope up with the environmental changes. Existing species evolve with the process of evolution and new species may take the place of existing species without disrupting the ecosystem. The range of variation and the range of activities they support are major factors in their resilience. When elements of biodiversity are lost, ecosystems become less resilient to abrupt pressures such as disease and climatic extremes.

The significance of wildlife has often been broadly overlooked or underrated in the earlier period by the global mass. There was inadequate aesthetic and tourism significance of wildlife. But such a situation has changed with time. The significance of wildlife to local people is nowadays internationally documented in community-based and shared or involved natural resources management programmes. The economic importance of wildlife not only relies on its consumptive and non-consumptive use, but also in its prospective dietary importance.

5.4.1 Consumptive value

The biological diversity provides most of the requirements that can be harvested and used directly for people's survival.

- The different varieties of cereals (rice, wheat, maize, corn, oats, and barley), pulses (moong, musur, tur), aquaculture (marine and freshwater fishes, oysters, lobsters, mussels), vegetables, beverages, spices etc comes directly from the diverse forms of wildlife that are their treasure and serve as food.

- We use timber as building materials.

- Dried biomass and the petrified products of coal, petroleum and natural gas that serve as fuel are all derived from biodiversity.

- Biodiesel can be prepared from sunflower and Jatropha.

5.4.2 Productive value

The products are harvested from the wild, marketed and sold directly at national and international markets either as raw or processed form.

- Natural crop varieties are bred and the new hybrid varieties are selected for better and improved qualities. Biodiversity provides the natural enzymes such as papain. The fruits act as rich sources of vitamins and minerals.

- A variety of dye such as red from madder (*Rubia tinctoria*), yellow from chamomile (*Anthemis tinctoria*) and blue from indigo (*Indigofera tinctoria*) is obtained.

- Fibres such as wool and cotton (*Gossypium* sp.) are used to make clothes. Sisal (*Agave sisalana*) leaves are used to make strings and ropes.

- Neem extracts are used as disinfectants.

- Gum Arabic is a product of *Acacia* trees.

- Trees provide shelter and food to many. Insects, birds, bats act as pollinators. Parasitism and predation helps in controlling the population of animals.

- Detritivores and decomposers help in forming compost and recycling nutrients. Green plants scavenge huge amounts of CO_2 and act as natural sink.

- Wetlands act as buffers between the land and water by filtering sediments, nutrients and contaminants from inflowing water.

- Indirect values assigned to biodiversity that may provide benefits to people are water quality, soil protection, recreation, education, scientific research, climate regulation etc.

A. Livestock rearing

- Livestock like poultry fowls, cattle, goat, and lamb is reared and harvested for meat, wool, egg and dairy products.

- The livestock section is globally the single largest human use of Earth.

- Animal fodder accounts for about one-third of cultivable land while grazing areas of the land comprises 26 per cent of the Earth (FAO, 2006).

- Increased agricultural yield is likely to exert its effect on the ecosystem and biodiversity services as it calls for land transformation for the purpose of crop production.

- The increasing livestock will directly contend with mankind for all sorts of natural resources including land and water.

- Livestock farming is not only the key to increased deforestation but also one of the biggest contributors of water pollution.

- 70 per cent of Amazon that was primarily a forest is currently utilized as a pasture land (FAO, 2006).

B. Wildlife farming

- 50 per cent of the farmed deer population of the world in over four thousand farms, which figures about 2 million, earns a revenue of NZ$ 200 million annually in New Zealand.

- Green iguana (*Iguana iguana*) ranching in Costa Rica is extensively practiced for the purpose of meat, leather/ hides, pet exports and tourist activities.

- In Latin America, the Capybara is of greatest interest for both meat and hides simultaneously.

C. Wildlife trade

- The legitimate wildlife trade comprises organisms of species that do not find place in any of the three **Convention on International Trade in Endangered Species (CITES)** appendix list.

- Organisms of species enlisted under the CITES may undergo international trade with suitable credentials. The credentials comprise permits and certificates issued by the individual country's CITES Management Authorities for exporting, importing or re-exporting.

- **TRAFFIC (Trade Records Analysis of Flora and Fauna in Commerce)**, is the wildlife trade monitoring network. It is an international association committed to ensure that trade in wild flora and fauna is not a menace to the protection of environment. TRAFFIC is a joint venture or partnership between **WCU** and **WWF**.

- South African auction of wildlife exemplifies the true financial value of big sized mammalians as is exhibited by the market price. R9 million (R stands for Rand) was earned in 1991 by selling 8,292 animals in nine sales. In 2000, in 48 sales the prices increased to R 62.9 million for 17,702 animals. On an average the auction price of roan antelope (*Hippotragus equinus)* rose to R 83,000 in 2000 from R17,000 in 1991 and that of sable antelope (*Hippotragus niger*) increased to R 53000 in 2000 from R 25,286 in 1991 (T. Eloof, personal communication).

- Not all wildlife trades are legal. According to CITES, around 25,000 African elephants were slaughtered in 2011 alone.

D. Medicines and healthcare

- Both traditional and processed drugs are obtained from biodiversity, such as penicillin, reserpine, quinine, atropine, morphine, vinblastin, vincristin etc.

- *Aloe vera* leaves are extensively used for healing wounds, burns etc. Biotechnology is finding new species to handle pollution and combat problem of incurable disease like cancer, AIDS etc.

- Newman and Cragg, 2007, opined that there is substantial connection between the biological diversity and modern day healthcare.

    i. Nearly half of all synthetic drugs have a natural origin. Out of 25 highest selling drugs, 10 owes its origin from nature.

    ii. Globally, 42 per cent of all anti-cancer drugs are natural and 34 per cent is semi-synthetic; for example Saracodictyin from *Saracodictyon roseum* and Paclitaxel from *Taxus brevifolia*.

    iii. The people of China uses 5,000 out of the 30,000 recorded plants of higher taxa for therapeutic purposes.

    iv. Nearly 75 per cent of the global population depends on conventional natural remedies.

    v. In 1997, the turnover of genetic resource based drugs was between US $ 75 billion and 150 billion in United States.

    vi. Substances highly effective against cardiovascular diseases from the gingko tree led to an annual income of US $ 360.

- The biodiversity-healthcare correlation also has a strong distributional equity dimension. There is often a disparity between the areas where benefits are produced, where their value is enjoyed, and where the opportunity costs for their conservation are borne.

Table 5.2: Few commonly used microbial products

Names	Products/ antibiotics	Application
Penicillium notatum chrysogenum	Penicillin	infections in skin, teeth, ear and respiratory tract, urinary tract and gonorrhea
Cephalosporium	Cephalosporin C	strep throat, pneumonia, bronchitis, tonsillitis, otitis, gonorrhea and skin infections
Streptomyces	Tetracyclines	malarial infections, chlamydia infections and syphilis lyme disease
Streptomyces venezuelae	Chloramphenicol	bacterial eye infections
Spectromyces grizeus	Aminoglycosides	pneumonia, typhus and other bacteria-causing illnesses

Table 5.3: Few commonly used plant products and their medicinal use

Plants	Scientific names	Parts used	Application
Amla	*Emblica officinalis*	Fruit	cough, diabetes, cold, laxative and hyper acidity
Aswagandha	*Withania somnifera*	Root, Leafs	restorative tonic, stress and nervous disorder
Guluchi	*Tinospora cordifolia*	Stem	gout, pile, general debility, fever and jaundice
Sandal Wood	*Santalum album*	Heartwood, oil	skin disorder, burning sensation, jaundice and cough
Tulsi	*Ocimum sanctum*	Leaves	cough, cold, bronchitis and expectorant
Periwinkle/ Nayantara	*Catharanthus roseus*	Whole plant	leukemia, hypotensive, antispasmodic and antidote
Neem	*Azardirchata indica*	Stem and leaves	sedative, analgesic, epilepsy and hypertensive

5.4.3 Social value

Understanding the social value of biodiversity is of extreme importance, especially in the current context, when unchecked exploitation of resources and unchecked development has already unleashed the twin menace of global warming and climate change. Often, what seems the logical step today, has far reaching adverse consequences in the future which is why the social value of such steps need to be calculated up front so that it may be evaluated in the correct perspective.

- Many of the plants like banyan, peepal, mango, bael, tulsi, etc., and animals like cow, snake and peacock are regarded as holy and sacred.

- Our lifestyle, songs, dance, scriptures and customs are closely related with wildlife and much more pronounced in the tribal lifestyle.

- Policy measures and resource utilization must therefore factor in the social value before setting into motion the wheels of change. Only after people know the social value that they may embrace sustainability.

- An example will suffice – while people know the cost of producing a ton of steel, and even what that one ton can do to the nation's economic agenda, they seldom bother about the 'social cost' of producing that ton of steel. What will be the impact on the biodiversity of the area where the plant will be set up? How many trees will be felled? From where and how will the raw materials be brought in, at what cost to the environment? How will the finished goods be shipped to the market? What will happen to the local population, the animals and the plants? What impact will the plant have on the region's ground water levels? What kind of pollution will the plant emit? If all these and other relevant costs are factored in, chances are, the proposed plant will be unviable. And that is precisely why calculating the social value is gaining in importance.

5.4.4 Ethical value

All species were created equal and have the moral right to live, procreate and grow.

- However, being at the top of the food chain, in the mad haste to control the forces of nature, human have played havoc with the fragile and interdependent ecosystems.

- Today, hoist as people are in their own petard of development, this intricate balance of life is becoming more and more clear to us, from which is emerging concepts like ethical value of biodiversity.

- Simply put, it is the concept that is urging people to understand and mould they moves as they take a holistic view of the consequences of their actions and do things that are sustainable, inclusive and honour the rights of every living organism that might be jeopardized by their wantonness.

5.4.5 Aesthetic value

- Wildlife is beautiful. It is a pleasure to see and smell flowers, to listen to birds singing, to watch the serenity and tranquility of forests etc. the sanctuaries, national parks and biosphere reserves are places attracting tourists.

- Tourists from all parts of the world spend a lot of money to visit the wilderness. They enjoy the tranquility, the natural and the aesthetic beauty of the forests and wildlife.

- Wildlife acts as the pillar of the tourism industry similar to the beaches that act as support to the beach tourism industries. Such type of tourism is basically rooted on nature and animal watching and is exclusively a division of service department.

- Plants and animals are often used as symbols in flags, paintings, sculptures, photographs, stamps. Their names are often used in songs and legends.

- Lion elephants, tulsi, bael, mango leaves are often associated with deities and rituals. Species like Asiatic Lion, panda are chosen as **flagship species** for their attractiveness and distinctiveness to represent an environmental cause.

5.4.6 Option value

It comprises the value of a species, its potential ability to provide economic benefit to human society in the future.

- It comprises protection of charismatic animals such as tigers, lion, elephants, panda, birds etc. in a direct way to contribute wealth to organizations.

- In addition, the complete range of wildlife behaviour generates revenue out of wildlife activities. Hence the gross national product (GNP) is also enhanced by contribution from the wildlife sector.

- This wildlife GNP can well be compared to the agricultural GNP and the national GNP.

A. Biodiversity and tourism

- In terms of exports wildlife based tourism ranks in the first or second place in Tanzania and Kenya. In Kenya, tourism is wildlife-based and is the leading foreign exchange earner. This industry earned US \$484 million in 1994, which represents approximately 35 per cent earnings of total foreign exchange per year.

- In 1989 the GNP varied from US \$ 30 million in Central African Republic to US \$131.7 million in Zimbabwe. Added to this wildlife also serves as a means of hard currency.

- In 1995, in South Africa, 90 per cent of the registered 1,052,000 tourists, visiting the national parks made a monetary change of R13 million.

- In Tanzania, the global annual income from wildlife tourism was around US \$570 million (E. Severre, 1999, personal communication). It is worth mentioning that Virunga National Park and Volcanoes National Park in Zaire and Rwanda respectively gain enormous revenue from tourism based on gorilla. Rwanda generated US \$10 million from the Volcanoes National Park in 1986.

- The popular Galapagos Islands National Park in Ecuador in Latin America generated US \$ 32.6 million in 1990 and US \$35 million in 1992 by tourism.

- The Yala and Udawalawe National parks in Sri Lanka receive about 2.5 lakh tourists annually and generate US \$60 lakh income.

- In 1998–99, Nepal had earned US \$0.75 million from Chitwan Royal National Park.

- Whale watching in Queensland can earn \$12 per year whereas the Kenyan elephant yields \$1 as revenue.

B. Bioprospecting

- Reid *et al.* (1993) defines bio-prospecting as: *'the exploration of biodiversity for commercially valuable genetic resources and biochemicals.'* That is, probing of plant or animal species for utilizing it as a source of commercially usable products, such as medicinal drugs.

- It includes a variety of business related activities comprising the pharmaceutical, biotechnological, seed and crop safeguarding, horticultural, medicinal, cosmetics and personal care, and foodstuff and beverage sectors.

- The early bio-prospecting literature produced a wide range of estimates of the value of conserving a species for pharmaceutical purposes, from \$23.7 million (Principe, 1989) to \$44 (Aylward, 1993) per untested species.

- The **Nagoya Protocol**, signed in October 2010, a supplementary agreement to the CBD, provides an international platform and clear legal outline for the effectual execution of the third objective of the CBD (i.e., fair and equitable sharing of benefits arising from genetic resources). It is to take actions to guarantee that only legally acquired genetic resources coupled with traditional information were used.

Case study 5.1: Merck–INBio Agreement of Costa Rica

In October 1991, the Costa Rican Asociation, Instituto Nacional de Biodiversidad (INBio), a private, non-profit, technical organization of Costa Rica and a US multinational pharmaceutical corporation, Merck, signed a two year contract. As per the deal, INBio would supply samples of flora, insects and microbes procured from Costa Rica's protected forests to Merck. Merck then would have the authority to use these samples to craft new pharmaceutical products. Merck being one of the world's leading company to export such products. Thus, the case involves issues of both goods and services, trade along with questions of Intellectual Property Rights (IPR) and biodiversity.

C. Biopiracy

- Biopiracy is an event where the traditional knowledge of the traditional people is utilized by others for revenue without the permission or compensation or acknowledgement of those native people.

- Such practices are very common between the developing countries richly endowed with biodiversity and the multinationals of the developed countries endowed with financial and technical expertise.

- The big international companies often patent such products (food and medicines) and deprive the rights of the native people to commercially exploit them.

- Traditional knowledge including both documented and non-documented information is an easily accessible wealth very vulnerable to misuse. The case studies 5.3–5.5 reflect use of indigenous knowledge and subsequently patenting it while the case study 5.6 reflects the vacuum in the benefit sharing mechanism.

Case study 5.2: Turmeric

Turmeric rhizomes have been in use since thousands of years as spices, medicines, cosmetics and dyes. It was granted US patent in 1995 based on its use in healing wounds. The CSIR challenged the case and filed re-examination. CSIR was able to defend its case by producing the evidence in its ancient use in Sanskrit literature and a paper publication in the *Journal of Indian Medical Association* in 1953. The US Patent Office (PTO) upheld the CSIR objections and cancelled the patent.

Case study 5.3: Neem

The use of neem extracts (*Azadirachta indica*) against fungi and other pests, its ability to cure cold, flu, malaria, meningitis and skin diseases has been known since long time. In 1994, the European patent office granted a patent to the US Corporation W. R. Grace Company and US department of Agriculture for a process of fungi control with the use of extracted neem oil. The farmers and NGOs filed a case against such patent in 1995 submitting evidence on such use in Indian agriculture from long ago.

Case study 5.4: Basmati rice

Rice Tec Inc. was awarded US Patent on basmati rice lines in 1997. The Government of India filed case on the grounds of violation of the WTO on the INBio. They claimed that rice varieties were similar to Indian Basmati Rice lines with their geographical range in the north, central and South America and Caribbean Islands. Evidence was backed up by germplasm collection of Directorate of Rice Research, Hyderabad since 1978 and the WTO bulletin. Based on the claims the various grain characteristics were evaluated by Central Food Technological Research Institute (CFTRI). Rice Tec even after filing for re-examination finally chose to withdraw its claim.

5.5 Biodiversity at Global, National and Local Levels

At global level: At present more than 1.2 million species have been formally identified and recorded out of the possibility of 1.5 to 20 billion species that have been estimated to be present. Roughly 80,000 species of snails and slugs inhabit the Earth. Around 5,000 species of frogs, 3,000 species of snakes and 10,000 species of aves have been documented. Earth provides shelter to at least 25,000 diverse types of pisces, almost 2,300 types of rodents as well as countless forms of viruses, bacteria, fungi, and insects. While mapping biodiversity across the globe, it is evident that there are some areas with high species diversity, though the trend is towards homogeneity.

- On Earth, biodiversity exists in eight broad realms with 193 biogeographical provinces. Each biogeographical province includes ecosystems, which are again constituted by communities of living species existing in an ecological region.

- Around 44 per cent of vascular plant species and 35 per cent species of four major vertebrate groups are confined to the former 25 biodiversity hotspots occupying 1.4 per cent of the land area (Myers *et al.*, 2000).

- Almost ten richest endemic countries covered 15.8 per cent of the world coral reefs (Roberts *et al.*, 2002).

- Most of the world's biodiversity rich countries are located in the south but the countries capable of exploiting these organisms are from the north block. These nations have low levels of biodiversity and therefore came to support the idea of considering biodiversity as a global resource. Countries like Brazil, Malaysia and Indonesia are richer in biodiversity than India.

- Over 500 types of mammals, 175 lizards, 300 reptiles, 30,00,000 insects and one-third of world's birds are found in the Amazon rainforest.

- About 500 bird species and 45 mammal species are found to occur in the African Savannah.

- The Great Barrier Reef harbours 1,500 types of fish, over 5,000 mollusks and 350 kinds of hard corals. The types of species occurring in these region are very different and hence we should conserve this biodiversity as a foremost economic source.

Conversely there is an urgent need to consider oil, uranium, intellectual and scientific expertise as global resource provided, biological diversity would have constituted a 'common property resource'. The significance of biologically rich areas is being progressively appreciated all around the world. The value is unbelievable and inconceivable.

At national level: India is immensely rich with biodiversity even only occupying 2.5 per cent of the global land mass, accounting for 7.8 per cent of recorded species of the world in an exemplary diversity of ecosystems ranging from forests, grasslands, wetlands, coasts to deserts.

- This includes over 46,000 species of flora and over 91,000 species of fauna.
- Its rich traditional knowledge is both coded as well as informal.
- The country is acknowledged as one of the eight Vavilovian centres for the evolution and variation of cultivable plants; still India possesses more than 300 wild relatives of the agricultural plants subjected to evolution. It has a repository of 50,000 varieties of rice, 5,000 varieties of sorghum and 1,000 varieties of mango.
- It has a rich assemblage of amphibians (61.2 per cent) and reptilians (47 per cent).
- Among the pisces there are two endemic families and 127 monotypic genera. It holds around 3,66,933 exclusive accessions of floral hereditary resources.
- Furthermore it is gifted with domestic faunal resources like cow, buffalo, goat and sheep, camel, horses, duck, goose, etc.

At local level: The Indian subcontinent is divided into 16 forest types with 251 sub-types.

- Over 50 per cent of the fishes, amphibians and reptiles are endemic in the Western Ghats.
- In the Himalayan range alone, nearly 979 bird species have been recorded.
- 10 states exhibit arid and semi-arid conditions.
- The chilly dry zone in Trans-Himalayan region covering 5.62 per cent of India's area is the strong hold of the lions, leopards and tigers.
- The desert represents 682 species of plants and six per cent flora of the Indian desert is endemic.
- Asiatic ibex (*Capra sibirica*), Tibetan argali (*Ovis ammon*), wild yak (*Bos mutus*), snow leopard (*Panthera uncia*) are all rare and endangered that inhabits the cold desert.

5.6 Mega-diversity Hotspots

The term mega-diversity hotspots are usually applied to the topmost and wealthiest nations in terms of biodiversity. This is to elevate countrywide awareness to conserve biological diversity especially, the countries with high variability and exclusive species. The concept was put forward in 1988 for the first time. A nation must possess at least 5,000 endemic plants in order to qualify to become a mega-diverse country. The mega-diverse country concept is based on four principles:

- Every nation's biodiversity is significantly important to that nation's survival, and must be a central component of any national or regional development policy.
- Biodiversity is unevenly distributed, greater variation lies in the tropical region than in other regions.
- Many of the wealthiest and most diversified countries are under severe threat to biodiversity loss.

■ To succeed maximum effect with limited resources, concentration should be on heavily on those countries rich in diversity and contains endemics and severely threatened species.

Box 5.3: The 17 mega-diverse nations

NORTH AMERICA – United States of America and Mexico

SOUTH AMERICA – Colombia, Ecuador, Peru, Venezuela and Brazil

AFRICA – Democratic Republic of Congo, South Africa and Madagascar

ASIA – India, China, Malaysia, Indonesia and Philippines

OCEANIA – Papua New Guinea and Australia

5.6.1 India as a Mega-diversity nation

About 70 Mya, a split in the Pangaea resulted in the creation of the northern and southern continents. India was a fragment of Gondwana along with Africa, Australia and Antarctica. Tectonic movement pushed India further northward across the Equator to unite with the Eurasian continent. With the closing of the intervening Tethys Sea, the flora and fauna that have originated and evolved in Europe and Far East, migrated and entered India before the mighty Himalayan formation have taken place. An ultimate entry of the Ethiopian species is from African country. These species underwent adaptations to suite the Savannahs and the semi-arid regions.

The location of India is at the intersection of three realms – Afro-tropical, Indo-Malayan and Paleo-Arctic. Consequently, it shares the characteristic features of each of them and such an assemblage contributes to richness and uniqueness to India. The geographic location of India is between 6^0 and 38^0N latitude and 69^0 and 97^0E longitude in South Asia. The Indian landmass has about 3,029 million hectares of total area. India is surrounded by the Himalayan Mountains in the north, Indian Ocean in the south, the Bay of Bengal in east and the Arabian Sea in the west. Globally it stands out as one out of seventeen mega diverse countries. (referred in World Bank Technical Paper no. 343). According to Miller Meier, 'India is remarkable in both species richness and endemism although it ranks tenth position'.

■ India occupies about 2.4 per cent of the Earth's land area and houses 11 per cent of the plants and 7.43 per cent of the world's animals.

■ India is divided in to 10 biogeographical zones and 26 biogeographical provinces. The climatic conditions are mixed – from moist tropical Western Ghats, the scorching deserts of Rajasthan, the cold deserts of Ladakh and the lofty Himalayas to the hot Indian peninsula. This is well reflected in the different types of ecosystems like grassland, forest, desert, wetland, estuary, marine etc.

■ India is placed amongst the leading 10 to 15 countries in terms of floral and faunal diversity. About 65 per cent of the total area of India was surveyed so far.

- Over 46,000 species of plants have been described so far, of which 130 species of ancient plants, 1,232 species of pteridophytes, 15,000 species of angiosperms or flowering plants, 65 species of gymnosperms, 40 species of insectivorous plants, 70 types of saprophytes, 14,500 types of fungi, 2,850 species of mosses, about 990 types of algae, 2,075 species of lichen, 1,082 species of orchids and over 850 species of bacteria and virus are found as reported by the Botanical Survey of India.

- 35 per cent of the angiosperms and 22 genera of monocotyledons are strictly endemic to India.

- The Zoological Survey of India has described over 81,000 species of animals. India is the home for 350 diverse mammalia (ranked eighth highest in the world), 1,200 avian species (ranked eighth in the world), 453 types of reptilian species (ranked fifth in the world), and 50,000 recognized species of insects, together with 13,000 butterflies and moths.

- The country is also one of the 12 prime centres of evolution of cultivated plants and domesticated animals. It is acknowledged to be the native soil of 167 major plant species of mueslis, pulses, millets, fruits, oilseeds, condiments, vegetables, fibre crops and 114 varieties of domesticated animals.

- Endemic species is predictable at 33 per cent with over 140 genera of endemic plants. At least 220 endemic species are recorded in the Andaman and Nicobar Island.

- Out of the 150 botanical sites considered to be important from the stand point of conservation, five places are in India. They are the Agastyamalai Hills, Silent Valley, New Amarambalam Reserve, Periyar National Park and the Eastern and Western Himalayas.

- Only 55 bird species and 44 mammal species exhibits endemism in India. Four endemic species are exclusive of the Western Ghats. They are the Nilgiri leaf monkey (*Trachypithecus johni*), Lion-tailed macaque (*Macaca silenus*), Nilgiri tahr (*Hemitragus hylocrius*) and Brown palm civet (*Paradoxurus jerdoni*).

- On the other hand, endemism in the Indian reptiles and amphibians are high. Approximately 187 reptiles and 110 amphibian species are endemic and occur in India. Eight genera of amphibians are unique and exclusively found in India.

Species diversity: India harbours a huge capital on biodiversity both on land and aquatic region. This richness can be demonstrated by the absolute numbers of species and the proportion of the world total they represent.

Table 5.4: Comparison between the number of species in India and the world

Group	Number of species in India (SI)	Number of species in the world (SW)	SI/SW (%)
Mammalia	350	4,629	7.6
Aves	1,224	9,702	12.6

Group	Number of species in India (SI)	Number of species in the world (SW)	SI/SW (%)
Reptilia	408	6,550	6.2
Amphibia	197	4,522	4.4
Pisces	2,546	21,730	11.7
Angiosperms	15,000	2,50,000	6

5.7 Hotspots of Biodiversity

The notion of 'biodiversity hotspots' was put forward by Dr Norman Myers to relate the biologically richest and most endangered eco-regions of the world. The criteria to designate an area as a hot spot goes like – when it contains at least 0.5 per cent or 1,500 species of plant species as endemic species and at least 70 per cent of the primary vegetation has to have lost. There are 34 such biodiversity hot spots of on a global level.

Biodiversity hotspots in India: There were two hotspots in India – the Indo-Burma and the Western Ghats and Sri Lanka. According to Conservation International, these hot spots covers around 2.3 per cent of the world's land surface and are found to possess about 50 per cent of the plants and 42 per cent of the terrestrial endemic vertebrate species. Presently there are four hotspots.

A. The Indo–Burma

- This hotspot includes the Asian countries of Cambodia, Vietnam, Laos, Thailand, Myanmar and Bhutan as well as parts of Nepal, far-off eastern India and extreme southern part of China. Additionally, it also covers several off-shore islands like Andaman and Nicobar Islands in Indian Ocean and Mainan Islands in the South China Sea. Indo-Burma is one of the most defenseless areas due to high rate of resource exploitation and loss of habitat.

- Six mammalian species Large-antlered muntjac (*Muntiacus vuquangenesis*), Annamite muntjac (*Muntiacus truongsonensis*), Grey-shanked douc (*Pygathrix cinerea)*, Annamite striped rabbit, Leaf deer (*Muntiacus putaoensis*) and the Saola (*Pseudoryx nghetinhensis*), are found here. It is the habitat for restricted number of monkeys, langurs and gibbons. Freshwater turtles found here are endemic. Practically 1,300 bird species including the endangered white-eared night-heron (*Gorsachius magnificus*), the leaf muntjac (*Muntiacus putaoensis*) and the orange-necked partridge (*Arborophila davidi*).

- Out of 13,500 plant species found here over 50 per cent endemic. Ginger is inherent to this region.

B. Western Ghats

- Western Ghats and Sri Lanka collectively known as 'Sahyadri Hills' embrace the mountainous forests in the south–western part of India as well as the Sri Lankan islands. The hotspot, which extended nearly 1, 82,500 km² once is currently 12,445 km² or 6.8 per cent owing to the exceedingly high population pressure.

- The series of hills on the western part of Indian peninsula comprises moist deciduous forests and rainforests.

- This region is typically similar to the Malayan region. Together with Sri Lanka, it also bears faunal similarities with Madagascar. The Purple Frog (*Nasikabatrachus sahyadrensis*) and Sri Lankan lizard genus *Nessia* are similar to the Madagascan genus *Acontias*. This is the home for Indian tigers, Asian elephant, and the endangered lion tailed macaque (*Macaca silenus*). Almost 77 per cent of the amphibians and 62 per cent of the reptile species are endemic.

- Over 2500 genera representing 6000 vascular plants are found here. More than 3000 plants are endemic. Spices such as cardamom and black pepper are native to this region.

- The species density is highest in the Agasthyamalai Hills, where more than 450 bird species, roughly 140 mammalian species, 260 reptiles and 175 amphibians are found. Over 60 per cent amphibians and reptiles are totally endemic to the hotspot.

C. Eastern Himalayas

- The region covering Bhutan, northeastern India, and southern, central, and eastern Nepal exhibits high altitudinal variation and includes two of the highest peaks, Mount Everest and K2. The region has more than 100 mountains with lofty heights.

- Around 163 threatened species including 45 mammalians, 50 aves, 17 reptilians, 12 amphibians, 3 invertebrates and 36 plant species are found here. One-horned Rhinoceros (*Rhinoceros unicornis*), Wild Asian Water Buffalo (*Bubalus bubalis)* and the Relict Dragonfly (*Epiophlebia laidlawi*) are amongst the animals that are found here. Himalayan Newt (*Tylototriton verrucosus*) is the only endemic salamander of India.

- Estimates states of 10,000 species of plants of which 33 per cent are endemic. Five families, Tetracentraceae, Hamamelidaceae, Circaesteraceae, Butomaceae and Stachyuraceae are totally exclusive here. The plant, *Ermania himalayensis* occurs even at a height of 6300 meters. Himalayan Quail, Cheer pheasant, Western tragopan are not only endangered but also endemic. Himalayan vulture and white-bellied heron, found here are, are quite endangered.

- 300 species of mammals live here of which 12 are endemic. Important names include Golden langur (*Trachipithecus geei*), Himalayan tahr (*Hemitragus jemlahicus*), pygmy hog (*Porcula salvania*), Langur (*Trachipithecus johnii*), Asiatic wild dog or dhole (*Cuon alpines*), Sloth bears (*Melursus ursinus*), Gaurs (*Bos gaurus*), Muntjac, Sambar (*Rusa unicolor*), Snow leopard, Black bear (*Ursus americanus*), Blue sheep (*Pseudois nayaur*), Takinor chamois or gnu goat (*Budorcas taxicolor*), the Gangetic dolphin (*Platanista ganetica*), Wild water buffalo (*Bubalus arnee*), Swamp deer (*Cervus duvaucelli*), etc. Namadapha flying squirrel (*Biswamoyopterus biswasi*) which is critically endangered and endemic are found here.

D. Sundaland

- The Nicobar group of islands is a part of this hotspot. The Nicobars are separated from the Andamans by about 160 km and often experience heavy tidal flow. Mount Thullier is the highest peak of the Nicobar group.

- The region has a volcanic origin with the coral reefs contributing towards raising the islands.

- The landmass to the north is Myanmar while the closest landmass to the south is Sumatra.

- Nicobar Tree shrew (*Tupaia nicobarica*), Nicobar Pigeon (*Caloenas nicobarica*), *Rhopaloblaste augusta* and Nicobar palm (*Bentinckia nicobarica*) are some of the important species.

Table 5.5: World's biodiversity hotspots and numbers of plant species endemic to each of the 34 hotspots as per Conservation International

Serial nos.	Biodiversity hotspots	Plant species	Endemic plant species	Endemics expressed as percentage of world total
1.	Atlantic forest	20,000	8,000	2.7%
2.	California floristic province	3,488	2,124	0.7%
3.	Cape floristic region	9,000	6,210	2.1%
4.	Caribbean Islands	13,000	6,550	2.2%
5.	Caucasus	6,400	1,600	0.5%
6.	Cerrado	10,000	4,400	1.5%
7.	Chilean winter rainfall – Valdivian	3,892	1,957	0.7%
8.	Coastal forests of eastern Africa	4,000	1,750	0.6%
9.	Guinean forests of west Africa	9,000	1,800	0.6%
10.	**Indo – Burma**	**13,500**	**7,000**	**2.3%**
11.	Madagascar and the Indian Ocean islands	13,000	11,600	3.9%

Serial nos.	Biodiversity hotspots	Plant species	Endemic plant species	Endemics expressed as percentage of world total
12.	Mediterranean basin	22,500	11,700	3.9%
13.	Mesoamerica	17,000	2,941	1.0%
14.	Mountains of southwest China	12,000	3,500	1.2%
15.	New Caledonia	3,270	2,432	0.8%
16.	New Zealand	2,300	1,865	0.6%
17.	Philippines	9,253	6,091	2.0%
18.	Polynesia – Micronesia	5,330	3,074	1.0%
19.	South west Australia	5,571	2,948	1.0%
20.	Succulent Karoo	6,356	2,439	0.8%
21.	Sunderland	2,5000	1,5000	5.0%
22.	Tropical Andes	30,000	15,000	5.0%
23.	Tumbes – choc	11,000	2,750	0.9%
24.	Wallacea	10,000	1,500	0.5%
25.	**Western Ghats and Sri Lanka**	**5,916**	**3,049**	**1.0%**
Serial nos.	New hotspots added to the existing ones			
1.	East Melanesian islands	8,000	3,000	1.0%
2.	**Eastern Himalayas**	**10,000**	**3,160**	**1.1%**
3.	Horn of Africa	5,000	2,750	0.9%
4.	Irano – Anatolian	6,000	2,500	0.8%
5.	Japan	5,600	1,950	0.7%
6.	Madrean pine – oak woodlands	5,300	3,975	1.3%
7.	Maputaland – Pondoland – Albany	8,100	1,900	0.6%
8.	Mountains of central Asia	5,500	1,500	0.5%
9.	Eastern Afromontane	7,598	2,356	0.8%

5.8 Threats to Biodiversity

Extinctions are normal and inevitable. Over the past 200 million years, the extinction rate is 1–2 species per year and 3–4 families per million years on an average. There are records of intermittent episodes of mass extinction, the period when many groups representing a widespread collection of life forms have undergone extinction in the similar blink of geological time. In the present times, species and ecosystems are threatened with obliteration mainly due to human activities. IUCN in 2008 had estimated that 70 per cent of the flora is in a state of jeopardy globally. Hundreds of medicinal plant varieties whose chemical constituents comprise the foundation of more than half of the approved drugs are facing danger with extinction, as per a recent report. Such a situation provoked the experts to seek action in order to 'secure the future of global healthcare' (Hawkins, 2008). As per CBD, 2010, there are five major threats to biodiversity – habitat change, over exploitation, invasive species, pollution and climate change.

A. Habitat loss and habitat fragmentation

- Loss of habitat, habitat degradation and habitat fragmentation represents significant causes of known extinctions. With continual increase of deforestation in the tropical forests, this becomes the chief cause of mass extinctions caused by human action.

- Fragmentation of landscape by construction of roads, seismic lines, and linear infrastructure also results in biodiversity loss. Another facet of habitat loss that is frequently not recognized is habitat fragmentation. The forests, deserts, pastures, or other habitats that are left are typically in small, inconspicuous, isolated bits rather than in vast expanses of whole and intact units. Each behaves as a tiny island that can sustain a very small population.

- Reduction in coyote populations in South California leads to increase in their prey populations. As because raccoons consume bird eggs, fewer coyotes caused more and more raccoons eating more song bird eggs, resulting in fewer number of song birds.

- Environmental fluctuations, disease outbreak, and other probable factors make such small isolated habitat highly susceptible to extinction. Species requiring a large home range, take for example, the grizzly bear, cannot survive in the small habitat, which is strongly affected and influenced by climate and other species.

- Habitat fragmentation thus leads to reduced genetic viability and ultimately **genetic erosion**.

B. Over exploitation

- Growing population accelerates the pace of transformation of the Earth and further results in loss of ecosystems and species.

- Seven billion of human population have degraded and destroyed about half of the world's forest canopy. The humans consume more than half of world's net primary productivity.

- As per WWF, 2008, there was a decline of terrestrial vertebrates by 33 per cent in between 1970–2005. 2 billion people depend on the drylands for livelihood and a change in dry woodland, savannah and desserts are noticed.

- There was a loss of 14 per cent of the marine species due to increase in temperature, pollution and fishing methods.

- The unsustainable mode of use and harvest of natural resources like plants, animals and microbes without replenishing them is a serious concern.

C. Exotic species

- The introduction of exotic or invasive and alien species is a significant reason of extinction.

- The great lakes of Africa like Lake Victoria, Tanganyika and Malawi are renowned for their native species of cichlid fishes. The introduction of Nile perch in Lake Victoria has made the way for extinction of most of the indigenous species.

- Invaders harmfully affect the native species by consuming them, infecting them, challenging with them, or even breeding with them.

D. Pollution

- Pollution from toxic chemicals undoubtedly proves to be a threat for the survival of species and sustenance of ecosystems. Pollution alters the physiochemical properties and degrades the quality of land, water and air.

- Species with a small range of tolerance is often threatened to extinction.

- The change in pH, acid precipitation, alteration of nutrient loads (such as nitrogen and phosphorous additions) and sediment loads, eutrophication, contamination from pesticides leading to bio-concentration and bio-magnifications, contamination of habitats and species from industrial byproducts, oil spills, pharmaceuticals, personal care products, ground level ozone, particulate matter, trans-boundary pollutants and photochemical smog all act as hindrances to the survival and perpetuation of species.

E. Climate change

- The distribution of species is largely governed by climatic and seasonal changes.

- Distribution of vegetation is dependent on climate, which in turn governs the distribution of animals. Changes in temperature, solar radiation, wildfire, rainfall, flood, drought, storms, snowfall, melting glaciers, and variable stream flow; are all limiting conditions to the distribution of plants and animals.

- Species populations may be lost if they are not capable of adapting to new conditions or relocating to new habitats.

- As per US Department of Commerce 2008, a small increase in the surface temperature of the sea in 1976 and 1998 ensued a series of global phenomenon that ultimately led in 1998, what was to be known as 'the year the world caught fire'. The lasting injury includes:

    i.   Burnt-out forests that seems not to recuperate within any significant timescale;

    ii.   An increase in the average temperature from 19°C to 25°C, of the surface waters of the central and western Pacific Ocean.

- In the past 30 years the coral reefs in Caribbean region have shrunken by 80 per cent. Earnings from diving based tourism was nearly 20 per cent of the total tourism earnings earlier but have presently declined so as to incur a loss up to US$ 300 per year. Fishery sector has also been severely affected (UNEP, 2008).

F. Hunting and poaching

- Both legal and illegal hunting are a major threat.
- Wildlife is continuously being hunted and poached for food, fashion, profit and make beliefs.
- The multibillion dollar illegal trade in protected species operates as one of the most profitable illicit markets in the world today.
- Notable examples are passenger pigeon and snowy egrets.
- Rhino horns are sold at an exuberant price, say each horn sold at $40,000 to $100,000; a thin hope is left about their sustenance in the wild.
- Many pets and ornamental plants are hunted and traded; yearly trade is projected to be a minimum of $5 billion, with possibly one-fourth to one-third of it being illegal.
- Hunting for sports or recreational, if properly managed and regulated, and may facilitate in bringing a species back from the verge of extinction.

G. Niche preference

- All species are limited by specific food and habitat requirements.
- More specific food prerequisites localize the habitat even more; greater will be the species vulnerability to loss and extinction.
- Only those species are likely to endure whose territories are highly protected.

H. Loss of forest land

- Conversion of forest land for agriculture and cultivation has a historical background. It started in China about 4,000 years ago and spread over Europe 400 years ago and USA about 200 years ago.
- The deciduous woods began to vanish after the establishment of Spanish and Portuguese colony in the new world tropics.
- It aggravated more in the twentieth century with population inflation, with inequitable land and income distribution.
- Satellite image shows the rapid decline of tropical forests. These forests possess at least half of the world's biodiversity. Originally the tropical rain forests stretched about 15 million

km^2 which at present is about 7.5–8 million km^2. The existing rate of loss is estimated at 2 per cent annually. It is predictd that in the future, the tropical forests will be depleted by 10– 25per cent of their actual cover by the end of twenty-first century.

■ Removal or alteration of riparian vegetation, logging to the stream bank, overgrazing, replanting different species after logging, fire, erosion, draining of wetlands, reservoir and dam construction, infrastructure development etc. all contribute to extinction.

I. Loss of keystone species

■ Islands are laboratories for evolution. Domino effects are likely to happen when two or more species greatly interdependent on each other or when one of the affected species, a 'keystone' species, exerts a strong control over many other species. Record says that amphibian and reptilian represent about 93 per cent of 30 extinct species and sub-species, 93 per cent of 176 species and sub-species are terrestrial and freshwater birds. Mammals represent only 27 per cent of 114 species and subspecies.

■ The black footed ferret was very abundant at one point of time in the western prairies. It usually preyed upon the prairie dogs and dig tunnels to live in. Poisoning has drastically reduced the prairie dog population and the black footed ferret is presently one of the rarest mammals in North America.

■ A small number of old Calvaria trees exist presently in the island of Mauritius. The seed coats of these fruits are tough and must pass the digestive tract of animals in order to germinate. The seeds while passing through gut undergo abrasion that facilitates germination. The Dodo bird was the key to this process. Hunting Dodo to extinction left the island only with very few old trees.

J. Tourism

■ Tourism in the protected areas creates vehicular pollution and casualties (roadkill).

■ Environmental conservation and restoration of scenic beauty becomes a challenge at times due to rush of tourists. Tourists require services, such as shops, hotels, restaurants, etc. Regulation of commercial activities becomes complicated.

■ Engagement of tourists in water sports disturbs the aquatic biota and activities like campfires may lead to wildfires.

■ Above all teasing and feeding is a persistent problem.

K. Disease outbreak

■ Spread of bacterial and viral infections like rabies, feline AIDS, feline leukemia, etc results in mass killing of wild animals.

L. Man-wildlife conflict

■ Leopards are territorial with strong social bonds. Leopards are extremely scared of humans and their first reaction is to run away from humans. Why then, should such an animal overcome a deep fear and kill humans? The two most common reasons are habitat destruction and lack of food. Undoubtedly, a human being killed by a leopard is a very grim issue. It does affect the lives of people who live in close proximity with these animals' habitats. Human welfare should not be compromised, either in the short term or the long term. As for food, a leopard has a home range of approximately 20 sq. km and all it needs is about a goat-sized animal per week. That means 52 dogs /goats in 20 sq. km in a year. Average livestock density in India is much higher and the leopard really does not need to kill humans for food.

- In Uttarakhand, humans have been attacked by leopards for decades; Himachal also have severe hunting pressures where leopards are killed in large numbers for the illegal wildlife trade. In the Himachal, when three people were killed in duration of a couple of weeks, fear grasped the local population.

- In the Sanjay Gandhi National Park, Mumbai between 2002 and 2004, as many as 84 people were attacked. Between 2006 and 2012 no confirmed human deaths occurred due to leopards.

- Most of the attacks were provoked either intentionally (e.g. because people were hunting the tigers) or unintentionally (as when a vehicle hit a cub and the mother responded by attacking people on the road). The pertinent question arises whether this killing is actually increasing the conflict and leading to loss of human lives.

- Forest officers in Palamu are responsible for protecting the wildlife as well as tasked with supporting villagers in the man-animal conflict. Poisoning of tigers by pesticides at Palamu Reserve Park is a common problem and authorities find themselves helpless in this conflict for survival. There is a history of tigers being poisoned to death by villagers for killing their cattle. As per official records, in 1984–85, two tigers were poisoned to death; one again in 1985–86; in 1996–96, one hyena was also killed. For instance, a tigress from the Palamu reserve park entering Dorami village created panic among villagers. The tigress confirmed a lavish dinner taking away two cows and a calf, belonging to a Chero farmer. The villagers thereafter paced up pressure on the administration for compensation. Early allotment of funds for payment of compensation is the only way to resist the growing resentment. Man-animal conflict has always affected the wildlife habitat in any tiger reserve in the country. Elephants are also the major destroyers of life and property in the PTR. This year, one elephant was found poisoned in the park.

Box 5.4: Threats to biodiversity

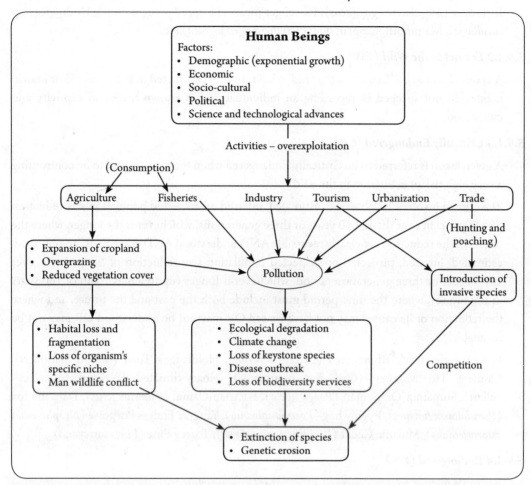

5.9 Red Data Book

The IUCN red list founded in 1963 serves as the world's most complete and inclusive record and inventory of the worldwide conservation status of floral and faunal species. The Red Data Book furnishes with a record of species that are placed under various threatened categories according to the extent or severity of threats faced by them. The list is compiled by the WCMC and the Species Survival Commission (SSC) of the International Union for the Conservation of Nature and Natural Resources (IUCN).

5.9.1 The IUCN Categories

5.9.1.1 Extinct (EX)

- A given taxon is referred to be 'Extinct' when surveys conducted in the expected or known habitat do not succeed in recording a single individual; there is hardly any uncertainty that the last individual has died.

- For example, Dinosaur, Tasmanian tiger (*Thylacinus cynocephalus*), Quagga (*Equus quagga*), Irish deer (*Megaloceros giganteus*), Passenger pigeon (*Ectopistes migratorius*), Dodo (*Raphus cucullatus*), Mammoth, Caspian tiger (*Panthera tigris vigrata*) etc.

5.9.1.2 Extinct in the Wild (EW)

- A taxon is said to be 'Extinct in the wild' whilst surveys conducted in the expected or known habitat do not succeed in recording an individual, but is known to live in captivity and cultivation.

5.9.1.3 Critically Endangered (CR)

- A given taxon is referred to be Critically Endangered when the taxa are said to be confronting a very high risk of extinction in the wild.

- The criteria involves 'an observed, estimated, inferred or suspected population size reduction of ≥ 90per cent over the last 10 years or three generations, whichever is the longer, where the causes of the reduction are clearly reversible AND understood AND ceased'; or, 'an observed, estimated, inferred, projected or suspected population size reduction of ≥80per cent over any 10 year or three generation period, whichever is longer (up to a maximum of 100 years in the future), where the time period must include both the past and the future, and where the reduction or its causes may not have ceased OR may not be understood OR may not be reversible'.

- For example, Black Rhinoceros (*Diceros bicornis*), Holdridge's Toad (*Incilius holdridgei*), Caquetá Tití Monkey (*Callicebus caquetensis*), Hoary-throated Spinetail (*Synallaxis kollari*), Sumatran Orangutan (*Pongo abelii*), Bactrian Camel (*Camelus ferus*), Darwin's fox (*Pseudalopex fulvipes*), Pygmy hog (*Porcula salvania*), Yangtze Finless Porpoise (*Neophocaena asiaeorientalis*), Mulanje Cedar (*Widdringtonia whytei*), Torrey Pine (*Pinus torreyana*).

5.9.1.4 Endangered (EN)

- A taxon is referred to as endangered once the best existing evidence indicates that it is confronting a high risk of moving to the critically endangered category if not conserved properly.

- The criteria involves 'an observed, estimated, inferred or suspected population size reduction of ≥ 70per cent over the last 10 years or three generations, whichever is the longer, where the causes of the reduction are clearly reversible AND understood AND ceased', or 'an observed, estimated, inferred or suspected population size reduction of ≥ 50per cent over the last 10 years or three generations, whichever is the longer, where the reduction or its causes may not have ceased OR may not be understood OR may not be reversible'.

- For example, Cayman Blue Iguana (*Cyclura lewesi*), Tiger (*Panthera tigris*), Indian elephant (*Elephas* maximus), Giant Panda (*Ailuropoda melanoleuca*), Red Panda (*Ailurus fulgens*), Philippine eagle, Albatross, Blue whale (*Balaenoptera musculus*) , African Wild Dog (*Lycaon pictus*), Snow Leopard (*Panthera uncia*), Pygmy Hippopotamus (*Choeropsis liberiensis*), Himalayan Musk deer (*Moschus leucogaster*), Mountain Gorilla (*Gorilla beringei*), epiphytic orchid (*Masdevallia atahualpa*), Monterey Pine (*Pinus radiata*), etc.

5.9.1.5 Vulnerable (VU)

- A taxon is Vulnerable whilst the finest existing evidence indicates that it is confronting a high risk of moving to the endangered category.

- The criteria involves 'an observed, estimated, inferred or suspected population size reduction of ≥ 50per cent over the last 10 years or three generations, whichever is the longer, where the causes of the reduction are clearly reversible AND understood AND ceased', 'an observed, estimated, inferred, projected or suspected population size reduction of ≥ 30per cent over any 10 year or three generation period, whichever is longer (up to a maximum of 100 years in the future), where the time period must include both the past and the future, AND where the reduction or its causes may not have ceased OR may not be understood OR may not be reversible'.

- For example, King Cobra (*Ophiophagus hannah*), Spiny Seahorse (*Hippocampus histrix*), Hippopotamus (*Hippopotamus amphibius*), Mountain Gazelle (*Gazella gazella*), Clouded Leopard (*Neofelis nebulosa*), African lion (*Panthera leo*), Komodo Monitor (*Varanus komodoensis*), Italian common name Lino di Katia (*Linum katiae*), sedge (*Bulbostylis pseudoperennis*), Hill Turmeric (Curcuma pseudomantana).

5.9.1.6 Near Threatened (NT)

- A taxon which does not meet the requirements for critically endangered, endangered or vulnerable at present, but is close to be eligible for or is likely to qualify for a threatened category in the proximate future.

- For example, Stellar Sea lion (*Eumetopias jubatus*), Jaguar (*Panthera onca*), Blue billed duck (*Oxyura australis*), Tiger shark (*Galeocerdo cuvier*), Solitary eagle (*Harpyhaliaetus solitaries*), Blackbuck (*Antilope cervicapra*), Günther's Vine Snake (*Ahaetulla dispar*).

5.9.1.7 Least Concern (LC)

- A taxon which is extensive and copious is included in this category.
- For example, Giraffe (*Giraffa camelopardalis*), Fallow deer (*Dama dama*), Chital (*Axis axis*), Andaman Wood pigeon (*Columba palumboides*), Wolverine (*Gulo gulo*), etc.

5.9.1.8 Data Deficient (DD)

- A taxon is referred to as Data Deficient whilst the information is insufficient or scarce so as to carry out a direct or indirect evaluation of its possibility of extinction.
- For example, Killer whale (*Orcinus orca*), Jewelled Tang (*Zebra somagemmatum*).

5.9.1.9 Not Evaluated (NE)

- A taxon which has so far not been evaluated or assessed against the criterion is NE.
- Siberian Tiger (*Panthera tigris altaica*), Sumatran Rhino, Asiatic lion (*Panthera leo persica*), Wild tailed mongoose etc.

5.10 Endangered and Endemic Species of India

Endemism: It refers to a species, which is only found in one given region or location and nowhere else in the world. The species may be 'site endemic' (for example, just found on Mountain), or a 'national endemic' (for example, found only in India) or a 'geographical range endemic' (for example, found in the Himalayan region). For example, Lion tailed Macaque (*Macaca silenus*), Malabar Parakeet (*Psittacula columboides*), Nilgiri leaf monkey (*Trachypithecus johni*), etc.

Table 5.6: List of some endemic birds in India

Megapodius nicobariensis (Nicobar Megapode)	*Psittacula columboides* (Malabar Parakeet)	*Gallus sonneratii* (Grey Jungle fowl)
Rhodonessa caryophyllacea (Pink-headed Duck)	*Heteroglaux blewitti* (Forest Owlet)	*Orphrysia superciliosa* (Himalayan Quail)

Table 5.7: List of some other vertebrate endemic organisms in India

	Scientific names	Common names
1.	*Macaca silenus*	Lion-tailed Macaque
2.	*Trachypithecus johnii*	Nilgiri Langur
3.	*Crocidura jenkinsi*	Jenkin's Andaman Spiny Shrew
4.	*Pteropus faunulus*	Nicobar Flying Fox
5.	*Biswamoyopterus biswasi*	Namdapha Flying Squirrel

Case Study 5.5: Olive Ridley Turtle conservation in India

The Olive Ridley Turtles are solitary creatures that flock together during the breeding season to seek out the beaches where they were born following an eternal cycle of life and birth. Their habitat being the warm sea waters, their nesting place is spread across the southern Atlantic, the Pacific and the Indian oceans.

The single most important breeding area for the Olive Ridley's in India is the Indian Ocean, along the Bay of Bengal in Orissa. The mouth of the Rushikulya River in Orissa is one such centre for large scale nesting. Scattered beaches around the Corromondal and Sri Lanka are also known to be the preferred habitats of the Olive Ridley's though they are far less significant.

The Olive Ridley Turtles are commercially valuable as their eggs are considered a delicacy, which leads to their poaching. While they face the threat of shrinking habitats because of human encroachment on their pasture, they are also increasingly poached upon because of their meat, shells and eggs. The local fishermen, who have been hunting these turtles from time immemorial have scant regard for the ecological importance of conserving the Olive Ridley's which today face a serious threat and may even be forced into extinction.

While most of the countries where the Olive Ridley's migrate to spawn and breed have laws protecting them, they are scantily and scarcely enforced, ensuring that the problem worsens with every breeding season. Conservation efforts that are inclusive in nature – efforts that involve the local community, spread awareness and enlist the help of the fishermen to act as custodians of the rights of the Ridley's to their age old breeding grounds and the safe passage of the hatched young back to the sea – have started yielding magical results and are widely being accepted as the logical step ahead.

Table 5.8: List of some endemic plants in India

Genera	Family
1. *Smithsonia 3 spp.*	(Orchidaceae)
2. *Xenikophyton 2 spp.*	(Orchidaceae)
3. *Deccania 1 spp.*	(Rubiaceae)
4. *Cynarospermum 1 spp.*	(Acanthaceae)
5. *Haplanthodes 4 spp.*	(Acanthaceae)

5.11 Conservation of Biodiversity

Conservation refers to the management of wildlife by the human beings to yield the greatest sustainable benefits. For catering conservation the Government of India has enacted the Wildlife Protection Act, 1972 that provides for the establishment of the posts of Chief wildlife wardens and wildlife wardens in state, state wildlife advisory boards. It also provides for the establishment of Protected Area Network (PAN). As per National Biodiversity Action Plan (NBAP), 2002–16, presently 4.90 per cent of the country's area extending over 1,61,221.57 sq. km is under protected area network in 668 Protected Area's (PA) as on (1 September 2011); but there is a plan of expanding it to 5.74 per cent. Protection of wildlife is in the concurrent list thereby enabling the union government to exercise some control on the state government. 'IUCN affirms that a goal of conservation is the maintenance of existing genetic diversity and viable populations of all taxa in the wild in order to maintain biological interactions, ecological processes, and function'. Conservation officers, administrators, supervisors and the decision-makers should approve and implement a practical and built–in approach towards conservation. The threats to biological diversity in situ persist to aggravate and inflate, and all the taxa have to endure in ever more human-modified situations. Threats may be in the form of loss of habitat, pollution, global warming and climate change, unsustainable use, introduction of exotic and pathogenic organisms, man-wildlife conflicts, which can be very hard to control. The reality of the present situation is that it will not be feasible to safeguard the subsistence of a cumulative number of threatened taxa without employing a varied range of balanced and harmonized conservation methodologies and integrated techniques.

5.11.1 In-situ conservation

In-situ conservation refers to the protection and maintenance of organisms in their natural habitat. In this type of practice isolation of organisms is not required but it requires elimination of the harmful factors threatening the very existence of the organisms. The natural area is protected and maintained to conserve all known or unknown species. It is one of the easy and cheap means to conserve biological diversity. The entire natural habitat is restored to ensure proper balance in the structure and function of the ecosystem. In the natural habitat the organisms are acted on by all the natural phenomena that provide the organisms to withstand stress and adapt well with the existing environmental conditions and evolve well. A drawback in such system lies in the fact that it requires a considerably large area for effective conservation. In-situ conservation is practiced in the form of Sanctuaries, National Parks and Biosphere Reserves. There are 515 sanctuaries, 102 national parks, 47 conservation reserves, 4 community reserves and 18 biosphere reserves in India.

Case Study 5.6: Keoladeo Ghana Bird Sanctuary in Bharatpur

The Keoladeo Ghana Bird Sanctuary – popularly known as Bharatpur – a protected Sanctuary and a World Heritage site, is home to about 350 plus species of migrating birds including the Saras Crane and is rich in flora and fauna that supports them. Once a grazing ground, its transformation onto a protected place had given rise to local anger which had even lead to the police firing on an irate mob. However, it is not the local ire that threatens Bharatpur today. Fed by the local dams, the erratic and scarce precipitation is leading to vast swathes of the wet lands drying up, which in turn is creating a scarcity of the fish and other small creatures on which the birds feed, forcing them to seek out greener pastures. Changing weather patterns are leading to lower rainfall in the entire catchment area and till date, no comprehensive system to bring adequate amount of water to maintain the sanctity of the sanctuary has been put into place. The result is an alarming flight that has taken place with most of the species of birds that used to migrate from distant shores during the winter months choosing to stay away.

5.11.1.1 Sanctuary

- A wildlife sanctuary is a protected area formed by the order of the state of central government but not bound by the state legislation. It focuses on the conservation of species.
- The ownership of a sanctuary could rest in the control either of a government or in any private organization or person.
- No forest could be cleared for agricultural purposes.
- Killing, poaching and hunting is prohibited but local people can gather flowers, fruits, firewood, medicinal plants, etc. in small amount.
- Public entry and roaming inside a wildlife sanctuary is allowed for research, educational, inspirational, and recreational purposes.

Examples Kedarnath (UP), Keladevi (Rajasthan), Askot (Uttarakhand), Mundanthurai (Tamil Nadu), and Pulicat (Andhra Pradesh) etc

5.11.1.2 National park

- The concept of national park was first introduced in 1969 to preserve ecosystems for conserving wildlife and natural objects without active human interference.

- The area of a national park is protected by statutory law.

- Private rights are non-existent.

- Photographs may be allowed and research and educational work can be carried out with prior permission.

- Human activities are prohibited and the park cannot be used for any reason, such as fruits, flowers and firewood or timber collection.

- National parks are more restricted.

- A sanctuary can be upgraded to a national park.

For example, Hailey National Park now known as Jim Corbett National Park, Bandhavgarh National Park (MP), Kanha National Park (MP), Bandipur National Park (Karnataka), Keoladeo Ghana National Park (Rajasthan), Dudhwa National Park (UP) and Gir National Park (Gujarat).

Table 5.9: List of some important national parks in India

Serial nos.	State	National parks	Year of establishment	Area (in sq. km)
1.	Uttarakhand	Corbett National Park	1936	520.82
2.	Madhya Pradesh	Kanha National Park	1955	940
3.	Assam	Kaziranga National Park	1974	471.71
4.	Karnataka	Bandipur National Park	1974	874.20
5.	Gujarat	Gir National Park	1975	258.71
6.	Uttar Pradesh	Dudhwa National Park	1977	490.29
7.	Rajasthan	Sariska National Park	1979	800
8.	Rajasthan	Ranthambore National Park	1980	392
9.	Madhya Pradesh	Bandhavgarh National Park	1982	448.85
10.	West Bengal	Sunderbans National Park	1984	1,330.10
11.	Karnataka	Nagarhole National Park	1988	643.39
12.	Haryana	Sultanpur National Park	1989	1.43
13.	Assam	Manas National Park	1990	500

Conservation Reserves are affirmed by the state governments in any region owned or possessed by the government.

Case Study 5.7: Kokkare Bellure, Karnataka, coexistence of humans and wildlife

The pelican (*Pelicanus sp.*), is one of the endangered avian species spawn in huge numbers at Kokkare, Bellure. Kokkare is one of the ten identified pelican breeding spots in India, giving rise in its wake an unique man-bird cohabitation example. Kokkare, Bellure is a small village in the state of Karnataka in South India. In December of each year, several hundred birds including spot-billed pelicans, painted storks, ibis etc. travel to this place in order to set up breeding colonies. They gather on the soaring tamarind trees of the village.

Instead of poaching the migratory birds as neighbouring villagers, around the world usually do, the local people have safeguarded and protected these birds through generations, having faith that they fetch good fortunes with respect to precipitation and crop yields.

The birds are believed to bless the villagers with a affluent supply of the natural manure that accumulates underneath the nests, the Guano. The droppings of these fish-eating birds are nitrate enriched and highly potent.

The people who own the trees on which the birds live, dig cavernous pits underneath the trees, into which the guano drops and gets collected. Sediments from neighbouring lakes and ponds are blended well with the guano and then applied in their fields. They assist in the process of cultivation and are hence sold as fertilizer. As a matter of fact it is a common practice of the locals to plant trees around their homes to encourage nesting enriching this unique interdependence and transfer of benefits.

This is in stark contrast to the acts of encroachment being played out around the edges of most of our protected areas. The pressure of humans, who depend on the forest for their livelihoods, is increasing by the day. The local villagers, fighting the animals for the same living space, often act in manners that go on to destroy the fragile ecosystem of whom they too are a part.

Community reserves can be affirmed by the state government in any private or community land, where a person or a group of people in the society have come forward to conserve nature and its wildlife.

5.11.1.3 Biosphere reserves

They are the designated areas of terrestrial and coastal ecosystems encouraging a way out to resolve the biodiversity conservation with its sustainable use. These are chosen by national governments, persist under sovereign jurisdiction of the states where they are located and are internationally recognized areas. The foundation of biosphere reserves owes to the 'Biosphere Conference' organized by UNESCO in 1968. The conference culminated in the launch of **Man and Biosphere (MAB)** programme, 1970 under UNESCO, to restore the totality of ecosystem, flora, fauna and the humans to ensure self-perpetuation and to cater unhindered evolution.

- The biosphere reserves (BRs) are unique entities (sites) for both public and the natural world and are living examples of the coexistence between mankind and nature although respecting each other's needs.

- These areas contain genetic materials that have evolved over millions of years and are crucial to future adaptations and survival. It is associated with high degree of diversity, presence of many endemic species, conventional agricultural practices and indigenous knowledge or information of the local populace.

- A biosphere reserves can serve as experimental sites or 'learning places' to discover and establish methods to conservation and sustainable development, providing lessons which can be widely applicable.

- A biosphere reserve fosters the development of relationship between man and environment, encourages economic and human development on ecologically sustainable principles. Such areas function as 'living laboratories' for experimenting and demonstration in combined management of land, water and biodiversity.

- It serves as the undisturbed natural areas for scientific research and study.

- Each biosphere reserve is envisioned to fulfill three elementary roles, which are harmonizing, complementary and mutually reinforcing:

  i. **A conservation purpose** – conservation of various landscapes and water bodies, maintaining ecosystem, species and genetic diversity; to observe the normal and anthropogenic changes in space and time; to endorse traditional resource use systems.

  ii. **A development purpose** – to nurture economic and human development that is socio–culturally as well as ecologically justifiable.

  iii. **A logistic or strategic purpose** – to afford provision for education, exploration, monitoring, and dissemination of information pertaining to local, national and global issues; to encourage community spirit in the natural resource management.

- A biosphere reserve is organized into three main zones:

  i. **Core zone** is legally protected with absolutely no disturbance. It contains centres of endemism and protects the wild lineages of economically viable species and also exemplifies significant hereditary reservoirs.

  ii. **Buffer zone** are areas adjoining core zone that meets for research and education. Their uses and activities include restoration, restricted recreation, and tourism, fishing and grazing.

  iii. **Transition zone** is the outermost part of a biosphere reserve. The areas include settlements, croplands, managed forests, wasteland reclamation, areas for rigorous amusement, and other fiscal uses and human settlement.

- Cooperatively, biosphere reserves forms the **World Network of Biosphere Reserves (WNBR)**. As per MoEF annual report 2012–13, Nilgiri, Gulf of Mannar, Pachmarhi, Simlipal, Sunderban, Nanda Devi, Nokrek and Achankmar-Amarkantak have been included under UNESCO WNBR. On 31st May Nicodar islands was also declared as part of WNBR.

- By 2007, there were 14 biosphere reserves in India with few more sites under consideration. There are at present 610 biosphere reserves spread in 117 countries including 12 trans-boundary spots and 18 biosphere reserves are in India.

Table 5.10: Biosphere reserves of India

Nos.	Year	State	Name
1.	1986	Tamil Nadu, Kerala and Karnataka	Nilgiri Biosphere Reserve
2.	1988	Uttarakhand	Nandadevi
3.	1988	Meghalaya	Nokrek
4.	1989	Tamil Nadu	Gulf of Mannar
5.	1989	West Bengal	Sunderbans
6.	1989	Assam	Manas
7.	1989	Andaman and Nicobar Islands	Great Nicobar Biosphere Reserve
8.	1994	Orissa	Simlipal
9.	1997	Assam	Dibru-Saikhowa
10.	1998	Arunachal Pradesh	Dihang-Dibang
11.	1999	Madhya Pradesh	Pachmarhi Biosphere Reserve
12.	2000	Sikkim	Khangchendzonga
13.	2001	Kerala, Tamil Nadu	Agasthyamalai
14.	2005	Madhya Pradesh, Chhattisgarh	Achanakamar-Amarkantak
15.	2008	Gujarat	Great Rann of Kutch
16.	2009	Himachal Pradesh	Cold Desert
17.	2010	Andhra Pradesh	Seshachalam Hills
18.	2011	Madhya Pradesh	Panna

BR is not envisioned for the replacement of prevailing protected areas; however it broadens the possibility of conservative protection method to further reinforce the PAN. Presently, legitimately protected areas such as national parks, wildlife sanctuary, tiger reserve and reserve/protected forests remain a part of the BR with no amendment in their official position.

UNESCO – World Heritage Convention (WHC): World Heritage Convention was negotiated to protect and improve such bio diverse areas. This Convention was adopted by UNESCO in 1972 to protect and preserve the sites for posterity. Such sites are known as World Heritage Sites. India being a party has declared five such areas as World Heritage Sites.

Box 5.5: Natural world heritage sites in India

Assam – Kaziranga National Park (1985)

Rajasthan – Keoladeo National Park (1985)

Assam – Manas Wildlife Sanctuary (1985)

Uttarakhand – Nanda Devi National Park (1988) and Valley of Flowers in 2005 as an extension of Nanda Devi

West Bengal – Sunderbans National Park (1987)

Project Tiger: One of the most successful ventures of the present day was the launching of Project Tiger in 1973–74. The project aims for the tiger conservation in specially declared and organized 'tiger reserves', which represents various biogeographical regions of India. Project Tiger is managed by the **National Tiger Conservation Authority (NCTA)** with a steering committee and a field director. At the turn of twentieth century the tiger population was believed to be 40,000. Nine subspecies of tiger was recorded – Siberian, Bengal, Indochinese, South China, Sumatran, Malayan, Javan, Bali and Caspian Tiger. The last three are extinct. A national ban on tiger hunting was imposed in 1970 followed in 1972 by the enactment of Wildlife Protection Act. The first all India tiger census was carried out in 1972 which showed the existence of only 1827 tigers and the first tiger reserve was launched at Palamau in 1973 on a 'core-buffer' strategy. The 2008 census shows the existence of only 1,411 tigers. The tiger census report released by the National Tiger Conservation Authority on 28 March 2011, estimates 1,706 tigers (i.e. varying from a minimum of 1,571 to a maximum of 1,875 tigers) with the latest record of 2,226 tigers in 2014. This result figures out from 17 states of India harboring tiger population. The project stresses on proper habitat management with scientific monitoring such as satellite imaging, Geographic Information System (GIS) modeling, radio tracking, etc. The core areas were unconstrained from all types of human intervention and the buffer areas were subjected to 'conservation oriented land use'.

Box 5.6: Important varieties of tiger, deer and antelopes

Subspecies of living tigers

Siberian tiger: *Panthera tigris altaica,* also known as Amur tiger, found in Primosky and Khabarovski Krais in Russia and Russia-China border.

Bengal tiger: *Panthera tigris tigris,* also known as Royal Bengal Tiger, found in India, Bangladesh, Nepal, Bhutan and Burma.

Indochinese tiger: *Panthera tigris corbetti,* found in Cambodia, Laos, Burma and Thailand.

South China tiger: *Panthera tigris amoyensis*

Sumatran tiger: *Panthera tigris sumatrae*

Malayan tiger: *Panthera tigris jacksoni,* found in Thailand and peninsular Malaysia.

Indian varieties of deer	Indian varieties of antelope
The antlers of the males are not permanent but shed annually.	The antlers are permanent, un-branched and never shed.
Swamp deer or Barasingha (*Cervus duvaucelli*) Sāmbhar (*Cervus unicolour)* Chital or spotted deer (*Axis axis*) Hog deer (*Axis porcinus*) Hangul or Kashmir Stag (*Cervus elephus hanglu*) Thamin or Brow antlered deer (*Cervus eldi*)	Black buck (*Antelope cervicapra*) Chinkara (*Gazella gazella*) Nilgai (*Boselaphus tragocamelus*) Chousingha (*Tetracerus quadricornis*)

Table 5.11: Important tiger reserves of India

Serial nos.	States	Name of the Project Tiger reserves	Year of establishment	Total area (sq.km)
1	Uttarakhand	Corbett	1973–74	1,318.54
2	Assam	Manas	1973–74	2,837
3	Maharashtra	Melghat	1973–74	1,676.49
4	Madhya Pradesh	Kanha	1973–74	1,945
5	Karnataka	Bandipur	1973–74	880

Project Lion: The **Asiatic lion or Indian Lion** (*Panthera leo persica*), exists as a distinct isolated population in the Gir forest of the state of Gujarat. It is listed as a threatened species by the IUCN because of its small population size. The lion populations, which had drastically fallen below 100 and were at the brink of extinction, have shown an increase to 411 individuals as of 2010.

Project Elephant: The Government of India launched Project Elephant (PE) in 1992 as a centrally sponsored scheme. 32 elephant reserves are so far been approved by the Government of India. As of now, various state governments have officially notified 26 elephant reserves out of in 11 proposed **Elephant Ranges (ERs)** that extends over about 60,000 sq. km. Project Elephant was executed in 18 states and union territories. They are Andhra Pradesh, Orissa, Arunachal Pradesh, Meghalaya, Nagaland, Assam, Bihar, Chhattisgarh, Jharkhand, Karnataka, Kerala, Maharashtra, Tamil Nadu, Tripura, Uttarakhand, Uttar Pradesh, West Bengal and Haryana. Elephant (*Elephas maximus*), the largest terrestrial mammal of India requires large areas. The requirement of food and water for elephants are enormous. Hence, the population can only be supported only by forest supplies that are under the most favourable conditions. The condition of elephant can serve as the best indicator of the status of the forests. The distribution of Asian elephants varies widely from Tigris–Euphrates, India, Sri Lanka, Java, Sumatra, Borneo and up to North China.

Current distribution of wild elephant in India is confined to West Bengal, Orissa, Jharkhand, Uttarakhand, Uttar Pradesh and South India. The Project is being primarily executed in 13 states. The objectives of Project Elephant are:

- to protect territory of the elephants as well as elephant corridors;
- to tackle man-animal dispute leading to conflicts; and
- to look after the well being of domesticated elephants.

The project has the following key pursuits like:

- natural refurbishment of current habitats and wandering course of elephants;
- developing management practices scientifically and strategically for elephant protection habitats and feasible populace;
- campaigning mitigation measures for man elephant conflict in critical habitats;

- intensification of actions in order to defend from hunters and smugglers and abnormal death;
- investigation and exploration on elephant management;
- community edification and responsiveness programmes;
- eco-developments; and
- veterinary concerns.

The all India enumeration of wild elephants is carried out after every five years. It was first done in 2005 and then in 2010.

Project Rhinoceros: The Indian rhinoceros (*Rhinoceros unicornis*), is also well-known as the **greater one-horned rhinoceros**. It is categorized as a vulnerable species by the IUCN. It is primarily found in northeastern India's Assam, Terai of Nepal; the populations mainly graze the riverine grasslands and the foothills of the Himalayas.

The Indian rhinoceros population fell drastically due to excessive hunting and reduced natural habitat. Presently, about 3,000 rhinos live in the wild, of which 2,000 are found in Assam alone.

Project Crocodile: A crocodile conservation programme was started in 1975 in a tripartite cooperation between the Government of India, state governments, the UNDP/FAO Crocodile Breeding and Management Project. In 1974, an investigation and analysis of the position of the three species of crocodiles was carried out. They are:

- gharial, *Gavialis gangeticus* – found in rivers of North India;
- estuarine crocodile, *Crocodylus porosus* – found in the deltaic areas of Orissa, Sunderbans and the Andamans; and
- mugger, *Crocodylus palustris* – formerly widespread and abundant but now restricted.

The aim of the project is to increase the reproductive output by collecting eggs, incubating and rearing until they could be released in the wild.

Setting up of wild life corridors: Wildlife corridors are areas connecting the wildlife populations separated by human activities. Such areas prevent the negative effects of inbreeding. These forest corridors play an important role in maintaining genetic variation and persistence of wildlife in this landscape. Corridors require proper legal status for effective implementation. **Wildlife Trust of India (WTI)** has identified 88 elephant corridors. Example – The **Siju-Rewak elephant corridor, 2007** in the Garo hills connecting Siju WLS and Rewak Reserve Forest in Meghalaya. The **Tirunelli-Kudrakote elephant corridor** in Kerala, connecting Tirunelli Reserve Forest and Kudrakote Reserve Forest was completed in 2012.

Elephant habitations in Valparai and Gudalur have become fragmented and that has left them with a discontinuous territory. In Coimbatore, around 70 per cent of the forest was in the gradients. This narrow corridor which links **Sathyamangalam Forest to Kerala's Silent Valley in Mannarkadu** is the usual migratory pathway for elephants. Any disturbance in this region ends in elephants venturing outside the forest.

Trans-boundary conservational initiatives such as creation of trans-boundary biological corridors, cross–border collaboration and supervision of illegal trade have promoted elevated increase in the tiger population of Nepal, along the Tarai plains of Nepal. It documented 63 per cent growths in the tiger populace in the last 4 years. **Bardiya National Park** (BNP) in western Nepal is linked with the **Indian Katarniaghat Wildlife Sanctuary** with the help of **Khata biological corridor** proves to be successful. Khata corridor is the only ecological corridor in the Tarai. BNP ranks second in number of big cats after Chitwan National Park (CNP) in the country. The Tarai areas are linked to Indian protected areas through wildlife corridors, which have helped restoring the fragmented habitats. **Chitwan National Park and Parsa are connected to the Valmiki Tiger Reserve** in India, **Banke National Park with the Sohelwa Wildlife Sanctuary** and **Shuklaphanta Wildlife Reserve with the Pilibhit Tiger Reserve**.

China is also striving for the protection of tigers. Of late China has resolved two agreements with Russia and India respectively to begin discussions in the effort to conserve tigers and other endangered organisms. The policy for combating illegal wildlife trade was adopted in the **International Workshop for Transboundary Conservation of Tigers** at Kunming. For the sake of survival of the Siberian tigers, China also called for trans-border cooperation with Russia in order to build 2 wildlife corridors between their shared borders. This will allow free movement of Siberian tigers.

In 1960, WWF had made a high alert for the Siberian tigers and reported that they are on the brink of extinction. Currently, there are nearly 450 such tigers with China being the home for about 20 along its boundary with Russia. Russia is planning to increase the number to 700 by 2022. China also extended an agreement with India to protect tiger habitats and prevent illegal wildlife trade. Tiger ranges in 13 countries. Currently, 3,200 to 3,500 wild tigers are surviving worldwide, and the 13 nations have decided to increase their number to 6,000 by 2022. In November 2010, the thirteen nations together decided to announce 29 July as the **Global Tiger Day** at the Tiger Summit in St Petersburg, Russia. Each year this day is being celebrated to raise responsiveness and to support the cause of tiger conservation in the wild.

Protection of keystone species: Keystone species (the term coined by Robert T. Paine) are those organisms that play a critical role in maintaining the ecosystem balance. Such species affect many other species in the ecosystem and assist in maintaining the numbers of other organisms of the community. They can be predators, mutualists or ecosystem engineers.

Examples – Grizzly bear feeds on salmon fish from the riverine system that is rich in nutrients. While eating they bring the partially eaten carcasses to the adjoining areas. By doing so it helps in recycling of nutrients between the land and water.

Bees are efficient pollinators that ensure the perpetuation of other plant species. Pacific salmon die after spawning and their bodies serve as food to many other species. Sea stars feeding on the mussels keep a balance in their numbers as the mussels usually have no other natural predators.

Protection of Evolutionarily Distinct Globally Endangered (EDGE) species: These species are extremely threatened and have unique evolutionary history. They are extremely unusual in the way they look, live or behave.

For example, Bumble bee bat (*Craseonycteris thonglongyai*), Yangtze River Dolphin (*Lipotes vexillifer*), Golden–rumped elephant shrew (*Rhynchocyon chrysopygus*), Elephant, Bactrian Camel (*Camelus bactrianus*), Pygmy Hippo (*Choeropsis liberiensis*), Purple frog (*Nasikabatrachus sahyadrensis*).

5.11.2 Ex-situ conservation

As per CBD, ex situ conservation may be referred to as, 'the conservation of components of biological diversity outside their natural habitats'. Ex situ collections consist of entire flora or faunal collections, botanic gardens and zoological gardens and parks, wildlife research facilities, and genes and germplasm assemblages of wild and domesticated taxa (zygotes, gametes and somatic tissue).

For some taxa, it includes usable practice of ex situ methods. If one waits to implement ex situ conservation until extinction is imperative, it will be too late and the taxa will be at permanent risk. Ex situ plant and animal conservation should aim at habitat management and restoration, harmonized demographic and genetic population management, long-term gene banking, reintroduction and translocation of wild populations, suitable sharing of benefits, capacity building, increasing public awareness, appropriate research, fundraising etc. Ex situ conservation should be recognized as a device in perpetuating wild population. Where ever and whenever possible, integration between the ex-situ and in situ approach should be sought for. Ex-situ methods can use a variety of techniques like captive breeding and reproductive propagation, gene banks and germplasm banks, cryopreservation, reinforcement and reintroduction of organisms into the wild and controlled environments. Ex-situ efforts are consolidated by setting up zoological gardens and botanical gardens. Breeding and multiplying organisms in captivity can be achieved with less difficulty, what appears a herculean task is their reintroduction in their natural habitat. Today a number of species exist due to ex-situ conservation efforts. Well-known examples are Przwalski's horse, Addax, Pere David's deer, Oryx, the douc langur, etc. Seed gene bank and field gene bank is used to store the germplasm.

Zoological gardens and Central Zoo Authority (CZA): *'Zoo' means a stationary or mobile establishment, where confined animals are kept for exhibition to the public and extends to circus and rescue centers but does not embrace an organization of a licensed dealer in captive animals.'* Over the years, zoos have got transformed into centres for wildlife conservation and education. If properly managed, it serves a useful role in the preservation of wild animals. In India, the Central Zoo Authority was established for supervision the functioning and development of zoos. Only recognized zoos are allowed to operate, as per the norms and standards prescribed by the Zoo Authority.

In about 800 zoos worldwide, more than 3,000 types of vertebrates are conserved. The Marble Palace Zoo in Kolkata, (1854), is the oldest existing zoo in India. The number of recognized zoo in India as on March 2013 is 192.

Box 5.7: Few examples of Indian zoos

1. Alipore Zoological Gardens, Kolkata, West Bengal
2. Nandankanan Zoo, Bhubaneswar, Orissa
3. Arignar Anna Zoological Park (Vandalur Zoo), Chennai, Tamil Nadu
4. Indira Gandhi Zoological Park, Visakhapatnam, Andhra Pradesh
5. Jawaharlal Nehru Biological Park, Bokaro Steel City

In the 1980s the enactment of new laws such as the National Wildlife Action Plan (NWAP), 1983 and the **National Forest Policy (NFP),** 1988 gave new impetus to conservation strategies.

According to NWAP, 1983, the objectives of botanical gardens are:

- to recover native and species in danger and their restorations to protected regions of their earlier habitation;
- commercial exploitation of plenteous species; and
- to sponsor wildlife edification to a wide array of target groups.

There are 71 Indian botanic gardens and over 1,500 botanical gardens worldwide. India also has botanical gardens for public and educational interest along with Government and University botanic gardens. The Empress Garden, Pune, Garden of Medicinal Plants, North Bengal University, West Bengal, Government Botanical Gardens, Ootacamund, Nilgiris in Tamil Nadu and the Institute of Forest Genetics and Tree Breeding (IFGTB), Coimbatore, Tamil Nadu are some well known examples.

The Botanical Survey of India, supported by the DoE, GOI supervises the botanic gardens throughout India. The Indian Botanical Garden has its Directorate of Survey at Howrah. There are nine regional circles and field stations in various parts of India. Out of this nine, six have experimental gardens. They are Northern Circle in Dehra Dun, Central Circle in Allahabad, Eastern Circle in Shillong, Western Circle in Pune, Southern Circle in Coimbatore and Andaman and Nicobar Circle in Port Blair. Arid Zone Circle in Jodhpur, Sikkim-Himalaya Circle in Gangtok, and the Arunachal Field Station in New Itanagar are the additional three.

India is a crucial centre of origin for 160 domesticated economically significant plants. Over 150 natural relatives and over 800 species of ethno-botanical interest can be found in India.

Two genera, 117 species, one subspecies and nine varieties have been discovered and added to Indian record by the BSI scientists recently. (MoEF, 2011–12)

Case Study 5.8: Beej Bachao Andolan (Save the Seeds Movement)

The movement was intiated in the lower regions of Himalayas led by Vijay Jardhari. The supporters of this movement collected seeds of miscellaneous crops and efficaciously conserved several hundreds of indigenous rice varieties, rajma, pulses, millets, vegetables, spices and herbs in Garhwal. Many such varieties were cultivated as a consequence of this initiative in local farmers' fields. The local women folk realized that these variations were of much superior quality than those delivered by the Green Revolution. On the contrary, money oriented men found it tough to welcome and appreciate the advantages of cultivating native and aboriginal varieties.

Way back, it was a practice to choose domestic animals and breed them for their capability to adjust to local conditions.

Customary Indian agro-pastoralists have selectively performed breeding programme for livestock for more than thousands of years. India has a variety of livestock, such as, 27 breeds of cattle, 40 breeds of sheep, 22 breeds of goats and 8 breeds of buffaloes. These indigenous strains need to be restored and nurtured for their genetic variation.

Techniques of ex-situ conservation include the following.

A. Captive breeding

It is the process of breeding animals outside the natural habitats of those organisms in restricted human controlled conditions. It might be practiced in the farms, zoological gardens or any other enclosure with adequate facilities. Hence the preference of animals and their breeding partners for such reason are manipulated by mankind.

The purposes of captive breeding are:

- commercial scale production of animals with economic value such as pets, food, hairs, medicines, etc.;

- generation of more and more animals for zoological gardens, aquariums, research laboratories, and other public services; and

- increasing the numbers of captive population of species that are on the edge of extermination or extinction.

Here are some examples:

- **Mysore Zoo** has started research work on synchronized captive breeding programme of **lion-tailed macaque.**

- Breeding programmes for the **Bengal tiger** of India was initialized in 1880. At **Alipore Zoological Garden of Calcutta** the tigers were subjected to breeding for the first time. The record of International Tiger Studbook of 1994 shows that the worldwide captive populace of the Bengal tigers to be more or less 333.

- **Captive lion breeding plan** initiated by the government as another substitute measure to relocate lions from the **Gir Wildlife Sanctuary** to Kunho Wildlife Sanctuary in MP and also to emphasize the planned Lion Safari in Uttar Pradesh. The Government of Gujarat had declined and did not accept to the shifting of lions from Gir in spite of their deaths and health intimidation. There are some 71 lions in different zoos.

■ As a response to the decline in the number of vulture population in an unprecedented speed and scale it becomes utmost essential to bring all critically endangered vultures of three resident Gyps species into captivity to ensuring the continued existence of these species. Removing diclofenac from the environment will permit the ultimate recovery of vulture populations. **BNHS** is heading the **vulture breeding programme** in India. The programme at the moment has 182 vultures in captivity at three centres in the states of Haryana, West Bengal and Assam.

■ As an emergency measure, the captive breeding programme of the world's most endangered cat species, **Iberian Lynx** is supported by **WWF.** It is under legal and illegal trafficking.

■ There are few good examples of captive breeding with its subsequent reintroduction. **Large blue Butterfly** was reintroduced in South west England, **wolves** are reintroduced in Yellowstone National Park and **Peregrine Falcon** is reintroduced in United States; introduced Golden–lion tamarin is slowly recovering in Brazil.

Captive breeding faces the greatest challenge when threatened species are bred for financial profit. For instance, utmost conservation risk is associated with tiger breeding in confinement in the commercial farms in Asia, for using them in traditional medicine. For tigers, which may possibly number between 3000– 5000, such practice will puts the lives of wild tigers at a deplorable risk.

B. Cryopreservation

Cryopreservation or cryoconservation is a technique where cells or whole tissues that are vulnerable to damage caused by chemical reactivity or time are preserved by cooling to sub-zero temperatures, preferably at -196^0C in liquid nitrogen. Cryopreservation aspires to facilitate stock of cells for storage and have all cell lines in culture in times of need. The additional benefits of cryopreservation are:

■ lower chance of microbe aided contagion;

■ low possibility of cross contamination with further cell lineages;

■ low likelihood of genetic flow and structural modifications;

■ work can be carried out using cells at a steady passage number; and

■ low expenses (consumables and staff time).

It is used for preserving both pre-implantation embryos and sperms that can be easily recovered to revive the line.

Comparison of two methods is provided below.

Table 5.12: Embryo cryopreservation

Advantages	Disadvantages
▪ Strong and dependable (gold standard for mouse cryopreservation).	▪ Costly to freeze: expensive upfront resources (needs plenty of female donors).
▪ Genotype of donor conserved.	▪ Supplementary per diems expenses linked with maintenance of homozygous males.
▪ Embyo in homozygous condition can be frozen.	▪ Longer queues: can extend over months to be completed.
▪ Revival or renewal of line is cost-effective and easy.	▪ Does not work proficiently for all lineages/strains.

Table 5.13: Sperm cryopreservation

Advantages	Disadvantages
▪ Easy and economical.	▪ Conserve only one half of the genome.
▪ Huge amount of material can be conserved (30 million sperm/male).	▪ More costly to recuperate a line (embryos must be created by In Vitro Fertilization or IVF/ICSI).
▪ Sperms may be secluded from unhealthful/ deceased male in urgent situation and lines recoverable by the Intra-Cytoplasmic Sperm Injection (ICSI).	▪ May not work proficiently for all lineages/strains.
▪ To a great extent quicker turnaround time (takes a few weeks for completion).	

C. Gene banks

They are a type of biorepository that conserve genetic materials. In plants, this can vary from frozen snips from the plants to seeds. In animals, this may be the sperm and egg freezing till they are needed. Pieces of coral can be subjected to storage in aqueous tanks under restricted conditions. Gene banks are utilized for storing and conserving the genetic resources of plants of major agricultural plants and their crop natural relatives. Gene banks are like future insurance policy, because they:

▪ safeguard the biodiversity;

▪ present resources for reproduction;

▪ offer food solutions in the period of calamity; and

▪ food security for coming generations.

The FAO established **World Information and Early Warning System (WIEWS)** on **Plant Genetic Resources for Food and Agriculture (PGRFA)**, as a global mechanism to encourage information exchange among member countries. At present, the FAO's WIEWS lists about 1,460 gene banks worldwide (465 in Europe, 468 in the Americas, and 298 in Asia). Major gene banks are in China, Russia, Japan, India, S. Korea, Germany, and Canada. Apart from these there are some gene banks executed by **Centers of the Consultative Group on International Agricultural Research** (CGIAR). The Food and Agriculture Organization of the United Nation enlists nearly 1400 collections, that vary in size from a single sample to the USA compilation of 464,000 diverse samples. The **Svalbard Global Seed Vault** on the Norwegian island of Spitsbergen is the most famous one.

In such a place the seeds are cleansed, dehydrated and put in a sealed package; maintained at 5°C for intermediate period of storage of 20 to 30 years and at -18 to -20°C for long period of storage for about 100 years.

- The Indian Council of Agricultural Research (ICAR), Government of India has established a number of gene banks for embryos, vegetable seeds and gametes at,

  i. **National Bureau of Plant Genetic Resources (NBPGR)** – The Indian National Gene Bank (INGB) will have the capacity to store 6 lakh seed samples under secure, longstanding storage, in vitro preservation and cryopreservation. Currently, it has gathered 92,046 native germplasm samples and has about 1.6 lakh accessions of agricultural species in long-term storage.

  ii. **National Bureau of Animal Genetic Resources (NBAGR)**

  iii. **National Bureau of Fish Genetic Resources (NBFGR)**

- Three National Gene Banks for Medicinal and Aromatic Plants were established at the

  i. Central Institute of Medicinal and Aromatic Plants (CIMAP), Lucknow,

  ii. NBPGR, New Delhi and

  iii. Tropical Botanical Garden and Research Institute, Trivandrum.

The most vital most gene bank assemblage of the globally important foodstuff and rummage crops was included under the **International Treaty on Plant Genetic Resources for Food and Agriculture**; it ensures that plant breeders, farmers and researchers will have right to use these plant genetic resources with applicable normal conditions and allocate the benefits that arise out of their use. The international agreement, by the FAO Conference in November 2001 was enforced on 29 June 2004.

D. Germplasm facilities

Additional germplasm amenities have been established to facilitate research, development and other services:

- More than 1,600 cultures are kept back in the **National Facility for Microbial Type Culture Collection** as reserve.

- The **National Facility for Blue Green Algal Collection** at the Indian Agricultural Research Institute possesses more than 500 strains. It has a number of pure cultures and soil supported cultures to be furnished to the farmers in order to produce bio-fertilizers.

- The **National Facility for Marine Cyanobacteria** at Bharatidasan University co–ordinates widespread survey on the southern coast of India.

- The **National Facility for Plant Tissue Culture Repository** (NFPTCR) at New Delhi, have shouldered the responsibility of in vitro germplasm conservation including seeds and pollen grains on a medium and long-term basis. This is specially initiated for the species for which usual methods are not enough.

- The requirement of quality animals needed for biomedical research is met by the **National Facility for Laboratory Animals** at the Central Drug Research Institute (CDRI), Lucknow and the **National Institute of Nutrition,** Hyderabad.

- The **National Facility for Animal Tissue and Cell Culture (NFATCC)** under DBT has a stockpile of 1,127 cultures including 594 various cell strains. This centre has delivered 401 consignments of cultures to 84 institutions across the country. It is also endowed with 50 vector and genomic libraries.

- For the northern region, three **National Gene Banks** were established for the medicinal and aromatic plants at the Central Institute of Medicinal and Aromatic Plants and National Facility for Plant Tissue Culture Repository; for the peninsular region, the TBGRI was set up at Trivandrum.

Table 5.14: Summary of biodiversity services

	Services	Nature of uses	Examples
1.	Ecosystem services	a. Safeguarding water resources b. Pedogenesis and soil protection c. Recycling of Nutrients and storing d. Degradation of Pollutants and its incorporation e. Restoring climate f. Preserving Ecosystem	• Crop residues feed cattle; Cattle egesta feeds the soil that nourish the crops; straw provides organic matter and fodder • Bacteria decomposes cellulose fibers of straw and return to the soil; • Flood control, soil erosion, toxicity removal • Amoebas feeding on bacteria make lignite fibers available to the plants; • Some algae serve as natural nitrogen fixers. • Burrowers like shrew, moles, earthworms are natural tillers of soil and increases the water holding capacity of the soil. Bees, other animals act as pollinators and thereby perpetuate plant species; There are 130,000 plants (from melons to pumpkins, raspberries) where bees act as pollinators Estimate says that implementing REDD might facilitate • halving the rate of deforestation by 2030, and

	Services	Nature of uses	Examples
		g. Resilience from unpredictable events	• cutting emissions by 1.5 Giga tons of CO2 annually. In terms of cost, it is estimated that • It would cost from US$ 17.2 – 33 billion annually • The expected gain in reduced climate change would be US$ 3.2 trillion. • Tsunamis, earthquakes, wildfires.
2.	Biological resources	Food	Cereals, pulses, oil, spices, fish, meat, milk, egg, butter, honey, fodder for cattle.
		Fibers	• cotton, jute, hemp.
		Clothing	• hides, silk, wool, fur.
		Gums, resin, tannins, lac, essential oils, dye	• Gum Arabic, Gum karaya, Douglas fir resin, rubber latex, tea, eucalyptus oil, ephedrine, reserpine.
		Source of medicines and drugs	• Vincristin, Penicillin.
		Timber products	• Teak, mahogany, rosewood.
		Paper and cardboards	
		Decorative plants	• Orchids, palms, cycads.
		Reservoirs for breeding and producing hybrids	• Domestic cattle breeds like Hariana, Kankrej, Tharparker; cross breeds like Jersey, Holstein Friesian, etc.
		Variability in genes, species and ecosystems; serves as a gene pool	
3.	Social, ethical, aesthetic and option benefits	a. Research, education and monitoring b. Recreation and tourism c. Business d. Cultural values	- Sanctuaries, national parks, zoos. - GMOs, trangenics, bioprospecting. - Protected areas, zoos, botanical gardens. - Cosmetics, pharmaceuticals, tanneries, food processing, organic farming, poultry. - Symbols of worship.

Summary

■ Biological diversity is the variability among living organisms from all sources and **encompasses ecosystem or community diversity, species diversity and genetic diversity.**

■ A biome is large unit of land or water containing a geographically distinct assemblage of species, natural communities and environmental conditions. The biogeographical classification of India uses four levels of planning and has 10 biogeographical zones.

■ Throughout the ages ecosystems are sufficient to cope up with the environmental changes. Existing species evolve with the process of evolution and new species may substitute the existing species without disrupting the ecosystem. When loss of biodiversity, ecosystems become less resilient to abrupt pressures such as disease and climatic extremes.

■ The worth of wildlife was broadly overlooked or underrated in the earlier period. This position has changed over times. The significance of wildlife to local people is at the moment internationally documented in community-based or participatory natural resources management programmes. The economic importance of wildlife not only lies in its potential of being consumed and non-consumption uses, but also in its possible dietary value. The productive value of biodiversity comprises harvesting the products and so is its social value rooted in the cultures and tradition. All species were created equal and have the moral right to live, procreate and grow. Option value is concerned with the value of a species, its potential ability to provide economic benefit to human society in the future.

■ At global level more than 1.2 million species have been formally identified and recorded out of 1.5 to 20 billion species that have been estimated to be present. At national level, India is immensely rich with biodiversity; even only occupying 2.5 per cent of the global land mass it accounts for 7.8 per cent of the world's recorded species in an exemplary diversity of ecosystems ranging from forests, grasslands, wetlands, coasts to deserts that includes over 46000 species of flora and over 91000 species of fauna. It is one of the mega-diverse countries and has four biodiversity hotspots. In terms of endemism India has innumerable species that are site endemic, national endemic or a geographical range endemic.

■ Extinctions are normal and inevitable but with an alarming rate of extinction the major threats to biodiversity are habitat change, over exploitation, invasive species, pollution and climate change.

■ In an effort to conserve, the IUCN red list serves as the world's most complete database of conservation status of floral and faunal species. It furnishes with a species record placed under various threatened categories according to the extent or severity of threats faced by them.

■ Conservation refers to the management of wildlife by the human beings to yield the greatest sustainable benefits. In situ conservation refers to the conservation of organisms in their natural habitat. Ex situ conservation is the conservation of components of biological diversity outside their natural habitats. Ex situ collections consist of entire plant or animal collections, genes and germplasm assemblages of wild and domesticated taxa.

Exercise

MCQs

Encircle the right option:

1. Match the following National Parks with their state of location:

   i.   Jim Corbett   a. Uttar Pradesh   A. ic, iia, iiid, ivb
   ii. Dudhwa   b. Arunachal Pradesh   B. ia, iic, iiid, ivb
   iii. Kanha   c. Uttarakhand   C. id, iic, iiib, iva
   iv. Namdhapa   d. Madhya Pradesh   D. ic, iia, iiib, ivd

   Ans:

2. The number of bio– geographical regions in India is:
   A. 10   B. 7   C. 26   D. 12

3. Match the following organisms to their current conservation status:

   i.   Tasmanian tiger   a. Critically endangered   A. ic, iia, iiid, ivb
   ii. Clouded leopard   b. Endangered   B. ia, iid, iiib, ivc
   iii. Black rhino   c. Vulnerable   C. id, iic, iiia, ivb
   iv. Blue whale   d. Extinct   D. ic, iia, iiib, ivd

   Ans:

4. Elimination of the _____ may result in extinction of other members of the community.
   A. Decomposers   B. Primary consumers   C. Keystone species   D. EDGE species

5. _____ is a biodiversity hotspot in India.
   A. Sunderban   B. Gangetic basin   C. Eastern Ghats   D. Western Ghats

6. Which one of these is an example of ex– situ conservation?
   A. National Park   B. Sacred groove   C. Wildlife Sanctuary   D. Seed Bank

7. The species with thin population confined within restricted area is known as:
   A. Rare   B. Threatened   C. Endangered   D. Foundation species

8. Identify the odd combination of Protected Area Network (PAN) and the specific animal concerned:
   A. Sunderban – Royal Bengal Tiger   B. Periyar – Elephant
   C. Rann of Kutch – Wild Ass   D. Dachigam National Park – Snow leopard

9. Namadapha flying squirrel is:
   A. Endangered   B. Endemic
   C. Critically endangered and endemic   D. None of these

10. Biodiversity of a geographical region represents:
    A. Genetic diversity of the region   B. Endemic species to the region
    C. Endangered species   D. Diversity in the organisms living in the region

Fill in the blanks.

1. _____ put forward the concept of biodiversity hotspots.
2. Species with restricted distribution are known as _____.

3. _____ is the anti–cancer drug obtained from periwinkle.
4. Kaziranga National Park is famous for _____.
5. Loss of habitat in scattered patches is known as _____ _____.

State whether the statements are true or false.

1. There are ten biogeographical regions in India. (T/F)
2. Domino effect refers to the extinction of weaker species by aggressive invasive species. (T/F)
3. Dodo is an extinct species. (T/F)
4. Quinine is obtained from peepal tree. (T/F)
5. Highest numbers of identified vertebrates are the fishes. (T/F)

Short questions.

1. What are the various types of biodiversity?
2. Differentiate between genetic diversity and species diversity.
3. Differentiate between alpha and beta diversity.
4. Name the biogeographical regions of India.
5. State the importance of Western Ghats in terms of biodiversity.
6. What are the direct values and indirect values of biodiversity?
7. Briefly explain the productive value of biodiversity.
8. What are the negotiations that regulate wildlife trade in a country?
9. Define bioprospecting. Give an example.
10. Define biopiracy. Give an example.

Essay type questions.

1. Compare and contrast biodiversity at global, national and local levels.
2. India is a mega-diverse nation. – justify.
3. What do you understand by man wildlife conflict? Give examples. What initiatives can you suggest to reduce such conflict?
4. Compare a Wildlife Sanctuary with a National Park. Give suitable examples to support your answer.
5. Explain with an example the instance of man and wildlife co –existence.

Environmental Pollution _____ 6

Learning objectives

- *To know about the different aspects of pollution.*

- *To identify the causes of pollution, to realize the impact of pollution and develop a plan for solution and to set up adequate control measures.*

- *To develop public awareness and to elicit collective and individual response to prevent pollution.*

- *To know about real life situations through case studies and arouse sensitivity.*

6.1 Introduction

6.1.1 Definition

Pollution can be defined as any undesirable change in the physical, chemical and biological characteristic of air, water or land that may or will harmfully affect human life or that of various species, the industrial processes, living conditions and cultural assets.

6.1.2 Classification of pollutants

Generally a pollutant is classified as biodegradable and non-biodegradable. Biodegradable pollutants are those which degrade easily such as sewage, whereas non-biodegradable pollutants are those which either do not degrade or degrade very slowly, such as plastic.

6.1.3 General source of pollutants

The sources of pollution caused by man-made actions can be grouped under four categories.

- Mobile transport: motor cars, motor bikes, aircrafts, ships, vapourization of gasoline.

- Stationary/immobile combustion: domestic, business sources, steam power-driven and thermal power plants.

- Factories and industry processes: metal and non-metal extractions, paper and pulp, oil refineries, synthetic chemicals, food processing industries.

- Disposal of solid wastes: domestic and mercantile refuse, refuse from coal combustion, burning of agricultural refuse.

6.1.4 State of pollution in India

A Comprehensive Environmental Pollution Index (CEPI) is used to typify the quality of the environment at a particular locality ensuing the source of pollution, their path or conduit and the receiver. CEPI is envisioned to function as an instrument of timely warning. Such a pollution index apprehends the well-being of environmental components like air, water and land in their various dimensions. The CPCB or SPCB or Pollution Control Committees in association with IIT Delhi, had applied Central Environmental Pollution index to 88 specific industrial clusters or areas. Out of 88, 43 were 'critically polluted' (with score more than 70 out of 100); 32 were 'severely polluted' with scores varying between 60 and 70. In the interim assessment report carried out jointly by CPCB and GPCB, 2012, Vapi ranked first in pollution with a CEPI score of 90.75 out of 100, Angul Talcher ranked second with 89.74, while Vatva and Ankleshwar rank sixth and seventh respectively.

As per WHO guidelines, the atmospheric PM levels should not be more than 20 $\mu g/m^3$ but PM levels in the city of Ludhiana are 251, whereas that of Delhi and Mumbai is 198 and 132, respectively.

Table 6.1: CEPI scores, 2009, for top 10 industrial areas/clusters as per CPCB.

No.	Industrial cluster/area	State	CEPI (air, water and land)
1	Ankleshwar	Gujarat	88.50
2	Vapi	Gujarat	88.09
3	Ghaziabad	UP	87.37
4	Chandrapur	Maharashtra	83.88
5	Korba	Chattisgarh	83.00
6	Bhiwadi	Rajasthan	82.91
7	Angul Talcher	Orissa	82.09
8	Vellore (North Arcot)	Tamil Nadu	81.79
9	Singrauli	UP	81.73
10	Ludhiana	Punjab	81.66

6.2 Air Pollution

It is any undesirable change in the quality of air that harmfully affects people's well-being. Atmosphere, the gaseous layer called 'air' is the life blanket of Earth, the essential ingredient for all living things. It is composed of mixture of gases (Refer Chapter 1). It makes up nearly 80 per cent of man's daily intake by weight. Humans breathe almost 22,000 times per day, taking in almost 16 kg of air. Therefore, preserving or maintaining the air in its pure form becomes necessary, failing which it (air) may profoundly affect the health and bring in other consequences. The atmosphere can cleanse itself of the impurities firstly by its vertical and horizontal mixing (dispersion – favoured by wind) and secondly washing out the pollutants to a certain extent by rain. But, the presence of high concentration of pollutants in air such as sulphur oxides (SO_x), nitrogen oxides (NO_x), hydrocarbons (HCs), carbon monoxide (CO) and particulates, affects the natural cleansing mechanism of the atmosphere, alters the percentage of concentration of gases naturally present in the atmosphere causing air pollution.

6.2.1 Classification of air pollutants

Agents or contaminants that cause or induce the formation of air pollution are called air pollutants. All air pollutants may be categorized according to origin, chemical compositions and state of matter.

A. Based on origin:

    i.   Primary pollutants are those pollutants that are emitted directly into the air and are found in the same chemical form in which they are released. For example, particulate matter, SO_2, NOx, CO, HC.

    ii.   Secondary pollutants are generated in the environment by interactions between two or more primary pollutants. For example, O_3, PAN, H_2SO_4, HNO_3.

B. Based on chemical composition:

    i.   Organic compounds: Organic compounds containing carbon and hydrogen are known as hydrocarbons. Aldehydes and ketones contain oxygen as well as carbon and hydrogen. For example, formaldehyde and acetone.

    ii.   Inorganic compounds: It includes carbon monoxide (CO), carbon dioxide (CO_2), carbonates, sulphur dioxides (SO_2), nitrogen oxides (NO_x), ozone (O_3), hydrogen fluoride and hydrogen chloride.

C. Based on state of matter:

    i.   Particulate pollutants/suspended particulate matter (SPM) are fine sized solids and liquids including dust, fumes and smoke, fly ash, mist and soot particles.

    ii.   Gaseous pollutants are formless fluids that completely fill up the space into which they are emitted. For example, Carbon dioxides, sulphur oxides, nitrogen oxides, hydrocarbons and oxidants.

6.2.2 Types of air pollution

■   Indoor air pollution: It is caused by burning of wood, animal dung, agricultural residues and coal used for food preparation and heating within the closed walls. Most of the victims of indoor pollution are women and girls, who owe the prime task of cooking and tending the home. Approximately 2.5 billion people, mostly in developing nations are exposed to high levels of indoor air pollution and consequently suffer.

- Outdoor air pollution: Industrial enterprises and automobiles are primary sources of atmospheric (outdoor) pollution. Increasing industrialization and urbanization has created growing demands to use the outdoor atmosphere as a medium of waste disposal. A pall of pollution entombs the thickly-populated and quickly growing cities such as Bangkok, Manila, Mexico City and New Delhi. The sources are mainly automobiles and uninhibited industrial emissions. Outdoor air pollution affects over 1,100 million people, predominantly in cities.

6.2.3 Sources of air pollution

A. Natural sources:

The atmosphere is polluted due to various natural processes, some of which are wind-blown dust, smoke, fly ash, gases from forest fires, gases and odours from swamps and marshes, pollen, fungi spores from microorganisms, salt spray from the ocean, fog in humid low-lying areas, anaerobic decomposition of organic matter, atmospheric reactions and natural terpene hazes from pine trees in mountainous region. Volcanic eruptions release great amounts of sulphur dioxides. Anaerobic decomposition of organic matter generates methane gas which upon oxidation in the air gives out carbon monoxide. Organic substances decay to produce stinking gases like H_2S too. Photooxidation of marine organic substance and biological oxidation by oceanic organisms generates carbon monoxide on the ocean surface which diffuses in the air. (See Table 6.2)

Table 6.2: Natural sources of air pollution

Process	Pollutants produced
Volcanoes	Sulphur dioxide, fly ash
Forest fire	
Biological decay	CO, fume, smoke
Storm wind	Methane(CH_4), ammonia (NH_3) Hydrogen sulphide (H_2S)
Ocean release	
Plants and micro-organism	Dust particles
	CO_2, salt spray
	Pollens, fungi spores

A. Anthropogenic or man-made sources:

Along with natural pollutants, there are pollutants of anthropogenic origin too. Bulk of the air pollution sources are linked to anthropogenic activities, an endowment of modern lifestyle. The multiple use of fossil fuels for transportation in vehicles, for industrial uses, for conservation of energy, for heating and for cooling as well as the treatment of industrial and municipal wastes, all add to air pollution.

The major anthropogenic sources of air pollution can also be grouped as:

- Domestic: Coal combustion generates enormous amount of, smoke, soot, dust, CO, SO_2, NO_x. However burning of LPG releases fewer amounts of pollutants comparatively.

■ Automobiles: Motor vehicles play an important role in air pollution (automobile emission). Automobile emissions have been identified as the major source of air pollution in Kolkata metropolitan region. It is contributing nearly 60–70 per cent of air pollution. India is the fifth leading car producer in the world in 2011. It is also the leading producer of three-wheelers (8,78,000 in 2011–12). The number of government-registered vehicles on roads in India is 142 million vehicles in the year 2011–12. (MoEF, Annual Report, 2012–13). The country vehicle population has grown in alarming proportions during the last decade. (Table 6.3)

Gasoline combustion produces carbon monoxide, various hydrocarbons such as aldehydes (HCHO), polycyclic aromatic hydrocarbons (PAH), NO_x, SO_x, various types of organic acids, ammonia, carbon particles and heavy metals like lead. Incomplete burning of fuel produces a hydrocarbon, 3, 4- benzpyrene.

Table 6.3: Allocation of various vehicular categories out of total vehicular population in India Annual Report, MoEF, 13 February 2012.

Serial nos.	Types of vehicles	Share (%)
1	Two wheelers	72
2	Cars and jeeps	14
3	Buses (inclusive of Omni buses)	1
4	Goods vehicles	5
5	Others	8

C. Industries:

   i.   **Fertilizer plants** – They generate sulphur oxides, nitrogen oxides, hydrocarbons, PM and fluorine.

   ii. **Thermal plants** – fly ash, soot, SO_2, CO, NO_x.

   iii. **Textile industries** – cotton fibres and dust, NO_x, chlorine gas, naphtha vapours, smoke and SO_2.

   iv. **Steel plants and metallurgical operations** – carbon monoxide, carbon dioxide, sulphur dioxide, phenol, fluorine, cyanide, particulate matter, copper, lead, zinc etc.

   v.   **Petroleum** – Fossil fuels include petroleum and coal; emissions are mainly sulphur dioxide. Additionally, carbon monoxide (CO), carbon dioxide (CO_2), nitrogen oxides, hydrocarbons, particulate matter and traces of metals are produced.

   vi. **Paper and pulp** – PM, SO_2, H_2S, methyl mercaptan.

   vii. **Food processing** – often releases dimethyl sulphide and various types of odour.

D. **Agriculture** – mainly pesticides and herbicides like chlorinated hydrocarbons, phosphates, nitrates etc.

6.2.4 Major air pollutants

Major air pollutants fall in six main categories of air pollutants:

■ Aerosols and vocs   ■ Other hydrocarbons   ■ Particulate matter (inorganic and organic)

■ Oxides of carbon   ■ Sulphur dioxide   ■ Oxides of nitrogen

Aerosols: Aerosols are tiny particles suspended in the air. Aerosol is produced when solid or liquid disperses in the air. Examples are smoke, oceanic haze, smog. On an average, globally, aerosols from human activities contribute about 10 per cent of the total amount of atmospheric aerosols. Most of this is concerted in the Northern Hemisphere, due to industrial development, slash-and-burn in croplands and overgrazed grasslands.

Natural aerosols in the air also arise from volcanoes, dust storms, forest fires, foliage and sea spray. Nonetheless, harmful aerosols are created by the particles of carbon. Anthropogenic activities, such as fossil fuel burning and the change of natural surface cover, also create aerosols.

Thick layer of tropospheric aerosols can have an effect on the climatic conditions by preventing the entry of the solar radiation. When aerosols are deposited on the leaves they interfere with the photosynthetic process. Aerosols are involved in the widespread scattering of the organic metallic pollutants. Sulfate aerosols exert a cooling effect on the climate and NOx form smog.

Volatile Organic Compounds (VOCs): The term VOCs refers to the class of organic chemical compounds with considerable vapour pressure. There are various types of VOCs present ubiquitously that can affect the environmental and human well-being. VOCs include both natural and man-made chemical compounds. It is the man-made VOCs that are controlled, especially for indoors environment where concentrations can be maximum. Commonly VOCs are used in household products like paints, paint strippers and other solvents, wood preservatives, aerosol sprays, cleansers and disinfectants, moth repellents, air fresheners stored fuels, automotive products and dry-cleaned clothing.

Common examples of VOCs are acetone, benzene, ethylene glycol, formaldehyde, methylene chloride, perchloroethylene, toluene, xylene, butadiene and an important class of compounds called terpenes, such as myrcene.

Exposure to VOCs is primarily through inhalation, which may affect the mucosa of the eyes, nose, esophagus and respiratory passage. Continued exposures to VOCs have shown to cause various types of malignancy. These chemicals also act as endocrine organ disrupters because they can mimic and interrupt the naturally occurring hormones actions in the human body.

POPs: POPs are toxic organic class of compounds that harmfully affect both the health and the environment throughout the world and has become a global concern. As they can be easily carried by currents of wind and water, most of the POPs produced in one country affect man and nature far-off from the place they are used and released. POPs including some pesticides, heavy metals evaporate from the soil in the equatorial and tropical countries; travel in the air towards comparatively cooler regions, where they condense with the falling of temperature. The processes repeat in 'hops' and are thus carried across thousands of kilometers away within few days. The more the volatility, the far they are carried and retained in the air. Such a mechanism of long range atmospheric transport and subsequent deposition of POPs is called '**grasshopper or global distillation process**'. The process goes on and on with highest concentrations in the circumpolar nations. In view of this, the Arctic Council was constituted by Denmark, Sweden, Iceland, Norway, Russia, Alaska, Greenland, Finland, etc., to monitor and assess the source and pathway of the POPs.

The POPs endure for an extended period in the environment and can amass and go into from one trophic level to the next through the food chain. In 1992, under United Nations (UN)/ECE (European Commission of Europe) the Convention of Long Range Transboundary Air Pollution (LRTAP) was constituted to identify the chemicals of potential concern. In 2001, the US with 90 other nations and EU signed the treaty in Stockholm, Sweden. Under this treaty the

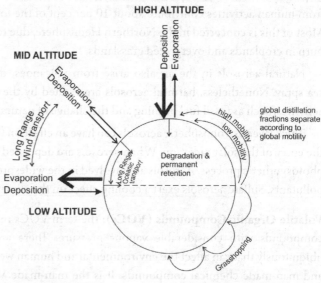

Figure 6.1: The global distillation and grasshopper movement of POPs

nations decided trimming down or eliminating the production, utilization, and/or discharge of 12 main culprits, known as 'dirty dozen'. (Refer Box 6.1)

Particulate matters: Particulates or particulate matter (PM) or fine particles is a generic term used to represent a subdivision of solid particles or liquid droplets suspended in a gas or liquid. PM is a complex mixture of airborne particles. They differ in origin, size, and their chemical composition, all of which are <10 μm in size. Their size ranges from coarse wind-blown dust to very fine particles from chemical reactions. They can be inorganic components like nitrates, sulphates, metals, carbon particles (soot) or organic compounds like aerosols, POPs etc. These fine particles are dispersed in air and can freely move over great distances. Most of particulates come from burning in fireplaces and wood stoves, automobile exhaust, etc.

Box 6.1: The dirty dozen

1. Aldrin – Pesticide, closely related to dielrin; extensively used on corn and cotton and for termite control.

2. Dieldrin – Pesticide widely used on corn and cotton pests. It is also a metabolite of aldrin.

3. Chlordane – Pesticide on crops, lawns and gardens and a fumigant for termite control.

4. Heptachlor – Insecticide for household and agricultural uses. It is also a component and a breakdown product of chlordane.

5. DDT– Pesticide; used for controlling malaria since Second World War; discovered by Paul Mueller.

6. Endrin – Used as insecticide, rodenticide and also to control birds.

7. Hexachlorobenzene (HCB) – used as pesticide and fungicide used on seeds, is also an industrial byproduct.

8. Mirex– used as insecticide and as flame retardant.

9. Toxaphene – Insecticide used on cotton pests.

10. PCBs or Polychlorinated biphenyls, widely used in electrical equipment.

11. Polychlorinated Dioxins

12. Polychlorinated Furans – Two notorious classes of 'unintentional' pollutants, produced as byproducts of incineration and industrial processes.

Particles larger than 100 µm diameter settle quickly. They can be dust, soot aggregates, sand particles and sea spray. Particles with aerodynamic diameter less than equal to 10 µm are known as PM10. Particles with aerodynamic diameter less than equal to 2.5 µm are known as PM2.5. The lesser the diameter, more will be the penetrating ability and greater will be the hazard. Evidence indicates that respirable particulate matter (RPM) or PM10 is linked to health hazards. Larger particles >10 µm are retained in the nasal passage. Particles between 5– <10 are arrested in the trachea. Further small particles may reach the trachea and ultimately settled in the alveoli.

The symptoms of the upper respiratory tract infections are stuffy nose, sinusitis, sore throat, hay fever, cough, and irritation of eyes. The symptoms of the lower respiratory tract infections are wheezing, phlegm, dysponea, asthma, pain in chest, emphysema, etc. RPM slows the ciliary beat and mucous flow inflammation of lung tissue. PM causes alterations in blood chemistry and can increase susceptibility to viral and bacterial pathogens. Particulates like arsenic, PAH, radioactive nuclei are carcinogens.

CFCs: CFCs are organic compounds comprising carbon, chlorine, and fluorine, produced as a capricious derivative of methane and ethane and commonly known as freons. A subclass of the CFCs is the hydrochlorofluorocarbons (HCFCs), which include hydrogen in addition; commonly they are identified by the DuPont trade name freon. One of the most recognizable examples is dichlorodifluoromethane (R-12 or Freon-12). CFCs had been extensively used as refrigerants, propellants and solvents. The production of CFC compounds is being phased out by the negotiation of Montreal Protocol since these add to diminution of ozone in the stratosphere. Presently, gases, such as helium, propane/isobutane mixtures are used as refrigerants.

Table 6.4: Summary of the various classes of air pollutants, their source and their impacts

Nos.	Pollutant	Source/Cause	Effect
1.	Carbon monoxide	Primarily produced by incomplete combustion; automobile exhaust, atmospheric photo- chemical reactions, biological oxidation by marine organism, etc	Affects respiration since haemoglobin has more affinity for carbon monoxide than for oxygen. CO simply combines with haemoglobin forming carboxyhaemoglobin (COHb);

Nos.	Pollutant	Source/Cause	Effect
			such formation reduces the oxygen–carrying capacity of blood. This in turn leads to fuzzy vision, headache, drowsiness, and finally death due to asphyxiation (lack of oxygen).
2.	Carbon dioxide (CO_2), though not a pollutant at normal level and is harmful in excess levels.	Fossil fuels combustion, forests diminution (forests take away excess carbon dioxide and help in maintaining the O_2 and CO_2 ratio).	Prime GHG that results in global warming.
3.	Sulphur dioxide (SO_2)	Industrial processes, fossil fuel combustion, wild fire, thermal power plants, smelters, industrial boilers, oil refineries and volcanic eruptions.	Respiratory ailments, rigorous headache, bronchitis, reduced plant production, yellowing, necrosis and corrosion to marble, spoiling of leather, increased corrosion of iron, steel, zinc and aluminium.
4.	Polynuclear Aromatic Compounds (PACs) and Polynuclear Aromatic Hydrocarbons (PAHs)	Vehicular exhaust and industries, leaky fuel tanks, leaching of noxious waste from dumping sites, coal tar lining of some water supply pipelines.	Potent carcinogens at high concentrations.
5.	Chloroflurocarbons (CFCs)	Refrigerators, air conditioning machines, foam shaving cream, spray cans and cleaning solvents.	When inserted in the troposphere they destroy stratospheric ozone layer and thereby allow the harmful UV radiation entry into the troposphere.
6.	Nitrogen Oxides	Vehicular exhaust, fossil fuel combustion, wild fires, thermal power plants, smelters, industrial boilers, oil refineries and volcanoes.	Formation of photochemical smog, acid rain and causes respiratory ailments such as inflammation, pneumonia, lung cancer; leaf damage at high concentration, influences photosynthesis.
7.	Nitrous oxide (N_2O)	Nitrogenous fertilizers, deforestation and biomass burning.	Greenhouse effect.
8.	Peroxy Acetyl Nitrate (PAN)	Chemical reaction of hydrocarbons and nitrogen oxides in presence of sunlight.	Irritation of eye, throat and trachea, damages clothing, paints and rubber, leaf damage and stomatal tissue damage in plants.
9.	Particulate matter lead halides (lead pollution)	Combustion of leaded gasoline products	Respiratory problems, bronchitis, emphysema, asthma: some are carcinogenic.
10.	Asbestos particles	Mining activities	Asbestosis
11.	Silicon dioxide	Cutting of stones, cutting, crushing and grinding, cement industries, making of pottery and glass.	Silicosis

Nos.	Pollutant	Source/Cause	Effect
12.	Biological matters like pollen grains, fungal spores, bacteria, and virus.	Flowers and microbes	Allergens and infections.
13.	Dioxins and furans	From combustion of chlorinated substance.	Carcinogen and possibly targets all the organ systems

6.2.5 Occupational health hazards

A. Berylliosis: It is a kind of pneumoconiosis; a systemic granulomatous disease, mainly affecting the lungs. This occupational health hazard can affect employees working in beryllium alloy industry, ceramic objects, foundry, cathode ray tubes, gas blanket, projectile, and nuclear reactors. Its incidence can happen in two forms: acute non-specific pneumonitis and chronic granulomatous disease with interstitial fibrosis, which may cause respiratory failure and ultimately death. Chronic disease may assume full form in 10 to 15 years after exposure. Most patients with chronic interstitial disease have only slight to moderate disability from impaired lung function and other symptoms. With each acute exacerbation, though, the prognosis worsens.

The list of signs and symptoms for berylliosis includes respiratory system symptoms in acute cases; coughing, difficulty in breathing, chest pain, tiredness, loss of body weight, allergic rashes all over the body, cyanosis; pulmonary granulomas, pulmonary nodular accumulations and pulmonary inflammatory cells, orthopnoea, haemoptysis, clubbed fingernails, corneal lesions in chronic cases with prolonged exposure.

B. Asbestosis: Asbestosis is very common. More than 3,000 products used currently include asbestos. Most of the products in heat and sound insulation, fire proofing, furnaces, floor tiles, electric wires, wall and ceiling panels, cements, contain asbestos. Workers working with asbestos products often develop asbestosis. The use of asbestos has been banned in goods such as pipe coverings, hairdryers, and artificial fireplace logs.

Asbestosis is labeled as a monosymptomatic disease as because the first symptom is only shortness of breath. Other symptoms are persistent cough, chest tightness, chest pain and appetite loss. Asbestosis is scarring and scraping of the lungs that leads to respiratory problems and cardiac failure.

C. Silicosis: Silicosis is a pulmonary disease caused by breathing in of silica dust. Silicosis is also known as grinder's asthma, grinder's rot, mason's disease, miner's asthma, miner's phthisis, potter's rot, rock tuberculosis, and stonemason's disease. Crystalline silica is a naturally occurring mineral, a main component in quartz, sand, flint, agate, granite, etc.

The most common early symptoms include cough, tiredness, appetite loss, chest pains, shortness of breath and bluish skin at the extremities. As crystalline silica exposure continues, the symptoms worsen. This is because the lung cell and tissues become more corroded and little efficient to carry out normal functions.

More rigorous symptoms of silicosis include all the above symptoms along with fever, loss of weight, night sweating, cyanosis and respiratory failure.

When one breathes in crystalline silica, it causes swelling of the lung tissue. This inflammation leads to the formation of scar tissue on the lungs. The scar tissue blocks the normal inflow of oxygen into the lungs and into the bloodstream. Silicosis also increases the susceptibility to bacterial or fungal infections. It can also cause lung cancer and tuberculosis. If left untreated, it is ultimately fatal.

D. **Anthracosis:** The earlier milder form of the illness from inhalation of coal dust is known as anthracosis (anthrac – coal, carbon). This is often asymptomatic and is found more or less in all city dwellers due to air pollution. Protracted exposure to considerable amounts of coal dust can result in more grave forms of the disease, the coal workers' pneumoconiosis.

E. **Black lung disease:** It is commonly known as 'coal worker's pneumoconiosis', a syndrome caused by the extended exposure to coal dust, which is progressively deposited in the pulmonary tissues. It is a chronic respiratory illness. Pneumoconiosis can be silicosis, asbestosis and the coal workers' pneumoconiosis or black lung disease. Silicosis and asbestosis, as stated earlier, are caused by the accumulation of silica and asbestos in the lungs. Black lung disease leads to black pigmentation of the lung and black salivation.

The main cause of black lung disease is the inhalation of man-made carbon particles. It is widespread among the coal miners, connected with the production of graphite and carbon black. The inhaled coal particles assemble in the lung and cause its discolouration. But ultimately, this transforms into small coal lumps or nodules, which gradually enlarge in size. The lumps are located in the lymph nodes and connective tissues of the lungs, and are impossible to be removed from the body. Finally they obstruct the air flow through the respiratory passages. Smoking, though not related, can add to the further deterioration of the lungs, and thereby exacerbating the situation. Usual symptoms of the disease are shortness of breath, constant cough and blockage of the air passages. However, sometimes the patients may be asymptomatic.

6.2.6 Global impacts of air pollution

6.2.6.1 Global warming

Major part of the Sun's ray that enters the Earth's atmosphere gets reverted back into the space. Yet some amount of heat is absorbed by the GHGs such as CO_2. The additional gases that contribute to this are water vapour, methane, CFCs and nitrous oxide. This helps to maintain the coziness and warmth very similar to the greenhouses in cold countries.

Greenhouse effect is an indispensable and normal phenomenon. Each year the temperature is rising on account of pollution and the concentration of GHGs is mounting too. This is referred to as global warming. At the present pace of increase, the average temperature of the Earth is predicted to increase by 3°C to 8°C in the coming century.

Global warming will exert subsequent effects on:

- changes in climatic pattern;
- changes in the distribution of flora and fauna;
- changes in the pattern of cultivation and crop yield; and
- snow caps will melt leading to increasing sea levels, this will lead to the submergence of the coastal cities like Kolkata, Mumbai, London, New York, etc.

Greenhouse gases and greenhouse effect: The term 'greenhouse gases' refers to the atmospheric gases that absorb the spectrum and emit them back within the thermal range of infrared radiation. These gases can significantly affect the global temperature. The temperature of the Earth's surface would have been almost $59°$ Fahrenheit colder than the present temperature in absence of these gases.

Everything is not good about these greenhouse gases. GHGs have a tendency to trap solar radiation, which ultimately leads to an increase in the surface temperature of the Earth. This is known as greenhouse effect or global warming. Gases are usually compared to one another on the basis of their Global Warming Potential (GWP); GWP of a GHG is its warming effect over a period of time compared to the same amount of carbon dioxide. Carbon dioxide is the universally accepted point of reference and its GWP is accepted as 1. It is the most significant greenhouse gas added by anthropogenic activities.

Table 6.5: GWPs and atmospheric lifetimes, IPCC, 2007

GHGs	Pre-1750 troposphere concentration	Latest tropsphere concentration	GWP (100 year time horiozon)	Lifetime in atmosphere (years)	Increased radiative forcing (W/m2)
Concentrations in parts per million (ppm)					
Carbon dioxide (CO_2)	280	392.6	1	~100	1.85
Concentrations in parts per billion (ppb)					
Methane (CH_4)	700	1874/1758	25	12	0.51
Nitrous oxide (N_2O)	270	324/323	298	114	0.18
Troposphere ozone (O_3)	25	34	Not applicable	hours-days	0.35
Concentrations in parts per trillion (ppt)					
CFC-11 (trichlorofluoromethane) (CCl_3F)	0	238/236	4,750	45	0.06

Water vapour, produced due to evaporation of water or through sublimation of ice, nearly constitutes 33 to 66 per cent of GHGs. Carbon dioxide is produced by respiration of flora and fauna (including human beings) and burning of fossil fuels. It constitutes 9 to 26 per cent of greenhouse gases.

Methane comprise 4 to 9 per cent of greenhouse gases, is produced by a process called methanogenesis and organic waste decomposition. When the permafrost region in the northern hemisphere begins to thaw with rising temperatures, it releases huge quantities of methane. Nitrous oxide used as an aerosol propellant and as an anesthetic, comprises around 5 per cent of the GHGs. Ozone is both natural as well as anthropogenic and constitutes approximately 3 to 7 per cent of greenhouse gases.

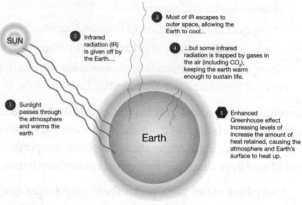

Figure 6.2: The Greenhouse effect

CFCs are the most popularly known haloalkanes, harmful for the environment. Methyl chloride is the only main naturally-produced organochlorine molecule made up about 0.6 ppb of that total; almost all others are anthropogenic. HFCs make up only a small portion but they are extremely potent greenhouse gases. They are used in refrigeration in airconditioning and as foam blowing agents, solvents, firefighting agents and aerosol propellants. PFCs, byproducts of aluminium smelting are also used in semi-conductor manufacture. Sulphur hexafluoride is the most potent greenhouse used in Nike Air shoes, making of tyres, electrical insulation, manufacture of semiconductors and in the magnesium industry. Other than these, gases such as nitrogen trifluoride and perfluorocarbons but owing to their presence in small quantities, these gases are not considered to be as important as others.

For ozone depletion, CFC and polar stratospheric clouds refer to Chapter 9.

Box 6.2: Common refrigerants and its alternatives

Common refrigerants used are of three types that are used in refrigeration and airconditioning systems. These refrigerants are 'halogenated,' means they contain chlorine, fluorine, bromine, astatine, or iodine. They are:

1. Chlorofluorocarbons (CFCs) – R-11, R-12, and R-114
2. Hydrochlorofluorocarbons (HCFCs) – R-22 or R-123
3. Hydrofluorocarbons (HFCs) – R-134a.

According to National Refrigeration Safety Code catalogues, refrigerants are grouped into three groups:

- Group I or safest of the refrigerants – R-12, R-22, and R-502
- Group II or toxic – R-40 (Methyl chloride) and R-764 (Sulfur dioxide)
- Group III or flammable refrigerants –R-170 (Ethane) and R-290 (Propane).
- Additional Refrigerants – R-717 Ammonia (NH3)

Eco-friendly approaches

HFC-152a can replace R-134a having similar operating features as R-134a but cools even better. An environmental advantage of HFC-152a is its global warming rating of 120, which is 10 times less than R-134a. This value is still higher than that of CO_2.

Greenfreeze uses a mixture of propane (R290) and isobutane (R600a), or pure isobutane gas as refrigerant to replace the ozone–destroying molecules.

Photochemical smog: The industrial revolution is the key to amplifying the atmospheric pollutants. Smog episodes in the nineteenth and twentieth centuries were named as 'pea-soupers'. Prior to 1950s, the bulk of pollution was due to coal combustion. Under suitable conditions smoke and sulphur dioxide (burning of coal) combined with fog to produce industrial smog. Industrial smog can be extremely poisonous to humans and other living organisms. The most notable incident that occurred in December, 1952 was the London smog episode when tranquil unclear weather conditions prevailed for five days and created a deadly atmosphere that claimed nearly 4,000 lives.

Presently, burning of fossil fuels like petroleum and diesel can produce additional air pollution dilemma recognized as photochemical smog. Photochemical smog is a situation occurring when the primary pollutants like NO_2 and VOCs react in presence of sun rays, forming a mix of many diverse and harmful chemicals known as secondary pollutants. Los Angeles, New York, Sydney, Vancouver and other cities habitually experience photochemical smog. High concentrations of NO_2 and VOC are coupled with industrialization and transportation. The timing of day serves as a significant factor in the aerial quantity of photochemical smog. Precipitation can ease the effect of photochemical smog as the contaminants are cleansed out of the air with it.

The prerequisites for photochemical smog formation are:

- solar radiation;
- NO_x;
- VOCs; and
- Temperatures should be over 18°C.

Ozone (O_3) and PAN are the two most overriding components of photochemical smog.

One of the following reactions produces nitrogen dioxide. It is to be noticed that the nitrogen oxide (NO) acts to eliminate ozone (O_3) from the air and such a phenomenon happens naturally in an uncontaminated condition.

$$O_3 + NO \longrightarrow NO_2 + O_2$$
$$NO + RO_2 \longrightarrow NO_2 + \text{other products}$$

Sunlight can degrade nitrogen dioxide

$$NO_2 + \text{sunlight} \longrightarrow NO + O$$

The atomic oxygen (O) then reacts with oxygen molecules producing ozone (O_3).

$$O + O_2 \longrightarrow O_3$$

Nitrogen dioxide (NO_2) reacts with the radicals produced from VOC in a sequence of reactions to form PAN.

$$NO_2 + R \longrightarrow PAN$$

Table 6.6: Important pollutants sources and environmental effects in photochemical smog

Toxic chemicals	Their sources	Environmental impacts
NO and NO_2	▪ automobiles and industrial combustion ▪ microbial action in soil ▪ wild fires ▪ volcanoes ▪ lightning	▪ reduced visibility due to brownish yellow colour of NO_2 ▪ contributes to cardiac and pulmonary complications ▪ NO_2 can retard plant growth and development ▪ increased susceptibility to infections
VOCs	▪ utilization as solvents ▪ incomplete fossil fuel burning ▪ natural compounds like terpenes from plants	▪ irritation of eyes and trachea ▪ carcinogen ▪ reduceed visibility
Ozone (O_3)	▪ photolysis of NO_2 ▪ ozone intrusion in Stratosphere	▪ constriction of bronchioles ▪ coughing, panting ▪ irritation of eyes and trachea ▪ reduced agricultural production ▪ retarded growth of plant ▪ damage to plastics ▪ cracking of rubber ▪ bad odour
Peroxy acetyl Nitrates (PAN)	▪ produced by the interaction of NO_2 with VOCs	▪ eye irritation and respiratory tract ▪ highly toxic to plants ▪ protein damage

Case Study 6.1: The London smog episode

In the winters of UK in the late nineteenth and twentieth centuries, on considerable occasions, the wind speed had fallen and temperature inversions were formed as stationary high pressure systems developed over western Europe. With increase in the concentration of pollutants, fogs became widespread. The condition was commonly referred to as smog. Initially the term smog was coined by Harold Des Voeux, the treasurer of Coal Smoke Abatement Society in 1905 to refer an odd combination of smoke plus fog.

Smog was formed from particulate emissions from the combustion of coal in the industrial furnaces, kilns and boilers, domestic grates and steam locomotives, canal ships. The visibility dropped, transport could not move, railways and air flights had to be closed and shops, stores, and institutions were shut down completely paralyzing the city. PM went up by 56 times the normal level and SO_2 level went up by seven times.

Some 8,000 young and old people were hit by bronchitis and heart disease. The number of deaths rose so high from this sulphur laden sooty condition that it was popularly known as Great Smog (Williamson and Murley, 2002). These profound events led to the enactment of Clean Air Act of 1956.

Case Study 6.2: Los Angeles smog episode

The sulphur-laden sooty smog in time, however, eventually gave way to the photochemical smog (the one we experience presently) with the development of internal combustion engine as a prime source of transportation. Such smogs are initiated by nitrogen dioxides. In presence of sunlight the free oxygen atoms react to form ozone. This was first evidenced in Los Angeles region of USA (Porteus, 1995). It irritated the eyes, damaged the plants, for example tobacco.

The major sources of nitrogen dioxides were vehicular emission, emissions from aircrafts, ships, trains, industries and houses. The symptoms are aching lungs, wheezing, coughing and headache. 'Smog complex' involves irritation of eyes and respiratory passages, chest pains, shortness of breath (dyspnea), nausea and headache. Lungs are ozone's primary target causing damages to cells in the airways, inflammation and swelling. It also reduces immunity. It poses a health risk to those people who already suffer from emphysema and chronic bronchitis.

Over millions of residents in the South Coast Basin including Los Angeles, Orange County, parts of San Bernadino and Riverside breathe foul air for over four months in a year. The level of ozone reaches twice the federal health standard. The level of ozone exceeded for 98 days at Basin locations, in the San Gabriel Valley in 1995.

6.2.7 Control of air pollution

A. At source: (in the industries)

- Raw material substitution such as replacing of high sulphur containing fuel with low-sulphur fuels), substituting toluene for benzene, titanium for lead and calcium phosphate for beryllium.

- Process modification: Pollution reduction can be achieved by new or modified processes. For example, the use of exhaust hoods and ducts over several types of industrial ovens has allowed the recovery of valuable solvents that could have otherwise become air pollutants. Dipping process instead of spraying can be implemented.

- Equipment alterations: Newer types of equipment that are less pollution-prone, can cut down air pollution.

- Removal of pollutants at source: Using of air pollution control devices such as cyclones, scrubbers, electrostatic precipitators, bag house fabric filters (discussed in the Box 6.3) play a vital role in controlling of air pollutants at source.

Box 6.3: Air pollution control equipments

A few control systems **at source** used by vehicles and industries are:

A. Systems to reduce particulate matter

  i. **Wet scrubbers** are used to remove pollutants (particles and gases) usually from furnace flue gas. The polluted stream of gas is let in by force through a scrubbing liquid. They are used in large thermal plants, asphalt plants, iron and steel plants, manure plants, etc.

ii. **Electrostatic precipitator (ESP)** is a particulate collecting device, which uses the strength generated by an induced electrostatic charge to remove PM from any flowing gas. ESPs are used in petroleum refineries, paper and pulp factories, etc.

iii. **Dust cyclones:** fluids and solids mixtures are separated by using centrifugal force in oil refineries, cement industry for the removal of particles 10 microns (μm) or larger.

iv. **Fabric filters or bag houses:** effective for very fine particles used in kilns, steel mills, foundries, fertilizer plants.

B. Systems to reduce NO$_x$

i. **Exhaust gas recirculation (EGR)** is the incoming air intermixed with the recirculated exhaust gas, results in dilution of the mixture with noble gas which in turn reduces the adiabatic flame temperature and also lowers the unnecessary oxygen in diesel engines. The peak burning temperature falls since the specific heat capacity of the mixture is raised by the exhaust gas.

ii. **Catalytic converter is** a device to reduce the toxicity of effluent gases that are produced by internal ignition engines.

C. Systems to decrease VOC:

i. **Gas flare:** When flammable, unusable waste gas and liquids are discharged by pressure relief valves, this device is used to burn them off. They are generally used in landfills.

ii. **Bio filters** are devices that use living matter to trap biologically degradable pollutants.

D. Dilution: Dilution is promoted by the self-cleansing mechanism of the environment. The pollutants released from the tall stocks or chimneys of the industry are dispersed into the atmosphere by strong wind, which helps in diluting the pollutants.

B. In vehicles:

- Discarding of old vehicles: The car companies should sell or dispose off technologically obsolete vehicles. For example, though the United States consumes 100 times more petrol than India, the vehicles in the two countries pump the same amount of benzene, a potent carcinogenic into the air. Therefore, the vehicles in India should meet the latest Euro standards like MPFi system.

- Catalytic converters: Catalytic converters are fitted between the engine and the tail pipe, and convert most of the exhaust into less damaging gases via a chemical reaction.

- Improving fuel quality: The quality of fuel used in automobiles is also an important factor in curbing air pollution. For example, reformulating or changing the composition of diesel or petrol in the refinery reduces pollution. Sulphur content in diesel supplied was brought down to 0.5 per cent in 1996 and it was further down to 0.25 per cent in 1997. It is necessary to have low sulphur content in diesel for complying with the emission norms ahead of EURO–II (from Euro–I to Euro–II norms, sulphur content in diesel is low by 0.3 per cent), the refineries are required to take initiatives for bringing down the sulphur content. Methyl tert–butyl ether, also known as methyl tertiary butyl ether and MTBE, is a volatile, flammable and colourless liquid immiscible with water. MTBE is an oxygenate and used as gasoline additive to raise the octane number, helping prevent engine knocking. Other compounds available as gasoline additives are ethanol and tert-amyl methyl ether (TAME).

- Unleaded petrol: Unleaded petrol is where lead has been removed from the petrol by adding aromatic compounds. (This produces another problem of releasing too much of cancer causing benzene, therefore alternative fuel is necessary). Introduction of unleaded petrol for new cars was introduced from April, 1995 and mandatory supply of unleaded petrol started from 1998. The lead content in the Delhi air has reduced by more than 60 per cent with the introduction of unleaded petrol.

- Alternative fuels: Alternative fuels like ethanol, methanol, CNG, LPG highly reduces air pollution. Even biofuels like biodiesel or soy diesel play a vital role in curbing air pollution. In December 2009, The GOI adopted National Bio-Fuel Policy that calls for blending of 20 per cent bio-ethanol in gasoline and 20 per cent bio-diesel in diesel by 2017.

- Efforts are made for developing and popularizing electric vehicles. For example, the Reva Motors.

- Fuel cells: The hope of the future are vehicles where hydrogen reacts with oxygen to produce electricity in a cell. It's only by-product is water vapour.

- Vehicular emission check: Continuous checking of vehicular emission for permissible limits and Pollution Under Control (PUC) certificates, strict enforcement of law and periodical survey of the emission control equipment is absolutely necessary for controlling air pollution.

Box 6.4: Emission norms and emission standards

Emission norms

Emission norms are prescribed levels of pollutants fixed by the government for the vehicle that they would release when running on roadways. They are the limits up to which the CO, HC and NOx gases are allowed. Every manufacturer is required to enforce and obey the same for vehicles that are manufactured since enforcement date.

The Automobile Research Institute, Pune is the certifying authority for the vehicles fulfilling the standards.

Prescribed permissible limits: Source – Society of Indian Automobile Manufacturers

Car

Year	Carbon monoxide (gm/km)	Hydrocarbons (gm/km)	Nitrogen oxides (gm/km)	Hydrocarbon + Nitrogen oxides (gm/km)
1991	14.30 – 27.10	2.00–2.90		
1996	8.68 – 12.40			3.00 – 4.36
1998*	4.34 – 6.20			1.50 – 2.18
2000 or EURO I	2.72 – 6.9			0.97 – 1.7
B.S II (2005) or EURO II	2.20 – 5.00			0.50 – 0.70
B.S III	2.30	0.2 – 0.29	0.15 – 0.21	–
B.S IV (2010)	1.0 – 2.27	0.1 – 0.16	0.08 – 0.11	–

Euro norms are the allowable and acceptable levels of emission for the vehicles that drive with petrol and diesel. It has been applied in Europe and the GOI has also approved the Euro norms as testing technique. In India, Euro I norms are referred to as INDIA 2000 because they have been effective since 4 January 2000. The regulations corresponding to Euro II, known as 2005 norms, have not yet been indicated by the Government of India. The Euro norms necessitate the makers to decrease the prevailing pollutant discharge levels in a further effective manner by adopting few technical alterations in their vehicles.

The manufacturers adopt the following changes to have a **Euro I compliant vehicles**.

- retuning of carburetor;
- secondary air intake;
- recirculation of exhaust gas;
- increase in catalyser capacity; and
- catalyser coated with trimetal.

In a **Euro II compliant vehicle** the carburetor is replaced by multipoint fuel injection system (MPFi).

Bharat stage emissions standards are emissions principles introduced by the Government of India to control the emission of atmospheric pollutants such as NO_x, CO, HC, SO_x and PM. In a lot of cases, they are related to European emissions standards. The first Indian emission regulations became effective in 1989. From 2000, India started accepting and implementing

European discharge principles and fuel control policies for four-wheeler light-duty and heavy duty vehicles. Emission policies framed by India still apply to two- and three-wheeled vehicles.

The implementation schedule of EU emission standards in India is summarized in the following table.

Indian emission standards (four wheeled vehicles)			
Standard	Reference	Date	Region
India 2000	Euro I	2000	Countrywide
Bharat Stage II	Euro II	2001	National Capital Region* (Delhi), Mumbai, Kolkata, Chennai
		04. 2003	National Capital Region* (Delhi), 12 Cities†
		04. 2005	Countrywide
Bharat Stage III	Euro III	04. 2005	National Capital Region* (Delhi), 12 Cities†
		04. 2010	Countrywide
Bharat Stage IV	Euro IV	04. 2010	National Capital Region* (Delhi), 12 Cities†
† Mumbai, Kolkata, Chennai, Bangalore, Hyderabad, Ahmedabad, Pune, Surat, Kanpur, Lucknow, Sholapur and Agra.			

C. **Legal and policy measures:**

- The government must impose taxes and levies on industrial units that contribute to maximum level of air pollution. The government has enacted the Air Act, 1981 and EPA, 1986.

The National Ambient Air Quality Standards (NAAQS)

- NAAQS were previously reported in 1994 under the Air (Pollution and Control) Act, 1981. The World Health Organization standards of 2005 and the revised NAAQS vary amongst each other. WHO recommended five parameters of which only four are to be monitored i.e., PM10/ PM 2.5, sulphur dioxide, nitrogen dioxide and ozone. The GoI prescribed 12 parameters, mainly, PM 10, PM 2.5, SO_2, NO_2, CO, NH_3, ozone, lead, benzene, benz (o) pyrene, arsenic and nickel.

- CPCB started the National Air Quality Monitoring Programme in 1984, in seven Indian cities later renamed as National Air Monitoring Programme (NAMP) in 1988–89. The Central Pollution Control Board monitors PM 2.5, ground level ozone, carbon monoxide, lead, hydrocarbons, ammonia, benzene, etc., at designated sites in few cities and sulphur dioxide, nitrogen dioxide and PM 10 at all places under National Air Monitoring Programme (NAMP).

National air quality monitoring programme

- There are over 542 monitoring stations in 223 cities all over the country as on 1 January 2013. The monitoring is undertaken with the aid of CPCB, SPCB, National Environmental Engineering Research Institute (NEERI), Nagpur and pollution control committees. The monitoring of weather parameters are also combined with the air quality data. They are wind speed, wind direction, air temperature, humidity, rainfall, solar radiation, air pressure etc. The observation is carried out with four hourly sampling for gases and eight hourly sampling for PM, twice a week so as to have 104 observations/year. The cheapest tool for monitoring air pollution is by its odour. Laboratory analysis of air pollutant is done by PM samplers, stack samplers and high volume air samplers.

High volume sampling: A high volume air sampler has an air drawing machine that sucks in the air through a special filter paper. The flow rate is adjusted while air sample is sucked in and the sample can be drawn in over a precise time period. This controls the volume of the air sample taken. Any measurable substance deposited on the filter paper is examined and evaluated further. High volume sampling yields a quantitative analysis of the air pollutants.

Bucket samplers: The sampler is a bucket made of plastic with a detachable bag placed inside it. The valve present on the nozzle of the pail is opened to draw in air. This operates by means of a pump that sucks in the air. Once filled with air sample, the bag can be separated and sent for analysis.

D. Green vegetation

- To reduce the spreading of air pollutants emanating from industrial sources, growing green vegetation around the industry has been recommended by scientists.

- The neem tree is reported to be helpful in checking atmospheric pollution caused by vehicular and industrial emission. Therefore, neem trees should be planted in the cities that face acute pollution problems.

Case Study 6.3: The Bhopal gas disaster episode

Commonly known as Bhopal gas tragedy, it is till date the most ghastly industrial disaster of the world. The incident took place on 2–3 December night, 1984 at the UCIL plant in Bhopal, Madhya Pradesh, meant for the production of Sevin – a pesticide.

Over 40 tons of methyl isocyanate (MIC) gas escaped out from the Union Carbide pesticide plant, which instantaneously killed nearly 3,800 inhabitants and caused considerable despondency and early death for several thousands. Estimates from various sources vary on death toll. Another agency claimed over 15,000 deaths. As per government sources in 2006, the leak caused 5,58,125 injuries with disabling injuries. The vent-gas scrubber, a safety device had been turned off three weeks prior. It becomes apparent that a defective valve permitted a load of water that was meant for clean-up of the internal pipelines, got mixed with 40 tons of MIC. When MIC is exposed to 200°C heat, it formed more deadly hydrogen cyanide (HCN) gas. The evidence gathered does reveal the temperature of the storage tank to reach that disastrous level. The cherry red colour of the victim's blood and their viscera were typical of acute cyanide poisoning.

Impacts on health from methyl isocyanate gas leak exposure in Bhopal (Environ Health, 2005; 4:6.)

Early effects (0–6 months)	
Opthalmic	Chemosis, redness, watering, ulcers, photophobia
Respiratory	Distress, pulmonary edema, pneumonitis, pneumothorax.
Gastrointestinal	Persistent diarrhoea, anorexia, persistent abdominal pain.
Genetic	Increased chromosomal abnormalities.
Psychological	Neuroses, anxiety states, adjustment reactions
Neurobehavioural	Impaired audio–visual memory, impaired alertness and reaction time, impaired analysis, interpretation and spatial ability, dysfunctional psychomotor coordination.
Delayed effects (6 months onwards)	
Visual	Constant watering, corneal opacities, chronic conjunctivitis
Respiratory	Obstructive and restrictive airway disease, decreased lung function.
Reproductive	Increased abortion, increased child mortality, reduced placental/ foetal weight
Genetic	Increased chromosomal abnormalities
Neurobehavioural	Impaired associate learning, motor speed, precision

The toxic plumes were hardly cleared when the first multi-billion dollar complaint was filed on 7 December by an US Attorney in the US court. This marked the beginning of legal intrigues in which the moral implications and its effect on the inhabitants of Bhopal were largely ignored. In March 1985, the Government of India passed the Bhopal Gas Leak Disaster Act as a pathway to ensure the speedy handling and equitable distribution of the claims arising from the accident. This Act also enabled the GOI to become the exclusive envoy of the Bhopal sufferers in matters of lawful proceedings in and out of the Indian Territory.

In 1989, in lieu of a settlement process mediated by the Supreme Court, the Union Carbide Corporation (UCC) acknowledged the ethical liability and decided to pay $ 470 million to the Government of India to be distributed to the lawful claimants as a complete and ultimate settlement.

Initially Indian Government-controlled banks and the Indian public held 49.1 per cent ownership share in this company. In 1994, the Supreme Court permitted UCC to sell its 50.9 per cent share. The Bhopal plant was then sold to McLeod Russel (India) Ltd. later renamed as Eveready Industries and then to Dow Chemical in 2001.

UCC chairman, CEO Warren Anderson was arrested after the incident and was released on bail of $ 2,100 by the Madhya Pradesh Police in Bhopal on 7 December 1984 and flown out on a government plane. Civil and criminal cases are pending in the US district court, Manhattan and the District Court of Bhopal involving UCC, UCIL staffs and Warren Anderson, the UCC CEO at the time of incident. In June 2010, seven former employees of UCIL (all Indian citizens and mostly in the age group of 70s) were convicted of causing death by negligence. Each of them was fined ₹ 1 lakh and was sentenced with two years of imprisonment. The eighth former staff, who was also convicted for the same died before the judgment was passed.

6.3 Water Pollution

Nearly 50 per cent of all marine pollution is caused by sewage and waste water discharge. Annually, about 400 billion tons of industrial waste are produced globally, bulk of which is discharged untreated into the streams, rivers, seas, oceans, and other water bodies. Water pollution means one or more substances building up in aquatic bodies to a limit that they cause trouble life forms. Water pollution also deals with amount of a polluting substance released and also the amount of water it is discharged into.

6.3.1 Types of water pollution

There are different perspectives of considering pollution.

A. **Based on water bodies:**

- **Surface water pollution** – The most evident type of water pollution affects surface waters like huge oceans, lakes, rivers and streams. For example, an oil slick from oil tanker can affect a huge area of the marine region.

- **Groundwater pollution** – All of Earth's water is not surface water. An immense amount of water is present in the underground rock structures or aquifers that are not visible to one and people hardly ever think of. Water that is stored in underground in aquifers is the groundwater. The rivers and streams are not only nourished by aquifers but they also supply drinking water. Fertilizers applied in the fields often seep into the soil and contaminate the waters. Groundwater pollution though less evident than surface water pollution is also a mounting problem.

B. **Based on source:**

- **Point source** – If pollutants are discharged from one place or spot, for e.g., an effluent pipe of a plant, oil spillage from a tanker, smoke stack discharge, etc. The pollution from such a source is known as point source pollution.

- **Non-point source** – A prodigious amount of water pollution also happens from variety of speckled sources. This is called non-point source pollution.

When point source contamination makes entry into the environment, the most affected place is usually the area adjoining the source. When an oil tanker has an accident the oil slick is mostly concentrated around the tanker itself and the pollution spreads further away from spot. This is unusual with non-point source pollution where the pollutants gain entry in to the environment from various places at a time.

C. **Based on chemical composition:**

- Soft water – water with little or no dissolved salts of magnesium and calcium.

- Hard water– In contrast to soft water, hard water has high mineral content. It primarily consists of calcium (Ca^{2+}), and magnesium (Mg^{2+}) metal cations, and sometimes other dissolved compounds such as bicarbonates and sulfates. Such water is unsuitable for use.

- Sometimes the pollution may enter the environment from one place and exert an outcome hundreds or even thousands of miles away from that place. Such pollution is known as **trans-boundary pollution**. Examples include the persistent organic pollutants and the radioactive wastes that may travel across the oceans and national boundaries.

6.3.2 Sources of water pollution

Approximately 33 per cent of water pollution is by domestic sources, followed by agriculture–livestock with 29 per cent, 27 per cent by industry and 11 per cent from other sources.

A. **Domestic:** Domestic sewage is wastewater generated from the household activities. It is 99.9 per cent pure water; remaining 0.1 per cent are pollutants which are both organic and inorganic materials. Organic materials are food and vegetable waste, excreta, faecal matter whereas inorganic materials such as phosphates and nitrates come from soaps and detergents.

In theory, sewage is an entirely natural matter that should degrade blandly in the environment. In reality, the sewage may contain a host of chemical substances, from drugs, papers, plastics, to the toilet flushes. Nearly everyone dispenses various chemicals of one type or another into their drainage system. Soaps and detergents used in washing machines and dishwashers are finally drained into the sewage.

In the urban and rural regions both, sewage comprising human and animal faecal matters emerges as one of the major pollutants. Sewage laden with organic matters augments the microbial growth that not only spread various diseases but also utilizes much of the dissolved oxygen in water resulting in oxygen depletion. Such waters creates ecological imbalance and aquatic organisms fail to survive under such conditions.

Usually people have the habit of dumping the household wastes in the nearby water source, which leads to water pollution. Sewage carries industrial contaminants and an increasing tonnage of paper and plastic refuse. The most common substance that washes up along with the waves is the plastics. They pose a great threat to the marine birds, fishes and other forms of life. The fishes are frequently choked up and strangulated in the left over or residual plastic fishing lines and this process is known as 'ghost fishing'. Scientific study reveals that one-fourth of all marine birds contain some type of plastic residue.

B. Agriculture: It is undoubted that extensive use of fertilizers and pesticides, collectively known as agro-chemicals, increases the agricultural output. These, mostly artificial chemicals, move into the water bodies along with rainfall and groundwater by leaching. Such chemicals are persistent and may often gain entry into the food chain causing numerous problems in the animals.

    i.   Pesticides: Any chemical utilized in controlling pests is a pesticide. The pest can be insects, microbes, fungus, snails and slugs, worms, weeds, etc. Consequently, pesticide can be in the form of insecticides, fungicides, herbicides, etc. Some of these pesticides can be contact pesticides, whereas others might have to be intaken to show effective actions. Application of pesticides needs to be based on its mode of action. Pesticides can cause a number of diseases like lungs malignancy, chronic liver damage, cirrhosis and chronic hepatitis, hormonal and gonadal disorders, immune-suppression, cytogenic effects, breast cancer, Non-Hodgkin's lymphoma, polyneuritis, etc.

Class of pesticides on the basis of their chemical nature:

    a.   Organochlorine compounds

This is the most hazardous of all pesticides. There are three classes of organochlorines –

-   **Dichlorophenylethanes** – DDT, methoxychlor. They are highly penetrable and soluble in fats.

-   **Cyclodienes** – Aldrin, endosulphan, endrin, heptachlor, toxaphene, chlordane. The metabolites of these products are more toxic.

-   **Hexachlorocyclohexanes** – Lindane or benzene hexachloride (BHC).

The best known representative of this class is DDT, made by the Swiss Scientist Paul Müller. He received the Nobel Prize for Physiology and Medicine in 1948 for his discovery. DDT was announced on the market in 1944. DDT when applied on the insects opens up the Na-channels in the neurons. Acute symptoms are parathesia, ataxia, dizziness, headache, nausea, restlessness. Chronic exposure results in anorexia, anemia, tremor, weakness and anxiety. They are potent carcinogens and damage the liver and endocrine organs.

    ii. Organophosphates: This subsequent class of pesticides acts by binding to the acetylcholinesterase and disrupts the propagation of nerve impulse, killing the pests and interferes with the ability to carry out normal function. They are frequently used in chemical warfare as nerve agents. Notable amongst them are sarin, tabun, soman, etc. that act in the same manner. Their toxic effects are additive; hence the toxicity amplifies with multiple exposure. The common organophosphates are malathion, parathion, diazinon and tetraethylpyrophosphate (TEPP).

Exposure to such compound may result in diarrhea, vomiting, sweating, blurred vision, abdominal cramps, slurred speech, slow heartbeat, muscle seizures etc.

iii. **Organocarbamates:** The mode of actions of this class is quite similar to that of organophosphates. Their actions last for smaller duration and are thus the least toxic amongst the three. These compounds inhibit acetylcholinesterase and hence disrupt the nerve impulse transmission from one cell to another. When the enzyme is inhibited, there is overstimulation and then paralysis of the secondary cell. Commonly used pesticides are carbaryl (sevin), propoxur (baygon), dimetilan. Exposure to such compounds may cause salivation, lacrimation and convulsions.

iv. **Pyrethroids:** A natural class of compound that mimics the insecticidal activity is pyrethrum, less intensely toxic than organophosphates and carbamates. These are non–persistent and act by modulating the sodium channels. They are usually used for household pest control.

v. **Neonicotinoids:** Neonicotinoids are broad spectrum, artificial analogous compounds resembling the natural pesticide nicotine. These compounds are nicotinic agonists of acetylcholine receptors and acts as systemic insecticides. They are sprayed for soil and seed treatments, frequently as a substitute of organophosphates and organocarbamates. The insects exhibit confused movement, tremors in legs, rapid movement of wings, stylet withdrawal, paralysis and death. Examples include acetamiprid, clothianidin, imidacloprid, sulfoxaflor, nitenpyram, nithiazine.

vi. **Biological insecticides:** More recently the task of reducing the use of broad spectrum pesticides has led to the introduction of biological insecticides back into trend. An example is amplified use of *Bacillus thuringiensis*. It is used to kill the larva against a diverse form of caterpillars. BT toxin from *Bacillus thuringiensis* could be successfully inoculated into the plants through genetic engineering. Products based on entomo-pathogenic fungi like *Metarhizium anisopliae,* nematode such as, *Steinernema feltiae* and virus, Cydiapomonella granulovirus are effective bio–insecticides.

C. **Industries:** Most of the industries are situated along the banks of the river for convenience and disposal of effluents. Often these effluents comprise acids, bases, dyes, paints and a range of other chemicals. Detergents form white foam. The industrial waste can be mercury, lead, cadmium, chlorides, fluorides, ammonia, etc. Such addition alters the pH of water and turns it into a condition that is fatal to aquatic forms. The industries regularly discharge heavy amount of superheated water used in the thermal power plants, oil refineries, nuclear power plants, etc. such high temperature injures and kills the aquatic flora and fauna. High temperature also reduces the amount of oxygen dissolved in water that may harm or even kill the aquatic life, the condition known as thermal pollution.

i. **Mining industries:** The process of extraction of minerals exposes heavy metals and suphur that were earlier inaccessible. Rainwater further leaches these chemicals out of the uncovered portions, causing AMD along with heavy metal pollution that lasts for a long period of time even after the closure of mining operations. The act of rainwater on the tailings contaminates the freshwater bodies. Cyanide is deliberately dispensed on mined piles in the gold mines for the extraction of gold. Some of cyanides used inevitably contaminate the close by water. Often, enormous amount of mining waste or 'slurry' are stored behind containment dams. In case of any accident or dam leakage or bursting, water pollution is assured.

ii. **Oil refineries:** Oil spill is a leading issue in the aquatic system. Oil seeps from the oil tankers and offshore oil refineries into the waters. Oil leakage from tanker accidents accounts to about 12 per cent; more than 70 per cent of oil pollution is from the usual shipping as well as from the oil that people drain out into the drainage system. Oil is lighter than water and thus floats on the surface of water. Such a film blocks the aerial oxygen from dissolving in water. Oil can coat and smother the body of aquatic animals fatally. Oil can also find entry into the body while drinking. The oily beaches disturb the ecosystem balance and rigorously affect tourism. The public bearing the pollution cost are not the people who were responsible for the problem. Yet, everyone using cars add up to pollution in one or the other way.

iii. **Radioactive waste:** High concentrations can be lethal, whereas low concentrations of radioactive substances can cause malignancy and various forms of sickness. Two most important contributors in Europe are Sellafield and Cap La Hague. These plants are engaged in reprocessing spent fuels from the nuclear plant and discharging the radioactive waste water into the sea that is subsequently spread throughout. Sellafield is located on the northwestern shore of Britain and Cap La Hague is situated on the northern shore of France. Norway, located downstream of Britain, gets considerable amount of such wastes from Sellfield. The Government of Norway has been recurrently complaining about increased radiation levels from Sellafield. Political pressure for the closure of these plants is constantly given both by the Irish and Norwegian governments.

D. **Rain drainage:** Rain drainage carries substances like highway debris, oil, chemicals from vehicular exhausts, sediments from public road and house construction, acids and radioactive wastes into the freshwater systems as well as into the marine systems. Animal waste from farms and feedlots, is also transported along with rain water. These pollutants impair rivers and streams, groundwater, and even the shoreline waters. Antibiotics, hormones, and other chemicals used to rear farm animals are components of such animal wastes. Pesticide and fertilizer residues from croplands also add to aquatic pollution via rain drainage.

E. **Other sources:** A cocktail of toxic chemicals covers the highways starting from spilled oil, brake fluids to the pieces of worn-out tyres and exhaust emissions from the automobiles. Sediments bring about yet another type of pollution that enters the sea from rivers. Fine particulates, soil may enter the neighbouring rivers and streams making the water turbid and murky. These sediments often clog the gills of aquatic organisms and kills them by suffocation.

Table 6.7: Major types of water pollutants

Nos.	Substances	Sources or application	Impact or effect
1.	Antimony - Sb	In electrical appliances, semiconductors, in expectorants, in manufacturing ammunitions.	In large doses it can cause poisoning, stomach ulcers, heart diseases.
2.	Mercury - Hg	In industrial wastes, dental fillings, fungicides, soldering, and various scientific instruments such as thermometers and barometers.	When ingested, mercury damages the Central Nervous System (CNS) leading to a disease called neuropathy. One of the famous cases is the Minamata incident in Japan in 1953. (See 5.3.5)
3.	Arsenic - As As_2O_3 and trivalent arsenic	As component of animal feed, in treated wood, ceramics, medicines, pesticides, paints, and fireworks.	70 mg of arsenic or arsenic compounds is considered lethal if ingested. Lower doses of arsenic can result in disturbance in peripheral circulation, black foot disease, damage to liver, kidney, various types of cancer, such as skin cancer, bladder cancer and lung cancer.
4.	Cadmium - Cd	In Ni-Cd batteries and part of various metallic alloys, electroplating, pigments, nuclear reactors, and the anticorrosion coatings of other metals, tyres.	Often deposited in the hepatic, renal, pancreatic and intestinal lining. Poisoning with cadmium causes headache, regurgitation, anaemia, pneumonia, diarrhoea, osseous deformation and renal necrosis. It results in bioaccumulation to interfere with the body's metabolism. It also leads to cancer, teratogenesis and itai itai disease.
5.	Chromium - Cr	Used in tanning leather, treating wood, storing data in magnetic tapes, making pigments, photography and in manufacturing iron-based alloys. The human body needs trace amounts of trivalent chromium metabolize fats and carbohydrates.	Hexavalent chromium is dangerous; it causes dermatitis, gastrointestinal ulcers, lung cancer, weak immune system, liver damage. It is also a teratogen.
6.	Lead - Pb	It is found in plastics, ceramics, glassware, paints, pigments, and batteries. Lead is also a crucial component in antiknock agents.	Lead is easily absorbed in the blood. It affects liver, kidneys, osseous system; central and peripheral nervous system; red blood cells leading to anaemia. This toxic heavy metal can cause brain damage, coma and even death.

Nos.	Substances	Sources or application	Impact or effect
7.	Nickel - Ni	Used in the steel industry, in making batteries, welding rods, wires, adding pigments to paints, desalination plants, and producing dental and surgical prostheses.	Overexposure may cause allergic reactions, asthma-like symptoms, lung embolism and various organ problems.
8.	Selenium - Se	Excellent photovoltaic and photoconductive property, hence widely used in electronic goods such as photocells, photo meters and solar cells; the glass industry, for animal feeds and food supplements, in photocopying and in the toning of photographs, in alloy making such as the lead plates used in storage batteries, in rectifiers for converting AC current in DC current; used to enhance the abrasion resistance in vulcanized rubbers, in antidandruff shampoos.	Toxicity causes giddiness, fatigue and irritations of the mucous membranes, fluid accumulation in the lungs and bronchitis occur; fragile hair and deformed nails, rashes, heat, swelling of the skin and severe pain; when it enters eyes it causes burning, irritation and tears.
9.	Copper - Cu	Used for electrical equipment, copper wires; construction, industrial machinery, copper alloys such as bronze, brass, gunmetal to make guns and cannons; copper and nickel used together, known as cupronickel, is the favoured metal for making low–denomination coins.	Prolonged exposure to copper can cause oro-nasal and eye irritation; it causes headaches, stomach pains, giddiness, vomiting and diarrhoea. High uptakes of copper may cause liver and kidney damage and become fatal.
10.	Beryllium - Be	Used mainly in defense and aerospace industries; used in the field of x-ray detection diagnostic and in the manufacture of a variety of computer equipments.	Breathing beryllium particulates is hazardous as it damages the pulmonary tissue causing pneumonia. The most common effect being berylliosis. (see section air pollution) Beryllium is an allergen to hypersensitive people; in severe conditions it causes a person to be seriously sick, a condition known as Chronic Beryllium Disease (CBD). The symptoms are weakness, fatigue and breathing problems; people suffering from CBD sometimes develop a

Nos.	Substances	Sources or application	Impact or effect
			norexia and blueness of hands and feet. In extreme condition CBD can cause their death.Beryllium enhances the probability of malignancy and DNA damage.
11.	Aluminum - Al	Used in aerospace industry and very important in transport and construction where light weight, durability, and strength are desirable. it forms alloys with many elements such as copper, zinc, magnesium, manganese and silicon. All present day mirrors and telescopic mirrors are prepared by using a thin reflective coating of	health effects, such as ▪ damage to the CNS ▪ dementia or loss of memory ▪ lethargy ▪ rigorous trembling ▪ lung problems
		aluminum on the back surface of a sheet of float glass. Other applications are electrical transmission lines, and packaging. Production of aluminum foam used in traffic tunnels and in space shuttles.	Inhalation of aluminium is reported to cause pulmonary damage in the form of pulmonary fibrosis. This effect, identified as Shaver's Disease, is complicated by the existence of inhaled silica and oxides of iron. Possibly it also contributes to Alzheimer's disease.
12.	Manganese -Mn	Iron and steel production. Manganese dioxide is also used as a catalyst. Used in decolorization of glass and craft violet coloured glass. Potassium permanganate is a strong oxidizing agent and used to disinfect water. Used in making fertilizers and ceramics.	Effects occur mainly in the respiratory passages and in the brains. Symptoms of manganese poisoning are hallucinations, lack of memory and neuronal damage. Excess manganese causes Parkinson's disease, lung embolism and bronchial infection. A syndrome that is caused by manganese toxicity exhibits symptoms such as schizophrenia, dullness, weak muscles, headaches and sleeplessness.

6.3.3 Global impacts of water pollution

6.3.3.1 Eutrophication

It refers to the enrichment of freshwater bodies by inorganic nutrients like nitrates, phosphates which may occur naturally but more readily as the result of human activity. It is predominantly apparent in sluggish rivers and shallow lakes.

Nutrient over amelioration of freshwater and coastal ecosystems is a rapidly growing environmental emergency. The number of coastal areas affected by eutrophication globally is over 500. High amounts of nitrates and phosphates are present in sewage, fertilizers and other organic matters. A eutrophic water body has high primary productivity due to excessive nutrients and hence favours the growth of algal blooms resulting in poor water quality. The waters at depth are usually deficient in oxygen, ranging from hypoxic to anoxic condition. The nutrients further augment the waters to promote algal growth. A restricted number of phytoplankons are found to be involved. A number of algal blooms are identified by discolouration of water bodies often due to the high density of pigmented cells. The water usually becomes greenish. This is known as 'algal bloom'. Rapid algal expansion leads to increased decomposers. All forms of aquatic life such as decomposers, other aquatic vegetation including the algae, aquatic animals including the fishes, consume the oxygen that is dissolved in the water for respiration. This in turn seeks a great requirement for oxygen and leads to oxygen depletion. Algal blooms can be found both in marine and freshwater environments.

Some algal blooms are harmful for instance the dinoflagellates belong to the genus *Alexandrium* and *Karenia*. Such bloom often assumes a red or brown colour and is conventionally known as red tides. An overgrowth of *Pfiesteria* and *Eicchornia* also results in eutrophication and covers up the entire surface. This decreases the penetration of light into the water underneath affecting productivity.

6.3.3.2 Bioaccumulation and biomagnification

Bioaccumulation is the accretion of substances like pesticides, metals and various organic compounds into the body of living organism over a period of time. This can occur either because the chemical gains entry faster than it can be utilized, or because the chemical cannot be metabolized by the organism. Thus, the longer the biological life of the substance the greater is the risk of chronic poisoning, even if its level in the environment is not very high.

Bioconcentration is a more precise term that refers to uptake and accumulation of a chemical substance from water alone. On the other hand, bioaccumulation relates to uptake from all sources such as water, food, air, etc.

Compounds like DDT and tetra-ethyl lead, being lipid solubles, are stored in the body's adipose tissues, which are used for energy production, the compounds on being released cause acute poisoning. Metals such as Strontium-90, radioactive fallout from atomic bombs, behave similarly and replace calcium that it is utilized in osteogenesis. The radiation from strontium can have long time damage.

Bio-magnification, also known as **bio-amplification** or **biological magnification**, on the other hand is the amplification in concentration of a substance up the trophic level. This occurs in a food chain as a result of

■ it being persistent;

■ energetics of food chain; and

■ low rate of internal metabolism and excretion of the substance.

Biological magnification is the process by which particular substances such as pesticides (DDT) or heavy metals (mercury) go up the trophic level. These chemicals are often released into rivers or lakes, gains access into phytoplanktons and zooplanktons; they are then consumed by aquatic organisms such as crustaceans, rotifers, snails, fish, which in turn are consumed by large birds, animals or humans. Each time, as they go up the trophic levels in the food chain, the substances become deposited and concentrate in tissues or internal organs. The contaminants are not easily excreted or metabolized by most organisms and therefore, they are accumulated in high levels.

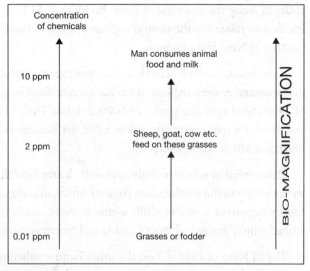

Figure 6.3: Bio-magnification along food chain

These contaminants remain in the fats and are not metabolized in the body. Over the years the quantity intensifies in the organism's body, a phenomenon called biomagnification. Pollutants like DDT may also enter the body of humans through dietary milk provided the bovine animals get the exposure of DDT contaminated grass and water. The consequences are severe blood and nervous system disorders. Substances that exhibit biomagnification are DDT, toxaphene, polychlorinated biphenyls, mercury, arsenic, cadmium, copper, lead, nickel, zinc, selenium, chromium, cyanide, etc.

6.3.3.3   El Nino, La Nina and southern oscillation

El Nino means 'Chirst child' and derives its name as it occurs sometime around Christmas. It lasts from few weeks to few months. The name, El Nino, is given to warm ocean current that flows along the coastline of Peru and Ecuador in South America.

Development of an El Nino is found to be related with a cycle of Pacific Ocean movement known as the Southern Oscillation. Normally, in parts of North Australia and Indonesia a low pressure develops in contrast to the development of high pressure along Peru's coast. Consequently, trade winds blow from east to west across the Pacific. This drives the warm waters towards the west which outcomes as storms in Australia and Indonesia. On the other hand, along Peru's coast there is upwelling of the cold bottom water to replace the warm water.

The year when El Nino occured the pressure fell over vast areas in Central Pacific and South American coast. This low pressure was counteracted by high pressure in western Pacific. The trade winds were consequently reduced and permitted the equatorial counter current to gather warm current along Peru and Ecuador coast. This warm water blocked the convection of the cold waters at depth along the coastline of Peru. Building of El Nino led to drought in the western parts of Pacific and rainfall in the central regions of Pacific. Weather conditions returned back to normal post the El Nino phenomenon.

Sometimes, the trade wind becomes stronger enough causing unusual accretion of cold waters in the central eastern regions of Pacific Ocean. Such incident is La Nina. La Nina in the middle of 1998 lasted upto the winter of 2000. Atlantic Ocean had an experience of hurricanes in 1988 and 1989. Of ten tropical storms in 1998, six became full scale hurricanes. In 100 years record hurricane Mitch is the strongest.

Other weather effects include aberrantly heavy rainfall in southeast Asia and India; colder and wetter winter in the southeastern parts of Africa, in India and Southeast Asia; damp weather in the eastern regions of Australia, chilly winter in west Canada and Northwest USA and an exceptionally soaked winter in southwestern Canada and Northwestern USA.

The El Niño of 1982–83 on the other hand resulted in tremendous heating of the equatorial Pacific. Surface temperature in certain parts of the Pacific Ocean rose by 6°C higher than usual temperature. Such warm current had destructive impact on oceanic life along the coast of Peru and Ecuador. Peruvian economy depends on fishing and particularly on selling of guano. The warm waters killed the teeming fishes and fish catches were 50 per cent lower than the previous year. Moreover, they failed to collect the bird dung guano.

Severe droughts occurred in Australia, Indonesia, India and southern Africa. Australia incurred a loss of $2 billion in harvest; sheep and cattle died in millions in the scorching heat. California, Ecuador and the Gulf of Mexico experienced heavy rains.

Figure 6.4: El Nino and its global climatic impacts

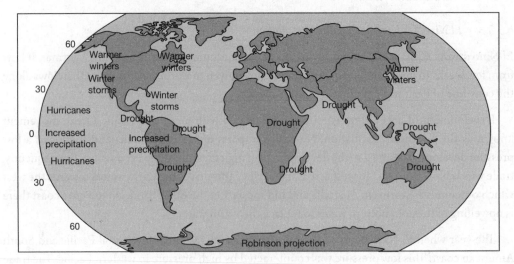

6.3.4 Biological pollution of water or water-borne diseases

Water-borne diseases are caused by pathogens. Microbes can also be the source of food-borne diseases through consumption of contaminated food. The most common water-borne diseases are summarized in the Tables 6.8, 6.9, 6.10 and 6.11.

Table 6.8: Few important protozoan infections

Disease	Pathogen	Common symptoms
Amoebiasis	*Entamoeba histolytica*	Discomfort in the abdomen, tiredness, loss of weight, diarrhoea, distension, pyrexia
Cryptosporidiosis	*Cryptosporidium parvum*	Flu-like manifestations, watery diarrhoea, loss of hunger, anorexia, distension, increased gas, queasiness
Cyclosporiasis	*Cyclospora cayetanensis*	Spasms, vomiting, muscle aches, pyrexia and tiredness
Giardiasis	*Giardia lamblia*	Diarrhoea, discomfort in the abdomen, distension and flatulence
Microsporidiosis	*Microsporidia sp.*	Diarrhoea

Table 6.9: Few important parasitic infections

Disease	Pathogen	Common symptoms
Schistosomiasis	*Schistosoma sp.*	Rash or scratchy skin, Pyrexia, Chillis, coughing and muscle aches
Dracunculiasis or Guinea Worm Disease	*Dracunculus medinensis*	Allergic response, urticaria rash, nausea, vomiting, diarrhea, asthma.
Taeniasis	*Taenia solium*	Intestinal disturbance, neurologic manifestations, loss of weight, cysticercosis
Fasciolopsiasis	*Fasciolopsis buski*	Gastrointestinal Tract (GIT) disturbance, diarrhoea, hepatic enlargement, cholangitis, cholecystitis, obstructive, jaundice
Hymenolepiasis	*Hymenolepis nana*	Abdominal pain, anorexia, itching around the anus.
Echinococcosis	*Echinococcus granulosus*	Liver enlargement, hydatid cysts present in the bile duct and blood vessels; if the cysts rupture they can cause anaphylactic shock.
Ascariasis	*Ascaris lumbricoides*	Mostly disease, is asymptomatic or accompanied by inflammation, fever, and diarrhoea. Severe cases produce Loffler's syndrome in lungs along with nausea, vomiting, malnutrition, and underdevelopment
Enterobiasis	*Enterobius vermicularis*	Peri-anal itching, nervous irritability, hyperactivity and insomnia

Table 6.10: Few important bacterial infections

Disease	Pathogen	Common symptoms
Botulism	*Clostridium botulinum*	Blurred vision, thirsty and dehydrated mouth, difficulty in swallowing, slurred speech, vomiting, diarrhoea, difficulty in breathing, death may be caused by respiratory failure
Dysentry	*Shigella dysenteriae*	Passing of faeces with blood and or mucus, in many a case vomiting blood
E. coli infection	*Escherichia coli*	Diarrhoea and dehydration
Cholera	*Vibrio cholerae*	Watery stool, nausea, cramps, nasal bleeding, rapid pulse, vomiting, haemarrhagic shock.
Salmonellosis	*Salmonella sp.*	Diarrhoea, fever, vomiting and abdominal cramps.
Legionellosis	*Legionella pneumophila*	Acute influenza, pyrexia, fever, chill, ataxia, anorexia, muscular cramps, pneumonia.
Typhoid	*Salmonella typhi*	Sustained fever up to 104 °F, profuse sweating, diarrhoea, rarely rashes, if untreated leads to sleenomegaly and hepatomegaly, progressive delirium and sometimes death.

Table 6.11: Few important viral infections

Disease	Pathogen	Common symptoms
Adenovirus infection	Adenovirus	Common cold, croup, bronchitis, pneumonia
Hepatitis A	Hepatitis A virus (HAV)	Fatigue, fever, abdominal pain, nausea, diarrhoea, loss of weight, itching, depression, jaundice
Gastroenteritis	Astro virus, Calici virus, Enteric adeno virus, Parvovirus.	Diarrhoea, nausea, vomiting, fever, malaise, abdominal pain.
SARS (Severe Acute Respiratory Syndrome)	Corona virus	Pyrexia, myalgia, lethargy, cough, sore throat, gastrointestinal disturbance
Poliomyelitis	Poliovirus	Headache, fever, delirium, sporadic seizures, spastic paralysis, seldom aseptic meningitis and death.

Pollution of water by organic matter is a prime cause for epidemic outbreaks like cholera, gastroenteritis in India. The microbes causing these ailments access the water bodies through the organic waste matter and then enter the bodies of healthy people causing diseases. In fact, a good indication of aquatic pollution is the existence of E. coli in the human intestines.

6.3.5 Occupational health hazards

A. Itai-itai disease: This was the famous case in Toyama Prefecture, Japan that was characterized by severe pain in joints and vertebral column. It was called as itai-itai disease by the local inhabitants and found to be produced by cadmium poisoning. Several mining companies discharged cadmium in to the rivers. The consequences are softening of the bones and kidney failure. It became one of the most prominent pollution-related diseases of Japan.

B. Chisso-Minamata disease: Also known as Minamata disease, the root cause of such disease is mercury poisoning. Primary symptoms are ataxia (loss of muscular coordination affecting speech, eyeball movements, swallowing, walking, etc.), lack of sensation in the palms and feet, weakness in the muscles, hearing impairment, reduced visual field, etc. Such a neurological syndrome, in extreme cases, can lead to insanity, paralysis, coma and death that might follow within few weeks of inception of the symptoms. Mercury may cross the placenta and affect the in-utero foetus.

C. Blackfoot Disease (BFD): BFD is an endemic disease confined to the southwestern coast of Taiwan and is caused due to arsenic toxicity. The disease is initiated with people complaining about the coldness or numbness in the appendicular extremities, especially in the feet along with irregular claudication (cramps and pain in lower leg while exercising and walking) making progress with course of time. This peripheral disease with severe systemic arteriosclerosis and gangrene like symptoms, affects the feet and occasionally the fingers. BFD are often grouped into two response classes histologically – arteriosclerosis obliterans (due to narrowing and gradual blockage of the artery) and thromboangitis obliterans (non-atherosclerotic vascular disease). Black pustules appearing typically on the feet, but can also be manifested on other parts of the body. The pustules are excruciating and if opened, ooze pus and blood along with a foul odour. Along with inflammation and itching, a blackened layer forms over the wounded pustule. If it is not treated the infected tissues decay.

D. Methaemoglobinemia – Excess release of nitrates from fertilizers enters the human body through water. When nitrates are ingested or intaken, they are transformed into nitrites in the alimentary system. The nitrites reacting with blood haemoglobin forms methaemoglobin. The heamoglobin molecule being preoccupied by nitrites cannot bind to the oxygen. The body is thus deprived of oxygen supply. This is fatal especially in the infants as they have very little amount of methaemoglobin reductase which could revert such effect. The syndrome is called blue baby syndrome or methaemoglobinemia. The symptoms are shortness of breath, vomiting and diarrhea. After confirming the test for the syndrome, an injection of required dose of methylene blue can help the baby's blood to return to normal.

6.3.6 Measurement of water pollution

Biological oxygen demand: Pollution in water can usually be measured by estimating the biological/biochemical oxygen demand (BOD) of water. Lower value of BOD indicates lesser pollution whereas higher BOD value indicates higher pollution.

It is one of the most useful parameters to indicate the organic strength of wastewater and can be defined as the amount of oxygen required by the aerobic bacteria to decompose the biodegradable matter in a given amount of water at 20°C over a period of five days. In reality it is an indirect measure to calculate the concentration of degradable matters present in waste. The BOD value of raw sewage may run to several hundred. Till 1971, the recommended BOD limit by World Health Organization (WHO) for potable water was 6 mg/l, there is no prescribed limit at present.

The concept of BOD test came from United Kingdom owing to the pollution of River Thames. It was said that a ferry carrying people tipped over in the Thames River; many of the people who fell in the water got sick and died. The cause of death was found to be river pollution and not due to drowning. In response, in 1908, the Royal Commission adopted the BOD_5 method. The test was carried out for five days since the sewage discharged in the Thames River usually took 5 days to reach the ocean.

Chemical oxygen demand (COD): This is the most common method of measuring the strength of industrial water, i.e., the amount of oxygen used for the chemical oxidation of the pollutants. It is expressed in mg/l and is the total measurement of all the chemicals in water that can be oxidized. Higher COD values indicate higher values of pollution in the wastewater sample. COD can be employed for waters too toxic for BOD test and usually takes few hours for completion, an added advantage over BOD test. The decomposition is brought out by adding and boiling with a powerful oxidant, usually potassium dichromate. COD value is usually higher than the BOD value.

6.3.7 Steps to prevent or control water pollution

It is not at all easy to find solution to the problem of aquatic pollution. In a broader perspective, three approaches might assist one to tackle such issue. They are education, legislation and economics that have to be treated synergistically.

6.3.7.1 Awareness

Creating mass awareness is the foremost step towards preventing and controlling water pollution. People walking along the world's most polluted beaches often express their resentment and often asemble together to organize themselves in beach cleaning sessions. Many NGOs have campaigned against over harvesting of fish and pleaded for tough penalties against the industries and factories dispensing effluents in the water bodies. More awareness can surely raise concern, bring responsibilities and contribute towards making a positive difference.

6.3.7.2 Legislation

The major problem with water pollution is mainly because of the flowing nature of water that results in trans-boundary pollution. Most of the rivers cross national boundaries whereas ocean and seas may span continents. The pollutants released by the factories in one nation with pitiable emission standards cause glitches in the neighbouring countries. In many a case, the countries that suffer has stringent environmental legislation and elevated environmental standards. The environmental laws make a situation difficult for the people to foul, but it is more difficult to make such laws practically operative as they need to be implemented across national and international borders. To handle such trans-boundary issues several international negotiations and laws were negotiated and agreed upon:

- The International Convention for the Prevention of Pollution of the Sea from oil tankers, London, 1954.
- The International Convention for the Prevention of Pollution from Ships, 1973, (MARPOL 73/78),
- UN Convention on the Law of the Sea, 1982 enforced in 1994.
- London Dumping Convention, 1972.
- Convention on the control of trans–boundary movement of hazardous wastes and their disposal, 1989, Basel.
- 1998 OSPAR Convention for the Protection of the Marine Environment of the North East Atlantic.
- Nairobi International Convention on the Removal of Wrecks, 2007.

The GOI has enacted the following legislations –

- The Water (Prevention and Control of Pollution) Act, 1974 further amended in 1988.
- The Water (Prevention and Control of Pollution) Cess Act, 1977, last amended in 2003.

6.3.7.3 Economics

Most environmentalists have the opinion of tackling pollution through the polluter pays principle as stated in Rio Declaration. It could mean that the oil company should have to take the responsibility for oil spill clean ups or customers have to bear the cost of the grocery bags made of plastics.

In 1993, WHO had given the guidelines for potable water standards; this acts as an international reference point for setting the standards and safety for potable water.

Table 6.12: Bureau of Indian Standards (BIS) drinking water – specifications for some of the important parameters IS 10500 – 2012 (second revision) as per Ministry of Drinking Water and Sanitation (MDWS), GOI

Serial no.	Characteristic	Unit	Requirement (Acceptable Limit)	Permissible limit in the absence of alternate source
1	Total Dissolved Solids (TDS)	Milligram/litre	500	2000
2	Colour	Hazen unit	5	15
3	Turbidity	Nephelometric Turbidity Unit (NTU)	1	5
4	Total Hardness	Milligram/litre	200	600
5	Ammonia	Milligram/litre	0.5	0.5

Serial no.	Characteristic	Unit	Requirement (Acceptable Limit)	Permissible limit in the absence of alternate source
6	Free Residual Chlorine	Milligram/litre	0.2	1.0
7	pH	--	6.5–8.5	6.5–8.5
8	Chloride	Milligram/litre	250	1000
9	Fluoride	Milligram/litre	1.0	1.5
10	Arsenic	Milligram/litre	0.01	0.05
11	Iron	Milligram/litre	0.3	0.3
12	Nitrate	Milligram/litre	45	45
13	Sulphate	Milligram/litre	200	400
14	Selenium	Milligram/litre	0.01	0.01
15	Zinc	Milligram/litre	5.0	15.0
16	Mercury	Milligram/litre	0.001	0.001
17	Lead	Milligram/litre	0.01	0.01
18	Cyanide	Milligram/litre	0.05	0.05
19	Copper	Milligram/litre	0.05	1.5
20	Chromium	Milligram/litre	0.05	0.05
21	Nickel	Milligram/litre	0.02	0.02
22	Cadmium	Milligram/litre	0.003	0.003
23	E. Coli or Thermo-tolerant coliforms	Number/ 100 ml	NIL	NIL

Case Study 6.4: Arsenicosis

Arsenic toxicity has become a universal issue. Arsenic is found in trivalent, pentavalent and as arsine state in nature. The main ores are arsenic-pyrites, orpiment, realger and arseno-pyrites. It is also found in the hot spring and sea water and exists in organic form in sea fishes. Arsenic in element state is water insoluble, but arsenic compounds have a varied range of solubility, basically dependent on the pH and other factors.

Sporadic instances of arsenicosis were reported early in the 1900s from the South American nations. The most affected nations were Chile, Mexico and Argentina with newer nations gradually adding up the list. Amongst Asian countries, India, Bangladesh and China comes to the forefront with news about arsenicosis pouring in from Pakistan and Afghanistan on one side and Myanmar, Cambodia on the other.

Widespread source of arsenic includes trace amount of arsenic associated with the pyrite ores. Weathering releases sulfate, arsenate and ferric hydroxide. While sulphates are washed out to sea, the charged arsenate gets adsorbed into ferric hydroxide and is deposited in the sediments. Presence of organic matters in these sediments creates a reducing environment, iron solubilizes and arsenates are released. The Pyrite Oxidation Theory and Iron Oxyhydroxide Reduction Theory explains the leaching of arsenic in ground water.

The affected region in India includes the Ganga-Brahmaputra-Meghna basin with a projected 25 million people and 6 million people were affected in Bangladesh and West Bengal respectively. Latest report indicates West Bengal to be the most awfully affected state with maximum numbers of arsenicosis deaths from the South 24 Parganas and huge number of cases being registered from Nadia. Nor surprising, Coochbehar, North Dinajpur and South Dinajpur are now listed as affected by arsenic pollution. Pyrite deposition in affected districts about 60,000 years old is perhaps responsible for localization of arsenic in these districts. Current report indicates the tendency to include more and more states in the river basin. The states of Jharkhand, Bihar, UP, AP and Chattisgarh are also announced to be in the grasp of this deadly disease.

The present scenario of ground water contamination by arsenic is bleak. In India, the first case of arsenic toxicity was identified by Dr K C Saha, Calcutta School of Tropical Medicine, in 1982. He found high concentrations of arsenic in the nails, hairs and skin scales of the patients who drank water from shallow contaminated tube wells. The affinity of culminating arsenical dermatosis (ASD) exhibits variation amongst the same population with analogous exposure.

- Certainly the link with malnutrition is very strong. Several studies disclose that poor nutrition usually increses the individual's vulnerability towards toxicity of arsenic.

- Another important determinant is the age. Most of the people with clinical symptoms are in their 30s to 40s. Clinical manifestations of the symptoms require a continual exposure of nearly 10 years. But that does not indicate that children are exempted from such incidence.

Arsenicosis has a variety of manifestations:

- Though no organ is left free from involvement, skin seems to be the chief target. Pigmentation (diffused and spotted melanosis), keratosis (palmo-planter and dorsal) are the specific skin lesions. Prolonged toxicity damages the melanocytes, its capacity to produce melanin and results in the manifestation of black and white spots (leucomelanosis). Pigmentation is also seen in the mucous membrane of tongue and lips.

When keratocytes are damaged carcinoma of skin may result.

Digestive, nervous, respiratory, circulatory and genitor–urinary systems are all affected due to prolonged exposure to arsenic. Weak liver conditions lead to portal fibrosis, portal hypertension and ascites. In case a major portion of liver is damaged, hepatic failure may result in jaundice, coma and ultimately death. Renal disorders, normocytic and

- megaloblastic anaemia are frequently reported.

- Symptoms of chronic lung disease were reported in West Bengal. There have been informations of liver cirrhosis after medication with compounds of inorganic arsenic.

- Minor symptoms may include weakness, myalgial burning body, hypothyroidism, ishcaemic gangrene, etc., to name a few.

- There are reports of an increase in tendency of foetal loss and premature delivery when a high concentration of arsenic enters the body.

- It is a devastating ailment with inclination for malignancy. Malignant arsenical skin lesions may be Bowen's disease, basal cell carcinoma, or squamous cell carcinoma. There is increased risk of urinary bladder cancer, lung cancer and liver cancer that been reported.

- In Taiwan, a form of peripheral vascular disease called BFD, has been reported.

- There are reports of peripheral neuropathy. The symptoms of peripheral neuritis are parathesia which includes tingling, numbness, weakness of limbs, etc. Such cases are found in almost 50 per cent of the patients of arsenicosis that is caused by consumption of water contaminated with As.

All the family members do not manifest clinical features. The exact cause for such difference in manifestation is a riddle. It is recent that arsenicosis is recognized as a public health problem and effective therapeutic measures are still unknown. Plausibly, nearly all of the existing intervention focuses on creating provisions for potable water free from arsenic.

Epidemiologically there are three levels of disease prevention:

A. Primary prevention basically dealing with pre-pathogenesis phase

- **To raise community awareness** about arsenicosis, its signs and symptoms through awareness campaigns.

- **Recognition of precarious sources of water** To identify and assess the current sources of water for levels of arsenic (WHO acceptable standard is <10 µg/l, while the Indian standard is <50 µg/l). For assessing both fields test kits and chemical analysis may be applied.

Methods of removing arsenic from arsenic free water; source substitution may be a better alternative. Surface water can be the alternative source, such as pond sand filters, rainwater harvesting or public supply through pipes; digging wells from safe aquifers or dug wells can be done under groundwater-based techniques. Wells from plio-pleistocene sediments contain low arsenic levels and luckily Kolkata draws water from these older sediments.

B. Secondary prevention dealing with identification and case supervision.

C. Tertiary prevention dealing with restraining of disability and rehabilitation.

Case Study 6.5: Fluorosis

Ground water has naturally more fluoride than surface water. High fluoride in groundwater is recurrent in parts of Africa, China, India, Ceylon, Turkey, Iraq, Iran, Afghanistan and Thailand. Fluoride is found mostly in the calcium deficient water of granite and gneiss aquifers in some of the sedimentary basins. The presence of fluoride in groundwater can be attributed to natural or man-made cause. The geogenic conditions such as the weathering of fluoride bearing rocks like apatite, fluorite, biotite, hornblende, contribute to most of the natural fluoride in ground water. In parts of Tanzania natural fluoride level is high as 95mg/l.

According to WHO, 1.5 mg/l is the maximum acceptable fluoride concentration, whereas the maximum level of fluoride permeability in potable water is 1.0 mg/l. WHO has also set up a target value between 0.8-1.2mg/l to obtain maximum benefits such as prevention of dental decay and strengthening of bones.

Seventeen out of 28 Indian states, are recognized as fluoride contaminated endemic regions. A projected 25 million people are said to be directly affected while 66 million is under jeopardy. To everybody's surprise the figure also includes 6 million children!

The most widespread cause of fluorosis in India is the high-fluoride containing well water obtained from borewells that are dug deep into the ground. Usually, three types of fluorosis are identified. Dental fluorosis is the most common of all; skeletal fluorosis and non-skeletal fluorosis being the other two. Dental fluorosis usually occurs at an exposure above 1.5mg/l whereas crippling skeletal fluorosis above 10mg/l. As per field survey report in India, fluoride levels below 1 ppm may result in skeletal fluorosis. Water

fluoridation is not generally practiced in our country. In 2004, both skeletal and dental fluorosis was found to be endemic in at least 20 states including Nalgonda, Uttarakhand, Jharkhand, Andhra Pradesh, Tamil Nadu, Karnataka, Gujarat, Rajasthan, Punjab, Bihar, Kerala and Chhattisgarh. Around 1 lakh people of Karbi Anglong in Assam were affected where the levels of fluoride vary between 5-23mg/l. the sufferers had anemia, painful and stiff joints, spotted teeth and renal failure. Till date, the maximum concentration in India is 48 mg/l as reported from the Rewari district, Haryana.

Yellow, brown or black streaks or spots appear on teeth in case of dental fluorosis; surface irregularities and pits may be noticeable. There is no cure for dental fluorosis. In early stage of skeletal fluorosis, this disease is less noticeable and extremely hard to diagnose. The beginning is often marked by stiff and painful joints rather hard to discriminate from the bone spurs, bearing remarkable similarity with other osseous diseases like, ankylosing spondylitis, osteoarthritis, renal osteodystrophy, osteopetrosis and diffuse idiopathic skeletal hyperostosis (DISH). Extreme and visible crippling effects on skeletons are produced in the later stages.

The Government of India has been compelled to set up fluoride removal plants with an array of technologies in order to reduce the level of fluoride in industrial waste and mineral deposits. The method can be chosen depending on local situations. Cartridge filled with bone charcoal can be set in to the domestic faucet to defluoridize water. The use of household filter candles using activate alumina, its cyclic adsorbtion–desorbtion studies point towards regeneration of activated alumina, thus, minimizing the expense of defluoridation. The Nalgonda technique, developed by National Environmental Engineering Research Institute (NEERI), Nagpur, in which pre-treated water is added with calculated amounts of alum, lime and bleaching and stirred thoroughly. On the other hand, defluoridation plants based on activated aluminae are very simple to operate. Reverse osmosis plants are widely used now. UNICEF has been working intimately with the Indian government and other NGOs to employ specific fluorosis mitigation programmes at the grassroot level.

6.4 Odour Pollution

In this modern and consumerist world, the market is loaded with cosmetics, deodorants, body sprays, talcum powder and air purifiers, all with one goal, to fight against the bad odour. One can witness large heaps of piled up garbage dumps in a metropolitan city like Kolkata. Organic decomposition of biodegradable matter emits large quantity of gases, or particulates which combining with the atmospheric air cause odour pollution.

6.4.1 Sources of odour pollution

A. **Sewage treatment plants (STP):**

   The human wastes (faeces, urine) from toilets and public urinals are treated in sewage treatment plants. Gases like ammonia and methane generated by this treatment process brings about foul odour. Public toilets and urinals in hospitals, cinema halls and even educational institutions generate bad odour.

B. **Garbage**

   While passing through the garbage dumps, one's hand goes to the nose inevitably to prevent nose irritation. The biodegradable parts like vegetable peels, fruit refuse, paper pieces, undergo decomposition and release gases like methane and affect the local atmosphere. Kolkata generates on an average about 3100 tonnes of solid waste per day and so proper solid waste disposal becomes mandatory to protect its citizens from odour pollution.

C. **Burning ghat/crematorium:**

   In India many dead bodies are burned in the open air which gives burnt flesh smell and thus spoils the atmospheric air, making it unfit for inhalation.

D. **Mortuary:**

   Most of the morgues near the hospitals store the corpse in the open grounds. The decomposed bodies leave a noxious smell, which pollutes the local atmosphere and give the passers-by uneasy feeling. Natural disasters like the Gujarat earthquake had buried many human bodies under the debris, which in turn generated bad smell hindering the rescue operations. One cannot forget the ghastly scenes of Orissa cyclone. Thousands of carcasses were left unattended and the stench from the decomposed bodies was unbearable.

E. **Domestic:**

   Poultry droppings, cow dung (from khatals) and pig food release harmful gases. For example, fermentation of cow dung produces methane and ammonia. Rotten egg generates hydrogen sulphide gas. These gases can be very harmful to human beings.

F. Drainage and unused wells:

When one lifts or opens the lid of a manhole (left unopened for many days) of the drainage it gives out pungent smell, which can cause asphyxiation (choking to death) in human beings. Unused wells due to algal bloom change the colour of the water and gives bad odour.

G. Tanneries:

Tanning process in leather industries is done with the help of barks of trees (vegetable tanning) or chemicals like chromium (chrome tanning). The effluent emitted out from the tanneries releases noxious gases in to the atmosphere. Pollutants generated during the leather dusting process can cause cancer in human beings.

H. Other sources:

Other industrial sources responsible for odour pollution are refineries and petrochemical industries, paper and pulp industries, plastic industry, chemical and textile injuries, fertilizer and pesticides manufacturing industries and food processing industries.

6.4.2 Causes of odour pollution

A. Fermentation:

Fermentation is a process by which breaking down of complex matter is carried out with the help of microorganism such as yeast or bacteria, usually in absence of oxygen (anaerobic conditions). For example, break down of sugar to ethyl alcohol in making beer, wines and spirits. This process gives out unpleasant smell, causing odour pollution.

B. Putrefaction/Decomposition:

Decaying or disintegration or rotting of degradable matter is carried out with the help of microorganisms. Thus the odour generated by these processes is very unpleasant, nauseating and affects public health.

6.4.3 Control measures

- Compost manure: The large amount of solid waste can be subjected to compost process by which large quantity of biofertilizers and manure can be obtained.
- Biogas generation: When solid waste is exposed to undergo anaerobic fermentation, biogas is generated. It can be used in domestic cooking.
- Avoidance of storing garbage in open dumps and regular or continuous disposal (at least twice a day) of collected waste will help in reducing odour pollution.
- Installing mortuary away from the residential area provides wholesome surrounding to the people.
- Setting up of common effluent treatment plant to treat tannery waste can reduce the odour pollution.
- Ozone-free deodorizers, disinfectants (phenol and dettol), improve d air purifiers can be used in public toilets to avoid foul smell.

6.5 Noise Pollution

In the late 1700s, the major cities of Europe were so quiet that the fire alarms could be shouted from the top of a watchtower in the centre of a town. Even in the early 1900s, a simple brass bell on top of a fire-truck was enough to clear a path through the street. In contrast, to be heard above ambient noise levels, fire vehicles in modern cities must use sirens that produce more than 120 decibel (dB) of sound – well past the pain threshold.

6.5.1 Definition of noise

Noise is considered to be any unwanted sound that may adversely affect the health and well-being of individuals and masses. Noise derives from the Latin word '*nausea*' – a feeling of sickness of the stomach with an urge to vomit. Any unwanted electromagnetic signal that produces a jarring or displeasing effect and which interferes with human communication, comfort and health can be called as noise pollution.

6.5.2 Types of noise

- Steady-state noise: Noise whose intensity and quality is practically constant (varying less than 5 decibels) over an appreciable period of time.
- Fluctuating noise: Noise whose intensity rises or falls more than 5 decibels (dB).
- Intermittent noise: It is discontinuous wave motion that prevails for a long period, remains silent and starts once again.
- Impulsive noise: Noise that is transient and somewhat sudden (e.g. gunshots, crackers). The impulse occurs within or less than 500 milliseconds (0.5 seconds) duration with a magnitude of at least 40 decibels within that time.

6.5.3 Characteristics of noise

Sound is a compression waveform. The maximum number of compressions per second is its **frequency**. The unit of frequency is Hertz (Hz). Infrasonic sounds are low frequency sounds usually ranging between 16 Hz to below 0.001 Hz while ultrasonic sounds are high frequency sounds ranging above 20000 Hz. The sound frequencies that can be heard are the audio frequencies. Human ears can hear between 20–20000 Hz. **Pitch** is the frequency of a sound as perceived by human ear.

The communication between elephants, whales, rhinoceros, hippopotamus, is by the infrasonic waves. The main application of such waves lies in the seismographs in order to detect tremors. Ultrasonic waves are used in sonograms to take pictures of the soft tissues, ultrasonic cleaners to clean lenses, medical instruments and jewellery or to find defects in the materials or to gauge their thickness.

The distance traversed by the wave in one cycle is its **wavelength** or it is the distance between two peaks of two consecutive waves.

Wavelength $(\lambda) = v/f$ where, v is the sound velocity, f is the frequency

Amplitude is the height of the wave from the undisturbed position to the top of the crest or the bottom of the trough.

6.5.4 Measurement of noise

Noise is usually measured either by its sound pressure or sound intensity. The sound intensity is measured in decibel (dB), which is tenth part of the longest unit 'Bel', named after Alexander Graham Bell (*decca* – 10; *bel* – name of the discoverer).

- Most commonly used instrument for measuring noise is known as Sound Level Meter (SLM) or Psophometer.

- The safe intensity level of sound as prescribed by the WHO is 45 to 50 decibels (dB).

6.5.5 Sources of noise

A. Natural sources:

The natural sources for noise pollution are thunder, earthquakes and loud volcanic eruptions.

B. Anthropogenic (man-made sources)

i. Industrial source: Various machines in different factories, industries and mills generate noise. For example, mechanical sows and pneumatic drill.

ii. Transport noise (traffic noise):

- Hooting of trains creates strain on ear because the noise level rises to 130 decibel.

- Air traffic noise caused by speeding jet aircrafts damage human hearing permanently (for example, sonic boom)

- Mechanized automobiles like motors, scooters, cars, motorcycles, buses, trucks, sirens of ambulance and fire engines produce tremendous amount of noise pollution.

iii. Neighbourhood noise:

Indians are famous for celebrating large number of religious festivals, and every occasion, function and sentiment is manifested in a noisy manner.

- Shrieking of loud speakers at the place of worship, blaring out film songs at marriage and birthday parties, bursting of fire crackers to celebrate Deewali, the festival of lights, political rallies and election campaigns, all involve noise that cause great damage to hearing.

- The reckless use of household gadgets like musical instruments, television, VCR, radio, telephones, washing machines, vacuum cleaners, air conditioners are also responsible for causing noise pollution.

Box 6.5: Noisiest city in India

According to the Institute of Road Traffic Education, Mumbai is the noisiest city in India followed by Delhi, Chennai and Bangalore.

Hindustan Times, 23 March 2011

6.5.6 Impacts of noise pollution

Noise is recognized as slow and insidious killer. According to Robert Koch, 'A day will come when a man has to fight with merciless noise as worst enemy of health.' The various impacts of noise pollution can be classified as:

A. Auditory impacts:

- Hearing Loss: The most immediate and acute effect of noise is the impairment of hearing. The damaging of hair cells and eardrums (tympanic membrane) occurs due to high and sudden noise and it leads to hearing loss.

- Acoustic Trauma: It occurs when a 130 decibel or louder blast of sound leads to mechanical destruction of the ear, rupture of tympanic membrane and crushing injury of hair cells (receptive organs for hearing). This type of trauma, often caused by firearms and explosions is immediate and permanent.

- Tinnitus: Workers suffering from noise-induced hearing loss experience continuous ringing in their ears and the condition is called as tinnitus.

B. Non-auditory impacts:

- Speech interference and sleep interference: Noise pollution interferes with speech communication. One has to strain the ears to listen to something. Noise intrusion causes difficulties in falling asleep (insomnia – sleeplessness) or awakens the ones in sleep, thus causing sense of fatigue and palpitation (throbbing or trembling) upon awakening.

- Human health impacts: Noise induces the physiological effects on human beings such as dilation of pupils, flow of saliva a gastric juices, pain, increased blood pressure leading to tension.

- Noise pollution causes contraction of blood vessels, makes the skin pale, and leads to excess secretion of adrenalin hormone into blood stream leading to high blood pressure.

- It causes nervous breakdown, tension and even insanity.

- It affects health efficiency and behaviour by causing damage to heart, brain kidney and liver.

- It affects the pregnant women by quickening of human foetus' heart rate and malformation of central nervous system.

- It causes bickering, loss of working efficiency, ill temper and violent behaviour.

C. Impacts on human properties

The non-living things such as buildings undergo physical damages such as cracks, broken windows, doors and glasses by sudden and explosive sound.

- Work place standards of noise pollution prescribed by Occupational Safety and Health Administration (OSHA) is given in Table 6.13.

Table 6.13: OSHA work place standards

Decibels (dB)	Permissible exposure
90	8 hours
95	4 hours
100	2 hours
105	1 hour
110	30 minutes
115	15 minutes

6.5.7 Control measures

A. Source level control:

i. Substitution of highly noise-generating machines with quieter machines, usage of vibrations isolated mountings (steel springs and elastomers), reduction of external surface of vibrating parts helps in noise reduction at source.

ii. Using muffler (a device used to deaden the sound) and dampening the vibrating surface subsides the noise pollution at source.

iii. Enclosing the noise-generating source in a box or in a glass cover can reduce the source generated noise transmission to its surrounding.

B. Path level control through modification:

Modification of the path along which the noise is propagated is highly effective in noise control. Tunnel construction with the help of barrier shields, hard boards and plywood to modify the path of noise transmission, and intense architectural planning by using acoustic materials on walls, ceilings and doors to absorb sound (for example, cinema hall walls in zigzag manner, and the roof with acoustic tiles) highly reduce the reverberation of noise.

C. Receiver level control:

In spite of the engineering methods some industrial noise cannot be silenced and so the receiver should wear 'ear-plugs' which can protect the receiver against noise pollution. Job rotation of the worker can restrict the length of exposure to potentially hazardous noise level.

D. Buffer zone by green cover:

Trees are capable of absorbing enormous amount of noise. Planting of Ashoka and Neem trees along the highways, around industries and jet ports play an important role in absorbing noise.

E. Miscellaneous control methods:

- Banning the use of horns, loudspeakers, bursting crackers in and around educational institutions, hospitals and wild life sanctuaries and national parks.

- Imposing huge fine on vehicles generating noise (Environmental Protection Act, 1986 and Motor Vehicles Act, 1988).

- Promoting community awareness by education and news media can help in noise control at the social gatherings and in usage of house appliances such as TV, VCR etc.

Table 6.14: Ambient air quality standards in respect of noise as per the Noise Pollution (regulation and control) Rules, 2000

Area codes	Class of area/ zone	Limits in dB(A) Leq*	
		Day time	Night time
(A)	Industrial area	75	70
(B)	Commercial area	65	55
(C)	Residential area	55	45
(D)	Silence Zone	50	40

Box 6.6: Noise pollution during Diwali

Seven cities are selected within which 35 locations are marked. They are Delhi, Mumbai, Kolkata, Chennai, Lucknow, Bangalore and Hyderabad. Real time ambient noise was monitored in 2011 and 2012 on the occasion of Diwali. There is no change in the trend of noise level in case of Mumbai, while Bangalore, Delhi, Kolkata and Hyderabad shows a decreasing trend in comparison to 2011. But the noise levels in Lucknow and Chennai showed insignificant variation.

Table 6.15: Few sources of sound and their sound intensity

No.	Source	dB
1	Rocket engine	180
2	Take off of jet aircraft	150
3	Rock music	130
4	Normal conversation	70
5	Air conditioners	60
6	Whisper	30
7	Rustling of leaf	10

6.6 Radioactive Pollution

Radioactive pollution is defined as a type of physical pollution of atmosphere, hydrosphere and lithosphere by emissions from radioactive materials. Certain materials possess the ability to emit the alpha, beta and gamma rays. Those materials are called radioactive elements. The sources of radiations can be natural or manmade. Radioactive wastes are those wastes containing radioactive material, usually byproducts of nuclear power plants, nuclear reaction or processes that involve radioactive substance such as research and medicine.

- Alpha particles have low penetrating power and can be blocked by a piece of thin paper and even human skin.

- Beta particles with medium penetrability can penetrate through skin, while they can be obstructed by glass and metal.

- Gamma rays are highly penetrable and can penetrate easily through the human skin and damage cells on its way through, reaching about 100 m, and can only be blocked by a very thick and massive piece of concrete.

The natural radiation, alternatively called the background radiation involves the cosmic rays. Natural radiation sources can be the high energy cosmic rays or the land radioactivity. Natural radiation occurs from the elements like radium, actinium, uranium, thorium, polonium, radon, strontium, potassium and carbon which are found in the rock, soil and water.

The man made radiations sources are the mining and refining of substances like uranium, plutonium and thorium, nuclear power plants and spent fuels. Nuclear weapons explosions result in radioactive fallouts.

6.6.1 Causes of radioactive pollution

Following is the summary of the major sources of the radioactive waste.

- cosmic rays;
- natural radiation for example, radon;
- nuclear fuel production;
- nuclear power plants;
- use of radionuclides in industries and domestic sources like television, tobacco, watches;
- nuclear weapon tests carried out by defense personnel;
- nuclear disposal;
- uranium, thorium and plutonium mining; and
- diagnostic purpose – iodine.

6.6.2 Impact of radioactive pollution

- The extent of damage depends upon the half-life period of the radioactive substance and the speed of absorption and excretion. The most sensitive regions appear to be actively dividing regions, such as the skin, gonads, and intestine.

- Reacting with the structural molecules it usually forms free radicals that can damage DNA and RNA, proteins etc. Such mechanism may cause cancer and congenital defects.

- In spite of the body cells possessing salvage and repair mechanisms, there is always some increase in the incidence of some types of cancer.

- Regarding the dose, scientists opine that there is no such threshold value and radiation at any dose poses a finite risk of causing some biological injury. The damage caused by very low doses radiation may be cumulative.

- Gases and particles produced by the radioactive materials can be carried by the wind and the rain as nuclear fallout. The Chernobyl nuclear accident is the notable example. Strontium has the ability to aggregate in the bones and cause bone cancer. Iodine may affect the WBC, bone marrow, spleen, lymph, and various damages.

- Radioactive materials contaminating the land and water adversely affect the aquatic animals. They are absorbed by the plants and ultimately enter the food chain.

6.6.2.1 Radioactive waste disposal and control

Radioactive waste is produced in small amounts and is hazardous as compared to that of power plants. Radioactivity decreases with time depending on the radionuclide in such wastes and their respective half-life periods. Those with longer half-life emits alpha and beta particles and those with short half–life emit highly penetrating gamma rays. Eventually all will form stable element lead. The radioactive waste can be categorized as:

- very low level waste (VLLW);
- low level waste (LLW);
- intermediate level waste (ILW); and
- high level waste (HLW).

The radioactive substance in VLLW is not considered as injurious to people and the environment. It comprises concrete, plasters, bricks, metals, valves, piping etc. generated during construction or dismantling operations. VLLW is produced in hospitals and industry and nuclear fuel cycle. It is often incinerated before disposal. It forms the bulk but contains only one per cent of radioactivity of all radioactive waste. ILW with higher amounts of radioactivity comprises resins and chemical sludge. HLW is produced when uranium used as fuel is burnt in a reactor and comprises fission products and trans-uranic elements produced in the core reactor.

Uranium oxide concentrate obtained during mine operations referred to as 'yellow cake' (U_3O_8) is not remarkably radioactive. This is refined at the beginning and then transformed into uranium hexafluoride gas (UF_6). The chief offshoot of enrichment process is depleted uranium (DU), principally the U-238 isotope. Around 1.2 million tons of DU is presently stockpiled. Organisation for Economic Co-operation and Development (OECD) countries produce about 81,000 m^3/year of radioactive wastes.

Preventive measures can be adopted to control radioactive pollution:

- explosion of nuclear devices underground;
- minimization in the generation of radio isotopes;
- utmost care to be taken in the disposal of industrial wastes that may contain radionuclides;
- efficient coolant system with highly pure gaseous coolants to be used in the nuclear reactors;
- fission reactions to be well regulated;
- in nuclear mines, wet drilling may be employed along with underground drainage; and
- use of nuclear medicines and radiation therapy absolutely when necessary in required doses.

6.7 Soil Pollution

When the natural soil is contaminated by xenobiotic (man-induced) chemicals, the result is soil pollution. Volcanic eruption, that spews lava with toxic elements robbing the land off its ability to regenerate or a destructive flood that washes in with it contaminants may also have the same effects, but considering their limited occurrence and destructive ability, the authors will concentrate on man-made causes and their effects.

The major contributors to soil pollution are:

A. **Coal ash:** The ash generated by burning coal, primarily in thermal power stations is one of the major causes of land pollution. Bourne by air, it creates a cloak of coal dust and tar within a radius forcing the soot bedecked vegetation to wither away to a dusty death. The lead and zinc that the coal slag carries with it, is in cases extremely hazardous and is known to have not only made barren vast tracts of land around power plants but has also caused untold miseries to the ecosystems that harboured them apart from exposing humans to serious health hazards.

B. **Sewerage:** The contamination of groundwater by untreated sewerage and the effect they have on the land is something that is of serious concern, especially in the developing and less developed countries, where the lack of resources stops communities from setting up treatment plants and the pressure on scarce land is further aggravated by the contamination by the discharge of sewerage. Treated sewerage, often termed as sludge is used as fertilizer, which too has a long-term detrimental effect on the land as even post treatment it carries a number of heavy metals and other contaminants.

C. **Pesticides and herbicides:** Intense cultivation to meet the growing needs of an ever hungry population has forced man to use more chemical fertilizers and bio–engineered, genetically modified inputs to increase the yield per unit of land. The chemicals that have seeped into the land have, in their wake, created the basic chemistries that were prevalent before their introduction leading to drastic falls in crop yields, the cropping patterns and ultimately to the carrying capacity of the land, making them barren.

An example would be the tea industry in India which in its mad haste to increase yields had embraced the chemical culture. As a result of this excessive use of chemicals today, most gardens have successfully terminated the earthworms which used to play a vital role in keeping the soil vital and regenerative. Similarly, the crop yields too have progressively fallen as the regime of chemicals which promises a bounty in the initial years, has started its regressive journey killing the ability of the garden eco–system to regenerate and revitalize itself the way it had over decades.

Today, forced by the failure of excessive chemical inputs, tea gardens are seeking to go back to their traditional roots by using organic inputs to nurse the scars caused by the greedy introduction of chemicals. However, this is proving to be easier said than done as it will be years, according to experts, before the adverse effects are addressed and mitigated and before growth and development can even be conceived of.

Case Study 6.6: Love canal waste dumping

It was an unfinished canal constructed in the city of Niagara Falls in New York in 1890s. From the 1930s to 1950s, the place was used as a chemical waste dump. The land was purchased by Hooker Chemical and Plastics Corporation.

Hooker Chemical turned that area near Niagra Falls into a municipal chemical disposal site of 352 million pounds with modern methods. A thick layer of impermeable red clay was used to seal the dumping so as to prevent leakage. In 1963, the place was sold to a city for $1 along with all warnings. Housing and schools were built. With time strange odours filled up the air, the red clay cap was damaged and seepage came to be noticed in the basements. Children fell ill, and miscarriage and birth defect problems aggravated. In 1978, a total of 21,000 tons of waste was dumped in the landfill with 248 different types of chemicals. 130 tons of Tetra chlorodiazo-p dioxin (TCDD) was discovered. More than 900 families moved away while 90 opted to stay. Hooker Company was sued by New York Supreme Court in 1983 and settled for $20 million. Even President Carter provided funds for shifting. In 1995, Occidental Chemical Corporation, the parent company of Hooker's Chemical Company agreed to pay $129 million to Environment Protection Agency as cleanup costs. The area was resealed, surrounding area was cleaned and the mother company paid additional $ 230 million for cleanup.

6.8 Marine Pollution

'The solution to pollution is dilution' is the catchphrase that sums up human concern about the oceans in general and marine pollution in particular. Conventional thought led us to believe that such is the huge expanse of the seas which surround us that irrespective of what is droped in the oceans, these will be insignificant and will not be to generate any serious concerns. This thought process, faulty as it may be on the face of it, led us to indiscriminately dump everything from sewerage to industrial effluents in the sea. The consequent catastrophe, of such irresponsibility being something whose real magnitude is only being understood now. A 1971 United Nations report defined ocean pollution as:

> 'The introduction by man, directly or indirectly, of substances or energy into the marine environment (including estuaries) resulting in such deleterious effects as harm to living resources, hazards to human health, hindrance to marine activities, including fishing, impairment of quality for use of sea water and reduction of amenities'.

6.8.1 Sources of marine pollution

Nearly 80 per cent of the marine debris comprises plastic, total mass of plastic can be as high as 100 million metric tons. Plastic bags, films, containers, fishing nets are often discarded in the marine waters.

Toxic organic and inorganic chemicals including heavy metals and radioactive wastes are discharged in to the oceans and seas from the industries and factories located along the coastal regions that may alter the pH and composition of the water at the point of discharge.

The sewage from the municipal wastes and agricultural runoff containing pesticides and fertilizers also gains entry into the marine waters.

Climate change, global warming and the windblown dust increase the crisis of the marine environment.

6.8.2 Impact

Discarded fishing nets drift for years, ensnaring fish and mammals. The fishing nets left or lost in the oceans, known as ghost nets often entangle the fishes, dolphins, sea turtles, iguana, sharks, dugongs, penguins, crabs, etc. Life on sea is threatened from ingestion of plastic which may get entangled in the intestine or cause suffocation. A sea turtle often consumes plastic thinking it to be jelly fish. Consuming flotsam blocks the passage of food in the gut. Plastics do not biodegrade but may photodegrade on prolonged exposure to sun. Plastic may also enter the food chain when it photodegrades down to the size of zoo planktons. In certain regions, ocean currents contain trillions of decomposing plastic items and other trash in gigantic, swirling garbage patches. One in the North Pacific, known as the Pacific Trash Vortex, is estimated to be the size of Texas. A new, massive patch was discovered in the Atlantic Ocean in early 2010. Substantial amounts of plastic accumulate on the Midway Atoll beach, the home to about two third of all Laysan Albatross.

The toxic chemicals from the industries, farmland and municipalities can leach out and bio-accumulate in the adipose tissues; over the time they biomagnify up the food chain and may act as endocrine disruptors. While the act of discharging pollutants into the oceans dates back in antiquity, environmental degradation has increased dramatically in the last century due to the discharge of industrial effluents, agricultural runoff and discharges from the coastal cities. It has also been found that medicines consumed by man, partially processed in the bodies, eventually end up in the fishes that people consume through a complex food chain. The surface runoff may comprise runoff from road construction, buildings, ports, harbours and highways. They are laden with carbon, hydrogen, nitrogen, phosphorous, etc. that may increase the tendency of algae to bloom and create hypoxic conditions. Fish death in North Carolina and Chesapeake Bay in the US is due to the profuse growth of Pfiesteria, a dinoflagellate, that flourished as a consequence of increased nitrogen and phosphorus from the wash-offs.

Oil spills have devastating effects and are extremely difficult to clean up. While oil spill hit the headlines, much of the oil in sea water comes from tankers discharging ballast water, leakage in the pipelines and engines. Apart from all these, discharge of cargo residues and containers are mounding up each year.

Oceans are huge carbon sinks that capture CO_2 from the atmosphere. With increasing atmospheric carbon dioxide, oceans too are becoming acidic. This raises concern about the calcium carbonate structures including the corals that becomes more vulnerable to dissolution. Rising oceanic temperatures reduce the ability of oceans to absorb the carbon as expressed in Monaco and Manado Declarations. Huge amounts of acidified water are seen upwelling within 4 miles of the Pacific continental shelf region of North America. The reserve of methyl clathrate has the potential to be released as well with global warming. Windblown dust and debris are found to blown long distances such as from the Sahara desert to the Carribean and Florida Dusts are also reported to be transported from Gobi and Taklamakan deserts to the Hawaiian Islands.

One of the major marine pollutants is the oil. Oil from cargo tankers is predicted to add about 3 million tonnes of oil annually to the sea. Collisions in the port may add 1 million tonnes of oil annually. Oil leakage may happen from the pipelines due to corrosion, cracks and punctures. Offshore drilling rigs also add to pollution. Vessel accidents and waste from tank washings cause oil spillage and extensive pollution of sea water. In 1984, more than 90,000 lites of oil were collected from Indra Dock Basin alone. Crude oil is a mixture of various hydrocarbons of varying complexities such as naphtha, kerosene, tar, wax, grease, asphalt etc. Oil pollution can be controlled by skimming of oil from the surface with a suction device, using adsorbents; spreading high-density powder which can sink the oil at the bottom, etc.

Apart from the obvious destruction pollutants in the sea waters also create havoc with the delicate coral reefs that have been formed over millions of years. While some pollutants destroy them outright, others ensure their mutation into grotesque new forms as they struggle to adapt and carry on with life. As these coral reefs are home to innumerable species of life, their decay and destruction have a direct impact not only on these life forms but also on the entire eco system that these coral reefs help create and foster.

Humans are just about to begin comprehending the narrow view of the philosophy of dilution. Innumerable legislations are existent at present that forbid dumping of harmful materials into the ocean, although enforcement can often be lax – being more on paper than anything else. Many pristine aquatic ecosystems are declared as sanctuaries. Even hardcore conservationists agree, there is a long way to go before one can stem the so-called tide.

Few things that one can do to prevent marine pollution:

- Eating sustainably; eat only what is abundantly available.
- Do not throw away plastic goods.
- Keep the beaches clean; do not litter.
- Ban all types of developmental activities along the coast.
- Run off from pollution sources should not be allowed to drain in sea water.
- Do not allow sewage effluents in the marine water.
- Do not dispose oil ballast in sea water.
- Flushing pet litter leads to harmful pathogens being carried to the seas while dumping aquarium fishes in the local water bodies can cause irreparable damage to the local ecosystems.
- People should collaborate and spread their hands of help to the various organizations that are working to reverse the trends of destruction. Spread awareness in one's community.
- Go for the most eco-friendly option that causes the least damage to the marine eco system.
- A number of response mechanisms are employed for controlling oil spill. They can be

mechanical containments that act as a primary line of defense, such a booms, skimmers and addition of sorbent materials. The sorbents can be natural (peat, moss, hay, feathers, sawdust, clay, vermiculite, glass wool) or synthetic (polyurethane, polypropylene) that can either act by absorption or adsorption. Booms control the spread of oil whereas skimmers can recover spilled oil for recycling or disposal. Under chemical methods, dispersants can be useful. They contain surfactants that can break the oil droplets into small particles so as to be subjected to natural process of breakdown. Gelling agents or solidifiers can react with oil and form rubber like solids. Biological agents like bacteria, fungi, also break down the complex oil compounds into simpler ones. To increase microbial degradation, either fertilizer may be added or a pool of microbes can be added to the native population. Lastly to protect birds and animals and keeping them away from the spilled site propane scare cans, helium filled balloons and floating dummies can be used.

Case Study 6.7: Exxon Valdez oil spill

On 24 March 1989, the tanker Exxon Valdez, on its voyage from Valdez, Alaska to Los Angeles, California, struck the Bligh Reef in Prince William Sound, Alaska. In order to avoid ice, the tanker was heading outside its normal shipping route. Immediately within 6 hours of hitting the reef, the tanker discharged about 10.9 million gallons of Prudhoe Bay oil. 8 tanks out of 11 on board had wrecked.

It emerged out as the largest oil spill in USA with the oil dispersing about 470 miles on the southwest of Chignik village in Alaska. Ultimately the spilled region had spread over 900 miles of shoreline. Clean up activities continued throughout the summer months of 1990 and 1991.

The predicted loss relied mainly on the sport fishing activities – like how many trips were conducted, their duration, the extent of fishing area, the varieties of fish, etc. The loss incurred was presumed to be in millions. After the incident the number of resident and non-resident vacation trips declined in the affected region which could be attributed to the shortage of accessible tourism services. This included accommodation, leased boats, air taxi, etc. There was acute shortage of labour in the tourism industries as most of the service industry workers went for clean-up jobs. Such jobs paid them higher amounts as compared to the tourism industry. The business turned out less than expected by 16 per cent and the cancellations went up by 59 per cent. They were all related to oil spill.

$4.9 to 7.2 billion dollars losses were caused by the spill. This included the cost for relocation and rehabilitation needed by the aquatic birds, marine and terrestrial animals that have been injured or damaged during the act of spill. Study of both long-term and short-term impacts reveal the loss of one million lives. This included 2.5 lakh seabirds, nearly 2,800 sea otters 12 riverine otters, 300 harbour seals, 247 bald eagles, 22 orcas and unknown number of salmons and herrings. Numerous sea otters, whales, sea lions, seals, bears, mink, deer, otters and eagles had to be relocated. It was predicted that 23 US gallons of Valdez oils accumulated in the sands and oils of Alaska. The rate of breakdown was estimated to be less than 4 per cent annually.

The fine imposed on Exxon was $ 150 million. It seems to be the biggest fine imposed for any environmental crime. In lieu of cooperation in the clean up operations from Exxon $125 million dollar was forgiven by the court.

Out of the remaining 25 million dollars, 12 million dollars was allotted to the North American Wetlands Conservation Fund and the remaining 13 million was given to the national victims of Crime Fund.

Case Study 6.8: Deepwater horizon oil spill

This oil spill, also known as Mexico oil spill or British Petroleum Oil spill is the largest marine oil spill till date. The incident that took place on 20 April 2010 was caused by the explosion on the Deepwater Horizon offshore platform. The platform was located about 50 miles to the southwest of Mississipi River delta.

Transocean Ltd. owned the nine year old submersible Deepwater Horizon offshore drilling unit. The accident happened in a position where British Petroleum (BP) was the developer of the Macondo Prospect Oilfield. The drilling unit could function in the waters at about 8,000 ft and the drilling process could go down to 30,000 ft. During the time of explosion it was drilling at around 5,000 feet in the Canyon Block 252 of the Gulf of Mexico, 66 km off the Lousiana coast. They were about to close the well so that they could start production later. In the mean time on 20 April 2010 methane gas spurged up the drilling column under tremendous pressure. With the release of pressure the gas expanded and ignited with explosion that engulfed the platform with fire. The US Government held BP responsible for the accident and British Petroleum accepted the accountability for oil spill as well as for the clean up costs.

Post accident most of the workers were evacuated safely. 11 workers went missing in spite of search and rescue operations. On 22 April 2010 the platform sank to about 5,000 ft. The search operation was ultimately stopped for missing workers as all were presumed to be dead. Crude oil gushed out of the 5,000 ft pipe as the rig platform sank. The pipe joins the well located in the bottom and the platform at the surface. The safety device called the blowout preventer. It failed to be activated and thus all attempts to shut down the oil flow failed. The oil spurted out from the broken well for more than 85 days.

The states of Lousiana, Mississippi and Alabama were heavily affected. The oil slick covered as much as 75,000 km. The slick changed its location almost daily because of the weather conditions. The NOAA had declared a ban on fishing activities. As of mid July 2010, 90–180 million gallon of oil spilled. Affected areas include Gulf Islands National Seashore, Mississippi and Alabama Barrier Islands, Florida Barrier islands, Galveston islands and approached Chaneleur islands, Delta National Wildlife Refuge and Breton Wildlife Refuge.

The fishing activities were stopped due to spill. Hotels and tourism business incurred huge monetary loss. The spill area hosted 8,332 sea species that include more than 1,200 fishes, 200 aves, 1,400 molluscans, 1,500 crustaceans, 4 sea turtles and 29 mammals. The spill threatened whale sharks to seagrass. The ocean floor was severely damaged which was the home range for the endangered Louisiana pancake batfish. A distinct link could be established between the death of the Gulf coral community and the oil spill. The oil and the dispersant (Corexit) mixture used for controlling the oil spill, including the PAHs, easily permeated the food chain through the zoo planktons.

Out of 1,746 birds collected, over 1,000 were noticeably oiled. Though 749 birds were caught alive, 997 birds died. Out of 528 sea turtles recovered, 400 were dead. Out of 51 mammals collected from the spill area, 47 were dead. The question of dolphin death due to oil spill is not yet proved.

British Petroleum resumed responsibility for the primary clean up and mitigation work. They engaged:

- response vessels;
- booms; more than 25 million gallons of oily water were recovered till date;
- dispersant; and
- skimmer ships, sand filled barricades.

They were able to recover 13.5 million gallons of oil till date. The US Government has set up a combined authority for the proper management of oil spills.

6.9 Thermal Pollution

Thermal pollution can be defined as the generation and discharge of waste heat into the water bodies thus resulting in undesirable effects in the aquatic environment. The major sources of such heat are the thermal power plants, nuclear power plants, iron and steel refineries, etc. The thermal plants utilize 33 per cent of the energy while rest of the energy is lost as heat. Such super heated water was earlier discharged in the water bodies which could raise the temperature of the water resulting in several impacts as stated below.

- It may immediately kill the flora and fauna at the point of discharge or in its vicinity.
- It alters the metabolic activities of several organisms.
- It hampers the migration, breeding and spawning of fishes.
- Temperature affects the physical, biological and chemical characteristics in a water body.
- It decrease the dissolved oxygen content of water.
- The elevated temperature may change the solubility of the pollutants in water.

Temperature of water can be brought down by taking the water to wet or dry cooling towers which pre-cools the heated water before its discharge. This can be achieved by the use of cooling ponds, spray ponds or cooling towers.

6.10 Role of Individuals in Pollution Prevention

A. The Power Saver Mode – all devices have in built 'power saver mode' that automatically gets turned off when left unused for a chosen period of time. Switch on to the mode and there would be an unbelievable and drastic reduction in energy consumption. Less energy means less burning of fossil fuels, which means lesser impact on the climate.

B. Unplug and switch off – once the charging is over, people simply jack off their gadgets, not switching off the power source which culminates into a huge loss of energy. Be responsible, switch off.

C. Turn off the tap – the amount of water we waste is humongous. By being water savvy, by turning on the tap for just the amount we need and by curbing wastage, we can all make a huge difference.

D. People should use their own tea or coffee mug – every time people have a cup of tea or coffee, they are either supporting the destruction of rainforests (paper cups are made from wood pulp sourced from trees) or are adding to heaps of plastic waste that promises to cover the Earth's surface one day and choke all forms of life to destruction. By using their own coffee mugs, people can stem the rot.

E. Hold that print – don't press that print button for every mail that one gets. Try to read from the screen. By printing that worthless chain mail, one is contributing to the denudation of trees, which are the best ways of combating global warming.

F. Pool resources – people should use public transportation as frequently as they can. It is better to use pool cars, which spread the use of fuel over many heads leading to better economics apart from having a lesser impact on the climate.

G. Walk that talk – don't just talk big. People should walk as much as they can. It is good for one's health and for the health of the planet. As more the people walk the less energy they use. And yes, if they can, they should drop that jacket and that neck tie. If people dress as though they are in Europe in the sweltering heat of Noida, they are going to need a battery of air conditioners to stop them from going insane. Instead, it is better to wear more tropical clothes and use less energy.

H. Plant a tree – Plant a tree and water it regularly. That's our ticket to a sustainable world.

I. Switch to energy saving devices – the shelves are full of energy compliant devices. They may cost a little bit more but effective over the long run. Switch over or perish.

J. Spread the word/embrace green products – if they can do nothing, then at the least they should spread the word. There are many products that are recycled, or are plain vanilla eco-friendly. Try embracing them.

Summary

- Pollution is any undesirable change in the physical, chemical and biological characteristic of air, water or land that may or will harmfully affect human life or that of various species, the industrial processes, living conditions and cultural assets.

- Generally a pollutant is classified as biodegradable and non-biodegradable. Biodegradable pollutants are those which degrade easily such as sewage, whereas non-biodegradable pollutants are those which do not degrade or degrade very slowly, such as plastic. The pollution sources can be mobile transportation, stationary combustion, industrial processes and solid waste disposal.

- In India, a CEPI is used to characterize the environmental quality at a given locality ensuing the system of source, pathway and receptor.

- The atmosphere may be polluted due to various natural processes or by anthropogenic activities. Bulk of the air pollution sources are linked to anthropogenic activities. Major air pollutants are aerosols and VOCs, other hydrocarbons, particulate matters (inorganic and organic), oxides of carbon, sulphur dioxide and oxides of nitrogen. POPs, particulate matter and CFCs have become a global concern. Associated occupational health hazards are berylliosis, asbestosis, silicosis, anthracosis and black lung disease. Global impacts of air pollution include global warming and green house effect, ozone depletion, photochemical smog, etc.

- Control of air pollution comprises raw material substitution, process modification, equipment alterations and removal of pollutants at source. Air pollution control equipments commonly used are wet scrubbers, electrostatic precipitators, dust cyclones, fabric filters and catalytic converters.

- Water pollution means one or more substances building up in aquatic bodies to a limit that they cause trouble life forms. Water pollution can be surface water pollution, ground water pollution, point source, non-point source, soft water or hard water pollution. Their sources can be domestic, agriculture, industries, rain drainage and others.

- Global impact of water pollution are eutrophication, bioaccumulation, biomagnifications, El Nino, La Nina and Southern Oscillation, acid rain, etc. Water pollution by organic matter is a prime cause for outbreak of epidemics like cholera, gastroenteritis and other water-borne diseases. Occupational health hazards are itai-itai disease, minamata disease, blackfoot disease, methaemoglobinemia etc.

- Odour pollution results from organic decomposition of biodegradable matter that emits huge quantity of gases, or particulates that diffuse in the air. Noise is considered to be any unwanted sound that may adversely affect the health and well-being of individuals or populations. Radioactive pollution is defined as a type of physical pollution of atmosphere, hydrosphere and lithosphere by emissions from radioactive materials. When the natural soil is contaminated by xenobiotic (man-induced) chemicals like pesticides and fertilizers, the result is soil pollution. Thermal pollution can be defined as the generation and discharge of waste heat into the water bodies, thus, resulting in undesirable effects in the aquatic environment. The major sources of such heat are the thermal power plants, nuclear power plants, iron and steel refineries, etc.

Exercise

MCQs

Encircle the right option:

1. Which of the following is an air pollutant?
    A. Nitrogen   B. Carbon dioxide   C. Carbon monoxide   D. Hydrogen

2. Which of the following is a secondary air pollutant?
    A. Ozone   B. Carbon dioxide   C. Carbon monoxide   D. Nitrogen dioxide

3. Depletion of Earth's ozone layer will cause
    A. Increased average temperature of earth
    B. Decreased oxygen content in the air
    C. Increased amount of UV radiation to reach Earth's surface
    D. Rise in sea levels

4. Acid rain is due to contribution from:
    A. Methane and ozone   B. Oxygen and nitrous oxide
    C. Methane and sulphur dioxide   D. Carbon dioxide and sulphur dioxide

5. Identify the primary health risks associated with greater UV radiation:
    A. Damage to digestive system   B. Increased liver cancer
    C. Neurological disorder   D. Increased skin cancer

6. Which of the following is not as a consequence of global warming?
    A. Rise in sea level   B. Increased agricultural production
    C. Deterioration in health   D. Increased storm frequency and intensity

7. The primary contributor to the greenhouse effect is by:
    A. Carbon dioxide   B. Carbon monoxide
    C. Chlorofluorocarbons   D. Methane gas

8. Presence of high coliform counts in water shows:
    A. Contamination by human wastes   B. Phosphorus pollution
    C. Decreased BOD   D. Hydrocarbon pollution

9. Enhanced organic matter in water _____ the BOD.
    A. Increases   B. Decreases   C. Remains unchanged   D. None of the above

10. Which of the following groups of plants can be used as indicators of SO_2 pollution of air?
    A. Epiphytic lichen   B. Ferns   C. Liverworts   D. Hornworts

Fill in the blanks.

1. Blue baby syndrome is also known as _____.

2. Los Angeles smog was caused by the emission of _____.

3. Surface run–off is an example of _____.

4. Twelve POPs that are banned collectively from being used are known as _____ _____.

5. _____ is the Indian city to have highest PM_{10} pollution.

State whether the statements are true or false.

1. PAN is a primary pollutant. (T/F)

2. Itai itai disease is caused due to cadmium poisoning. (T/F)

3. Pheromones are third generation pesticides. (T/F)

4. SO_2 is not emitted during burning of coal. (T/F)

5. BOD value is always greater than COD value. (T/F)

Short questions.

1. Define pollution.

2. What are the different types of pollutants?

3. What is air pollution?

4. Name the major indoor and outdoor pollutants.

5. What are secondary pollutants?

6. What is the health impact of carbon monoxide poisoning?

7. What is the source of sulphur dioxide?

8. Classify the sources of pollutants.

9. What is particulate matter?

10. State the effects of PM on human health and the environment.

Essay type questions.

1. Distinguish between primary and secondary pollutants. Give examples.

2. Global warming is a myth or reality. Justify.

3. What are photochemical smogs? How is it formed? Briefly describe its health impacts.

4. Describe the Los Angeles and London smog episodes.

5. Suggest a few control measures for air pollution.

Waste Management 7

Learning objectives

- *To develop an overall concept of waste management.*
- *To know about the ways of waste water treatment.*
- *To be familiar with the drinking water standards and permissible limits of pollutants in water.*
- *To know the 3 R strategies in solid waste management towards resource conservation.*
- *To learn the various steps in waste management so that we can apply them in our daily living.*
- *To know about the recycling of specific materials like glass, paper and plastic.*

7.1 Definition

Waste refers to eliminated or discarded substances that are no longer useful or required after the completion of a process. According to the Basel Convention, *'wastes are substances or objects, which are disposed of or are intended to be disposed of or are required to be disposed of by the provisions of national law'*. The term usually applies to substances generated as a result of human activity.

Figure 7.1: Waste management hierarchy

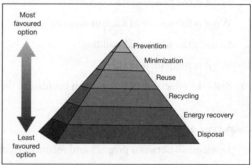

Waste management refers to the collection, transport, processing/treatment or disposal and monitoring of waste materials. Such process is generally undertaken to reduce their impact on health, environment or aesthetics. Waste management is based on 3 R strategies: **reduce, reuse and recycle**.

7.2 Wastewater Treatment (WWT)

Wastewater is that liquid waste that has been adversely affected in terms of quality by human interference. Usually it is discharged by the residences, industries, agricultural practices and often contains various contaminants. Based on its origin wastewater can be classed as sanitary, commercial, industrial or agricultural runoff. In this context, it is worth mentioning that sewage is a subset of wastewater contaminated with faeces or urine.

Wastewater treatment is a process by which the physical (large, coarse, suspended solids), chemical (dissolved minerals, gases, chemicals and heavy metals) and biological impurities (disease-causing pathogens) present in the wastewater/sewage/effluent can be reduced to an acceptable level by a series of treatment process. Wastewater treatment plant is also called an **Sewage Treatment Plant (STP)** in the municipal circle and **Effluent Treatment Plant (ETP)** in the industrial circle. This treatment makes the treated water fit for human consumption and recreational purposes. In other words, wastewater treatment means the various steps undertaken in the process of transforming wastewater into a harmless liquid that suffices to meet the requirement of sanitation, health and decency. Thus, wastewater treatment system is composed of a combination of **unit operation** and **unit process** designed to reduce the various contaminants in wastewater to an acceptance level.

- **Unit operation:** It involves contaminant removal by physical forces; for example skimming and sedimentation.

- **Unit process:** It involves contaminant removal by addition of certain chemicals like alum and microorganisms like bacteria; for example coagulation, chlorination, filtration and activated sludge process.

Each step of unit operation and unit process is explained in details under the section of wastewater treatment methods.

7.2.1 Wastewater sources and components

Wastewater sources mainly include:

- spent water from bathroom, lavatory and basins of residences;

- process water from manufacturing concerns i.e., industrial wastewater let out as effluent after the manufacturing of products or after cooling of machines; and

- semi-liquid waste like human and animal excreta and dry refuse like house and street sweepings; thus wastewater from these sources is composed of large volume of suspended and floating debris, dissolved impurities of gases and chemicals, biological impurities like disease causing pathogens.

Box 7.1: Impurities in wastewater

A. Human wastes

B. Surface run-off comprising
  - detergents;
  - disinfection by-products;
  - insecticides and herbicides;
  - petroleum hydrocarbons;
  - ammonia from food processing waste; and
  - fertilizers containing nutrients (nitrates and phosphates).

C. Metals and non-metals from mining and processing

D. Heated water from thermal power plants

E. Radioactive wastes from nuclear power plants

F. Deep sea mining

G. Oils

7.2.2 Necessity for wastewater treatment

Wastewater arising from the above mentioned sources, if left untreated,

- gets accumulated and creates a condition offensive to sight or smell;
- the impurities present in wastewater affect the water quality parameters;
- affects the water colour and taste, for example effluent from textile and dyeing industry make the receiving water highly turbid, silts from flood or surface runoff make water highly murky, thus preventing sunlight penetration required for photosynthesis of aquatic plants;
- sand silts clog the gills of small fishes and thus destroys the valuable aquatic life;
- high quantity of minerals with corrosive properties present in wastewater interferes with water supply appurtenances and thus cause blockage to the public water distribution systems;
- wastewater acts as vehicles of disease carriers for numerous pathogens like bacteria, virus responsible for intestinal diseases like typhoid, cholera, gastroenteritis and hepatitis (jaundice).

Therefore, wastewater treatment becomes highly necessary to make water pathogen-free, odor-free, potable, safe for human consumption, pleasing to senses and suitable for ordinary domestic and industrial uses.

7.2.3 Objectives of wastewater treatment

Wastewater treatment aims:

- to avert contamination of the receiving waters;
- to kill or disinfect all the pathogenic germs in polluted water;
- to make water free from offensive odour and unpleasant taste;
- to remove murkiness, so that water becomes clear in appearance and enriches aquatic life;
- to reduce corrosive properties of water which in turn enhances the long life for pipe conduits; and
- to make water safe for drinking and suitable for domestic purposes like cooking and washing, industrial purposes like brewing, dyeing and cooling of machines, agricultural purposes like irrigation and aquaculture as well as recreational purposes like boating and other water games.

7.2.4 Wastewater treatment methods

A. Natural or self-purification process:

When untreated wastewater is directly discharged into natural water bodies like river or lakes, it undergoes self-purification by the following five actions of dilution, oxidation, reduction, sedimentation and sunlight.

- **Dilution**: It is often said that the solution to pollution is dilution. The wastewater in the form of sewage or effluent if released in small quantity gets diluted or dispersed into larger volume of flowing water. Thus, dilution helps in reducing the potential nuisance of wastewater and the impurities become harmless when mixed with the larger volume of receiving water bodies.

- **Oxidation:** It is a process by which the dissolved oxygen (DO) present in the natural water bodies, oxidizes or breaks down or decomposes the organic matter let out by the wastewater. The aerobic microorganisms (bacteria living in presence of oxygen) present in the natural water bodies eat up or decompose the organic matter and thus reduce the organic load of the flowing natural water. Dissolved oxygen content of the natural water is increased by the mixing up of the atmospheric oxygen by the action of wind, water current (continuous turbulence of water) and by photosynthesis of microorganisms like phytoplankton and algae.

- **Reduction:** Anaerobic microorganisms (bacteria capable of living in absence of oxygen) and the chemicals present in the flowing natural water bodies split the highly complex organic constituents of the wastewater and releases dissolved gases like carbon dioxide and methane. These gases in turn undergo the above oxidation process to generate oxygen beneficial to aquatic microorganisms.

- **Sedimentation:** It is a process by which suspended solids and particles heavier than water settle down due the force of gravity. Thus, this process plays an important role in the reduction of settle-able solids.

- **Sunlight:** Sunlight plays a vital role in self-purification by acting as a bleaching agent, i.e., it kills the harmful pathogens present in water. Sunlight also helps out in photosynthesis process of aquatic microorganisms and consequently increases the dissolved oxygen content of the water.

All these five steps of natural purification process invariably occur uninterrupted in the natural water bodies. When the pollution load of the wastewater is high and goes beyond the natural cleansing capacity of the receiving water body, it is treated by engineered process of wastewater treatment plants. Here the pollutants are reduced to an acceptance level and then finally released into the natural water.

B. Artificial or engineered methods

The treatment of wastewater under engineered methods is broadly classified in to three categories (refer Figure 7.2).

i.   primary or physical treatment (unit operation);

ii. secondary or chemical and biological treatment (unit process); and

iii. disinfection or tertiary treatment.

Wastewater from domestic origin/sewage is transported from the place of origin to the treatment facility or the site of disposal. Collection system comprises **sewers** and is so designed that the sewage reaches the treatment site soon after it enters the sewerage system. As far as possible, sewers are laid in straight lines. Corners or sharp bends may reduce the velocity of flow; permit clogging as a result making cleaning difficult. Since sewer does not flow under pressure, a slope is needed to keep it flowing. Sometimes pumps may be necessary to give adequate pressure. Sewer may be made of vitrified clay, concrete, cement asbestos or bituminous impregnated fibre. The mains or the crossing are generally made of cast iron or cement asbestos.

Primary or physical treatment (unit operation): Primary treatment can also be called as physical treatment, because all the physical impurities (for example, floating bodies, larger solids) are removed by employing physical forces alone, i.e., no chemicals or biological organisms are used in this step. Primary treatment includes skimming and sedimentation.

Secondary treatment or chemical and biological treatment (unit process): Secondary treatment or chemical and biological treatment involves using of chemicals or biological organisms to remove the dissolved minerals and colloidal particles. The organic matter is used as nutrient for the biological microorganisms (for example, activated sludge process). Fine suspended matter is filtered (for example, trickling filter). The purification process is almost complete after the secondary treatment and the wastewater is nearly purified by these above mentioned processes, except that it contains harmful pathogens and so it has to be disinfected.

Disinfection or tertiary treatment: The process by which killing of harmful pathogens is carried out is known as disinfection. Chlorination, ultraviolet ray treatment or ozone treatment are some of the methods used in disinfection.

The various stages involved in primary, secondary and tertiary treatment of wastewater is summarized in Table 7.1.

Figure 7.2: Schematic representation of a typical wastewater treatment plant

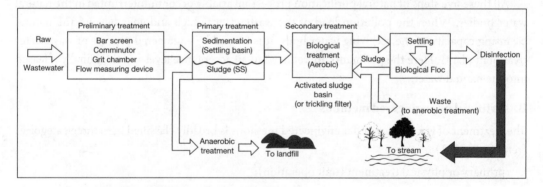

7.2.5 Major types of wastewater and their treatment

A. Municipal wastewater and STP

STP is a combination of processes arranged in a sequence in such a way that each step would support the performance of the downstream processes as the wastewater progresses through the following stages. Irrespective of the size and complexity of STP, it comprises the following steps:

Preliminary treatment ⟶ Primary treatment ⟶ Secondary treatment ⟶

Tertiary treatment ⟶ Sludge treatment

B. Industrial wastewaters and ETP

ETP is the most cost effective way to remove the unwanted hazardous chemicals from the effluents in order to meet the statutory pollution control requirements, especially applicable to chemicals, pharmaceuticals and electroplating wastewater. ETP covers mechanisms to treat water that has been contaminated by industrial and commercial activities prior to its release or reuse.

Figure 7.3: A typical STP producing wastewater sludge

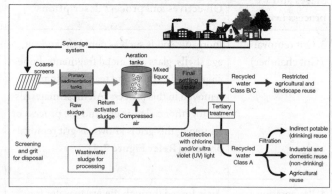

Table 7.1: Treatment of wastewater

Stages and process involved	Objectives and process descriptions	Advantages
A. Preliminary treatment	It is an operation performed before primary treatment.	▪ Protects Pumps and mechanical equipment from clogging and thus the efficiency of the water treatment plant. ▪ Removes large suspended solids like gravels, sand, paper, metals, glass plastics, etc.
1. Screening	The screens that have openings of uniform size remove large, coarse solids like dead animals, tree branches, food wastes, kitchen refuse, hair, paper and fibres, rags and faecal matters. Bar screens are a grating of steel bars spaced about 2–4 cm apart. The waste water is then passed into a basin so that sand and gravel settle down rapidly. **(Refer Figure 7.4)**	▪ Prevents clogging of valves and other appurtenances in wastewater treatment plants.
2. Comminuting/ grinding	The floating solid is grinded or shredded (cut) into uniform size of 8mm to 1/4 inch by a shredding device called comminutor. This method helps to hasten the screening process. **(Refer Figure 7.6)**	▪ Supplements screening by reducing the size of the particles. ▪ Prevents clogging.

Stages and process involved	Objectives and process descriptions	Advantages
3. Grit removal (Grit chamber)	Inorganic solids such as pebbles, sand silt, egg shells, glass and metal fragments, bone chips, seeds, coffee and tea grounds are removed in this process. All these materials are together called as grit and the process of removing this grit is known as grit removal process. (Refer Figure 7.5)	▪ Helps to speed up the screening process. ▪ Inorganic solid load in the wastewater is highly reduced. ▪ Removal of sand silts benefits the fish , i.e. it prevents the clogging of fish gills.
4. Pre-aeration tanks	Air is forced through the waste water for about half an hour.	▪ Enhances the efficiency in removing suspended solids and grease in the subsequent steps.
5. Pre-chlorination	Wastewater is mixed with chlorine.	▪ Prevents further decomposition of wastewater. ▪ Controls odour. ▪ Helps in sedimentation.
B. Primary treatment	The first step in the wastewater treatment process with the incoming wastewater.	▪ Removes settle-able inorganic and organic solids.
1. Skimming	By this process removing of grease, oil and other floating fat bodies are carried out.	▪ Removes up to 65 per cent of oil and grease. ▪ Helps in sunlight penetration on the treated water.
2 . Sedimentation	It removes large settle-able solids by settling it at the bottom of large tanks by the force of gravity. Organic matter slightly heavier than water settles down slowly by gravity. This settle-able matter is known as sludge. The sludge solids are pumped away. A longer period results in depletion of oxygen and anaerobic conditions. The sedimentation tanks may function intermittently or continuously. Settling of solids depend on the velocity of flow, the viscosity of water and the dimension and specific gravity of the waste particle. (Refer Figure 7.7)	▪ Reduces organic load. ▪ High removal of sediments/ sludge. ▪ Removal ranges from 50–65 per cent. ▪ BOD reduces by 30–40 per cent. ▪ Increases the efficiency of the disinfecting process.

Stages and process involved	Objectives and process descriptions	Advantages
3. Coagulation and chemical sedimentation	A process by which the finer suspended and colloidal particles (0.0001 mm) are converted into masses of large bulk by the action of coagulants like alum and ferric sulphate is known as coagulation. The coagulants such as aluminium sulphate (alum), ferrous sulphate (copperas), ferric sulphate and lime, all help in aggregation or flocculation of wastewater (sewage) particles, make them heavier to settle down fast under the force of gravity. Since the chemicals like alum induce this process of sedimentation, it is called as chemical sedimentation. $Al_2(SO4)_3.18H_2O + 3Ca(HCO_3)_2 = 2Al(OH)_3 \text{ (floc)} + 3CaSO_4 + CO_2 + 18H_2O$	■ Finer and colloidal particles are removed rapidly. ■ Turbidity is less and the water colour becomes clear. ■ Bacterial load is reduced by 5 per cent.
4. Filtration a. Rapid sand filter	Sand size of 0.35–6 mm is used as a strainer or filter medium to remove colloidal and suspended matter remaining after sedimentation. The pathogens attach themselves to the surface area of the sand particles and feed on the organic matter present in the wastewater/effluent. The filtered water is collected and then sent for disinfection.	■ Suitable for large cities. ■ Filtration is faster, the filtrate is to be sent for coagulation.
b. Slow sand filter	Though the filtration process is similar to rapid sand filter process, the sand size is smaller (0.2–0.3 mm) and the filtration process is very slow. This process is highly efficient in reducing the pathogen load.	■ Suitable for small towns. ■ It reduces bacterial load by about 98 per cent.

Note: Sludge digesters have a temperature between 30–35 °C where the sludge is pumped in. It might take 20–30 days or even more in during winter. Only well-digested sludge should be withdrawn and continuous addition of raw sludge is necessary. Dried sludge is then placed on drying beds to drain out the liquid or to evaporate it. The dried cake can then be used as fertilizer base.

Stages and process involved	Objectives and process descriptions	Advantages
C. Secondary treatment	The process involving the removal of organic matter and BOD in presence of air. Such treatment often involves a biological process. Water after secondary treatment will still possess nitrogen, phosphorus, heavy metals and microbes.	
1. Trickling filter	It is also known as percolating or sprinkling filters; it is a rigid bed biological filter operating in absence of oxygen. Wastewater is allowed to sprinkle or spray through the nozzles (sprayers) over a bed of coarse crushed stones or pebbles or gravel or rock wool of filter media, covered by biomass. The water is acted upon by the microorganisms and then collected by the under drainage system. A material with surface area between 30–900 m² is desirable. The filter has a depth of 1–3 m; sometime it can be up to 12 m. The particle preferably should be of uniform size; almost 95 per cent of the particles should vary between 7–10 cm sizes. The stones are coated with zoogloea. A bacterial film or bio-film layer is created around the filter media; air is circulated by convection and the aerobic bacteria carry out the oxidation of organic matter releasing carbon dioxide and water. Biological act mostly happens in the uppermost 50 cm of the filter bed. When the slime layer becomes very thick it sloughs off the rocks. A secondary basin or secondary settling tank is required to clarify the effluent by chlorination for a 30 minutes contact period. **(Refer Figure 7.9)**	▪ Highly decreases the organic load. ▪ BOD is reduced by 80–90 per cent. ▪ Colloidal matter is reduced by 70 per cent.
2. Activated sludge process	Activated sludge process is carried out in a multi-chamber reactor unit. The operation is similar to that of trickling filter. It uses bar screens, comminutors, grit chambers, primary and secondary settling tanks. It requires a good supply of oxygen to maintain a suspension of active mass and makes use of aerobic bacteria. The sludge, which settles down after the wastewater has been freely aerated (supplied with air) and agitated for a certain period of time in the multi chamber reactor. **Aeration** in aeration tank removes the dissolved gases and hence offensive odour is reduced in the wastewater. This activated sludge, which is rich nutrients and bacteria, performs the following function when mixed with the raw wastewater; 1. Oxidizes (decomposes) the organic solids. 2. Promotes coagulation and converts the colloidal and suspended solids into settle-able solids resulting in sedimentation.	▪ Give clear, sparkling treated water. ▪ Treated water is free from bad odour. ▪ Sludge obtained is used as fertilizers. To meet specific effluent goals of BOD, nitrogen and phosphorus, the design of the activated sludge is to be modified.

Stages and process involved	Objectives and process descriptions	Advantages
	Activated sludge process consists of the following stages. A. **Primary settling tank** where raw wastewater is treated with activated sludge and then sent to aeration tank. B. **Aeration tank** where the wastewater mixed with activated sludge is aerated for 4–10 hrs and a biological **floc** is formed. Bad odour is removed. This provides not only oxygen but turbulence too so that the waste might come in bacterial contact. The microbe oxidizes organic carbon in water and generates CO_2, water and new cells. Apart from aerobic bacteria, anaerobes and nitrifying bacteria are also present. The floc is transferred to the clarifier to settle down. C. **Secondary or final settling tank**, where the colloidal and suspended solids settles down to form activated sludge. Supernatant is let out for subsequent treatment and some amount of sludge is returned back into the aeration tank.	
	D. The sludge obtained is highly rich in nutrients and after being treated in sludge digestion tanks (Gives methane gas of high calorific value) it is used as manure or returned to the aeration tank to reseed with wastewater again. (**Refer Figure 7.10**)	
3. Oxidation ponds	Large and shallow with a depth ranging between 1–2.5 m, used singly, parallel or in series. The ponds comprises microorganisms that devour on the organic matter received from primary effluent. Algae are crucial in the oxidation pond system. Algae convey a steady flow of oxygen to the bacteria.	▪ Input of energy is small. ▪ Nitrogen and phosphorus degraded. ▪ Can be operated easily. ▪ Aerobic condition maintained easily as the products formed encourage algal growth as well as of other vegetation.
4. Rotating biological contractor system (RBC)	Use of bar screens, comminutors, grit chambers, primary and secondary settling tanks in a similar manner as that of trickling filter. It also makes use of bar screens, comminutors, grit chambers, primary settling tanks, secondary settling tanks, digesters that operate in the same way as in trickling filters.	▪ Simple and effective method of secondary wastewater treatment.

Stages and process involved	Objectives and process descriptions	Advantages
	It comprises biomass-coated plastic partly kept in the wastewater. With rotation, it raises wastewater film into the air with a rotation speed of 1–2 RPM. The wastewater is acted upon by the living biomass. Sloughing of the surplus biomass takes place with the gradual rotation of the medium. (**Refer Figure 7.11**)	▪ Prevents clogging of media. ▪ High degree of organic removal. ▪ Maintains a stable microbe population.
5. Imhoff tank	It also operates in the same manner as the trickling filters. A combined sedimentation and digestion tank; the upper compartment is meant for settling of solids and the lower compartment for anaerobic digestion. The upper compartment forms a channel with a 20 cm slot at the underneath. Accumulated solid are removed periodically through the draw off pipe located approximately. 30 cm above the tank bottom. (**Refer Figure 7.12**)	▪ The sides are provided with overlapping, horizontal and vertical slopes to prevent the escape of gas formed in the digester.

Figure 7.4: Bar screen

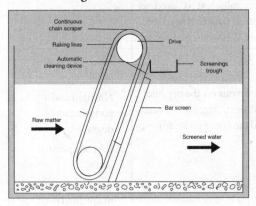

Figure 7.5: Grit chamber

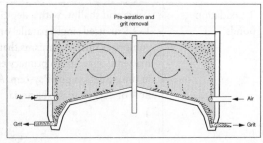

Figure 7.6: Comminutor

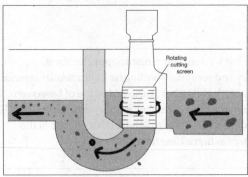

Figure 7.7: Primary settling tank

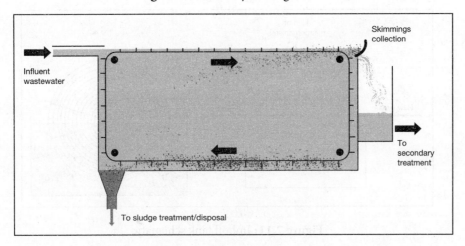

Figure 7.8: Trickling filter

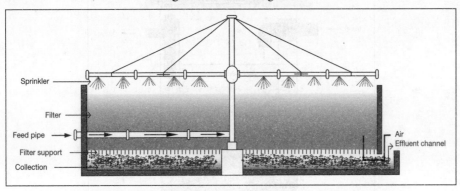

Figure 7.9: Activated sludge system

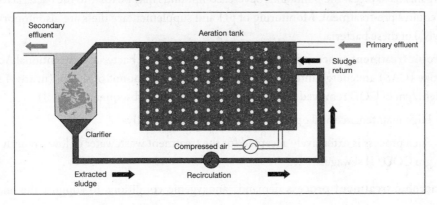

Figure 7.10: Rotating biological contactor (RBC)

Figure 7.11: Imhoff tank schematic

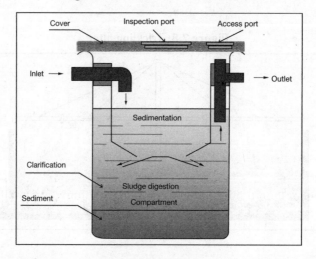

Biological treatment process should be carefully managed as they involve the use of living microorganisms to digest the pollutants. Several compounds may be toxic to the bacteria itself and hence requires pre-treatment. Monitoring of pH and supplementary diets are also important for the survival of these bacteria.

A. **Aerobic treatment process** demands oxygen by the aerobic bacteria. Conventional Activated Sludge (CAS) are energy-intensive and produce soaring amount of sludge (nearly 4 mg dry weight/gm of COD removed which needs treatment and subsequent disposal).

  ▪ High maintenance and operation cost occur subsequently.

  ▪ Such process is extensively accepted for the treatment wastewater of low strength (< 100 gm COD/ l) sewage.

B. **Anaerobic treatment process** demands anoxygenic conditions to support the anaerobic bacteria.

- Anaerobic process is more cost-effective especially in the tropics and sub-tropics with warm climatic conditions during the year.

- They are effective in removing high organic load and pathogen.

- The process involves four phases: hydrolysis, acidogenesis, acetogenesis and methanogenesis.

- The microbes may use a substrate which is formed as a product by another group of microbes along with the production of methane and carbon dioxide from the organic matter.

- Sludge production is low with methane gas production and energy consumption is less.

- Sewage treatment plant in anaerobic systems has restricted relevance. The reason is that municipal sewage is too weak to hold higher content biomass in the reactor.

- The operations in anerobic digestor usually take place in mesophilic condition nearly, 35°C.

- Nevertheless, in most of the cases, the wastewater is subjected to 18°C.

- In many a case there is heat application to wastewater before other treatment. About 30 per cent of the energy produced may be consumed in this process; reduction in expenditure of sewage treatment and minimization of the excessive sludge production is the chief intention.

- Examples include Upflow Anaerobic Sludge Blanket (UASB), the Anaerobic Sequencing Batch Reactor (AnSBR) and the anaerobic filter (AN) that can successfully treat high-strength industrial wastewater.

Box 7.2: The UASB

- It is usually a single tank process that evolved from anaerobic clarigester.

- Wastewater is made to stream upward through the blanket, when it is broken down by the anaerobic microbes.

- A blanket of regular sludge is formed which remains in suspension. Suspension of blanket with the assistance of flocculants is possible due to the upstream flow that is coupled with settling action of gravity.

- No support matrix; the microorganisms are attached to each other, capable of surviving and multiplying in this formed selective environment. They amass forming dense granules.

Figure 7.12: Schematic representation of UASB

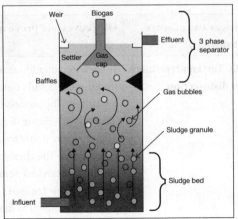

- The granules are 0.5 to 2 mm in diameter in the blanket that helps in degradation of organic wastes. Large granules of sludge are formed within few weeks that in turn acts as a filter for smaller particles as the effluent water tries to rise up through the sludge cushion.

- A 0.6 to 0.9 m/hour upflow velocity should be maintained in order to maintain the blanket suspended.

- Blanket may reach maturity within three months time.

- Biogas produced contain high amount of methane and carbon dioxide that can be trapped for generation of electricity.

Advantages of UASB:

- Reduction in organic waste is high.

- Able to endure high levels of hydraulic loading rates as well as high organic loading rates (up to 10 kg BOD/m^3/day).

- Low production of sludge.

- Biogas formed can be used in production of energy.

Disadvantages of UASB:

- The upflow and settling rates should be balanced properly which may otherwise be difficult to maintain proper hydraulic conditions.

- Time to start up is more.

- Treatment often unstable with variable hydraulic and organic loads.

- A continuous supply of electricity is necessary.

- The parts and materials are not available locally.

- Requires expert designing and supervision during construction.

Stages and process involved	Objectives and process descriptions	Advantages
D. Tertiary treatment or disinfection	Wastewater after filtration and activated sludge process still contains bacterial impurities. The process by which the harmful bacteria are destroyed or killed in order to make it safe for drinking is called disinfection. The chemicals (for example, chlorine, potassium permanganate bleaching powder, lime) used in this process are known as disinfectants.	

Stages and process involved	Objectives and process descriptions	Advantages
1. Chlorination	The process of applying chlorine to water to kill the pathogenic germs is called chlorination. Chlorination is done by a chlorinator and can be added in for gas, liquid or solid. N.B. The exact dose of chlorine for chlorination is very important, for if chlorine is less the bacteria don't die and if the chlorine dosage is high, it is highly toxic and adds bitter taste to the water.	▪ The most commonly followed method by any public water distribution system. ▪ Easy to operate and economically cheap method.
2. UV ray treatment	Ultraviolet rays (1000 to 4000 µm) are powerful disinfectants. They can be generated by passing electricity through mercury enclosed in quartz bulbs. When the treated water is allowed to pass through the UV rays given by the mercury vapour lamps, it penetrates the microbial cell wall and damages their genetic material thereby killing the pathogens. The process is physical and leaves no trace of chemical. For example, Aquaguard machines used in homes use this method of disinfection.	▪ Though expensive, it is highly efficient in killing the germs.
3. Ozone gas treatment	Ozone gas, produced from high voltage current trough airstram in a closed chamber, is also used as disinfectant in destroying the disease-causing virus and bacteria. It quickly breaks down to oxygen and nascent oxygen. This nascent oxygen oxidizes the microbes in water. It is uneconomical due to high energy costs.	▪ Efficient in destroying pathogens.
4. Reverse osmosis	Reverse osmosis (RO) is a technology using a semi-permeable membrane. A force is applied in order to overcome osmotic pressure, which is compelled by chemical potential.	▪ The solutes are retained on the pressurized side of the membrane allowing the pure solvent to pass through.

7.2.6 Removal of nitrogen and phosphorous

A. Removal of nitrogen

- Nitrogenous compound may result from the proteins and urea of the body.
- Hence, nitrogen may be present in multiple forms that are not removed by the secondary treatment.

- The organic nitrogen in water gets converted into ammonia and ammonium ions exerting oxygen demand or promoting profuse algal growth.

- Ammonia is oxidized to nitrite and then to nitrate in presence of molecular oxygen by the nitrifying bacteria present in the wastewater plant.

- The process is sufficient enough to convert ammonia into non-toxic nitrates or else air stripping method may be followed.

- If nitrates must be entirely removed an additional biological process can be extended to the conversion of nitrate to nitrogen through de-nitritification. The effluent is placed in an anaerobic tank with addition of carbon containing chemicals such as methanol. Under such environmental conditions the bacteria reduces nitrate to release molecular nitrogen.

B. Removal of phosphorous

- Phosphorus also stimulates the growth of algae. So reduction of phosphorus is needed to prevent the formation of algal bloom in the water.

- Polyphosphates comprise the bulk of synthetic detergents used in a wide variety of purpose. Polyphosphates are hydrolyzed into orthophosphates.

- Phosphorus may be removed by chemical precipitation where they combine with trivalent aluminium or iron in an acidic pH.

- Biological removal of phosphorus can be backed up with chemical precipitation using either lime or alum. This latter option would change the quantity and nature of the produced sludge requiring final disposal. Quantities can increase substantially and the sludge may have a high content of lime or aluminum depending on the coagulant used.

Biological processes like biological nutrient removal (BNR) can be employed for removal of these nutrients.

7.2.7 Impacts of wastewater

A. Impending benefits: Wastewater may prove to be a valuable resource

- Wastewater proves to be a consistent source of water for agricultural production.

- It minimizes the requirement of any synthetic fertilizers and thus help in conserving nutrients.

- Returns from agriculture are good as crop production is high.

- It acts as a supplementary income source by its use in aquaculture.

- A low-priced method for the sanitary disposal of municipal wastewater.

B. Impending costs: Harmful impacts of wastewater

- Its employment in agriculture could increase its exposure to the farmers and local communities to contagious diseases.

- May percolate and lead to groundwater contamination.

- Long-term usage of wastewater can have negative impact on soil resources such as build-up of salts and accumulation of heavy metals; this may reduce soil productivity in the long run.

■ Have negative impact on property values nearby.

■ Have other negative impact on socio-ecological systems.

7.3 Solid Waste Management

Solid waste refers to all types of garbage, refuse or sludge from water treatment or water supply plant as a result of household, industrial, commercial, mining or agricultural activities. In solid wastes, garbage refers to putrecible or rapidly decomposing waste such as waste from food, slaughter houses, etc. Rubbish applies to non-putrecible waste both combustible and non-combustible. Combustible rubbish comprises wood paper rubber leather while non-combustible wastes are the glass, metals, ceramics, stones, etc. Trash, on the other hand, is dry waste such as papers, plastics, glass, wood, etc.

Solid waste management refers to collection, transport, processing, recycling, disposal and monitoring of solid wastes.

7.3.1 Types of solid waste (SW)

On the basis of its source, SW can be grouped into three categories:-

A. Municipal solid waste refers to the household wastes, construction debris, sanitation residue, and street wastes. Household wastes include paints, solvents, pesticides, electronics, aerosols, refrigerants, batteries, radioactive wastes etc. MSWs are governed by Municipal Solid Waste Rules, 2000.

B. Industrial or hazardous wastes are mainly the residues from various industries and the left over raw materials. Their management is governed by Hazardous Wastes Rules, 1989.

C. Biomedical waste or hospital wastes comprise the sharps, anatomical wastes, medicines, excreta, blood syringes, swabs, bandages, catheters, gowns, masks, pipettes, slides, culture media etc. These wastes are governed by Biomedical Waste Rules, 1998.

Bio-medical waste means any solid and/or liquid waste including its container and any intermediate product, which is generated during the diagnosis, treatment or immunization of human beings or animals or in research pertaining thereto or in the production or testing thereof.

7.3.2 Cause of increase in solid waste

Production of solid waste commences with the retrieval of raw materials at each step as it progress towards the transformation of product for the purpose of consumption.

The amount of total waste generated over four billion tonnes per year (Veolia, 2009) within which nearly 45 per cent believed to be MSW and the remaining as industrial waste including hazardous one. The generation of urban food waste has risen by 44 per cent from 2005 to 2025.

According to Wilson *et al.,* (2012), the per capita waste generation increases with the level of development and level of income. It is a fact that increased international trade has helped in reducing poverty, raised living standards and augmented the purchasing power of people in developing countries, although no one can deny the footprint it leaves on the waste management. Pant, 2012 mentioned that the distant, inaccessible and insignificant territories have gained immense importance on account of globalization.

Besides offering precious minerals, energy, timber, etc., these places have become prospective dumping sites which nobody wants in their backyard. Globalization facilitates to extend the existing technology, helps in the emergence of newer techniques that have more comprehensive resource extraction options to spur green techniques. Globalization shrinks both time and space and puts emphasis on the current style of global control. Megacities are products of incessant urbanization; usually metropolitan areas having more than 10 million people. Megacities may comprise a single, two or more metropolitan areas. These cities exhibit rapid growth, newer forms of population density in space, formal and informal economic conditions, paucity, and crime with soaring societal fragmentation and **enormous amount of waste generation**. The number of megacities increased from two in 1950, four in 1975 to 21 in 2003. By 2015, there will be 33 megacities of which 27 will be in the developing nations. 283 million people were living in the megacities in 2003 with 207 in the developing countries. The number is likely to become 359 million by 2015. According to Wilson *et al.* (2012), the composition of waste varies from high organic, thick with more moisture content with low calorific values in medium to low income cities to light with lower organic content in the high income nations.

Municipal solid waste management comprises collection, transportation and disposal methods. Waste is generated during both technological and consumptive processes.

Table 7.2: Source and main components of SW mainly from urban areas

Sources	Producers	Composition
Domestic	Residences and accommodations.	Kitchen residues, paper and cartons, plastics and rubbers, leather, glass, garbage, ashes and metals.
Industrial	Construction site wastes, demolition debris, food processing industries, slaughter houses, breweries, tanneries, carpet industries and garments factories, chemical plants and tourist facilities.	Earth, stones, sand and wood, packaging materials, food wastes, fibres, bones, feathers and body residues, hazardous wastes and old machine parts.
Commercial	Stores, tea stalls, business premises, stores, restaurants, markets, fruit vendors, hotels and motor repair shops.	Paper and cartons, glass, waste from food preparation, hair, ashes, spoiled, discarded spare parts and organic wastes.
Agricultural	Dairies, poultry farms, livestock and other agricultural wastes including vegetable and fruit cultivation.	Bio-degradable components.

| Institutional | Hospitals, religious places, schools, banks, offices. | Paper and cartons, food wastes, glass, plastic wrappers, flowers and leaves, papers, pens and other stationeries, hazardous and pathological wastes |
| Natural | Trees, plants and animals. | Leaves, tree branches, seeds and animal carcasses. |

7.3.3 Environmental impacts of solid waste disposal

- Groundwater contamination by leaching.
- Offensive odour and outbreak of vectors, carriers, pests, rodents etc.
- Bird crowding at dumping sites may be a threat to the aircraft.
- Erosion problems in places with slopes.
- Change of pH, like acidity.
- Possibility of fires.
- Production of inflammable gases like methane.
- Surface water contamination by the run–off.
- Epidemics through stray animals.
- Greenhouse gas emission.

7.3.4 Objectives of solid waste management

Solid waste of late has mounted to an environmental threat and a means of losing energy and precious resources that cannot be recovered or reused. The very purpose of solid waste management (SWM) is to trim down the disposal of solid waste along with salvaging material and energy from solid waste.

Global recycling markets contribute to huge quantities of recovery of materials. At the same time there is increase in waste trafficking activities. There is an urgent need to fight waste trafficking in order to establish a sound global system of waste management. Of all wastes, plastic trade is increasing worldwide. Over 50 per cent of plastics, paper and scrap metals are send to Southeast Asia in the name of export and the tendency is set to boost up. In 2010, over 7.4 million tons of plastic wastes, 28 million tons of paper wastes and 5.8 million tons of steel scrap were imported by China. 20 per cent was shipped for disposal and the remaining 80 per cent was shipped for recovery.

7.3.5 Principles of solid waste management

This comprises solicitation of the norms of **Integrated Solid Waste Management (ISWM)** in order to accomplish the identical objectives of waste minimization, operative and effectual waste management that is still there even after reduction of waste. The latent energy in the organic fraction of the waste can be recovered for gainful utilization by adopting suitable waste treatment methods. The quantity of waste reduction can vary from 60 to 90 per cent. Waste management has become an exceedingly important requisite for healthy living with increase in demand of land with rising population, increasing cost of the transportation of wastes and an overall decrease in environmental pollution. ISWM can be defined as the strategic approach towards sustainable solid waste management covering all the sources, generation, sorting, transport, treatment operations, recovery and disposal in such a manner so as to emphasize the maximum efficiency of resource use. It helps in clean surroundings, minimizing waste management costs, good business opportunities, economic growth and participation of local people.

7.3.6 Effective management of solid waste

Effectual solid waste management systems are desired to safeguard improved healthiness and well-being of mankind and to prevent the spread of diseases. Additionally, solid waste management is ought to t be both ecologically and financially supportable.

- **Environmentally sustainable:** Waste management should be able to reduce the environmental effects of waste generation and disposal.

- **Economically sustainable:** Waste management should operate at a cost that the society will accept.

A balance is needed to trim down the general environmental impact of waste within a tolerable level of cost. An effective solid waste management system comprises the following options:

- Collection of waste followed by its transport.

- Resource recovery through categorization and recycling such as paper, glass, metals through various methods of segregation.

- Resource recovery by compost processing or energy recovery through other biological, thermal means.

- Waste makeover or reduction of volume and toxicity by physical and chemical properties.

- Ecologically harmless along with sustainable disposal in the landfills.

Fundamental elements of MSW management comprise the following steps:

A. **Waste generation:** These are the materials that are identified to be no longer valuable.

B. **Waste handling, categorization, storage, and processing at the source:** This involves the actions related with management when the wastes are first kept in storage containers for the purpose of collection. Waste handling also comprises the transit of truckload to the collection venue. Categorization of the waste ingredients is the most important step towards handling and stockpiling of waste at source. In fact, the best place to segregate or sort out waste in order to reuse and recycle is the point of waste generation. Households are becoming more conscious about segregating newspapers, cardboards, plastics, bottles, glass, and organic kitchen residues, ferrous and non ferrous metals. Recycling of solid waste is primarily carried out by the informal sector in the low and middle income based developing nations; they also offer basic collection services (Medina, 2008). The work may comprise the associations structured into co-operatives doing door to door waste collection that are segregated at source or mixed wastes to the people individually rummaging the dumping places, transit stations and societal bins (Wilson, 2006).

The sorted waste can be:

- dried up recyclables such as glass, chapter, plastics, cans etc;

- biological and garden waste;

- huge and bulky waste;

- hazardous substances in domestic waste; and
- structural and demolition waste.

Household waste reach curbside collection bins (dhalaos) also through the rag pickers. Often waste processing at source comprises backyard composting. People who work in informal sectors have a precarious and difficult lifestyle. Contradictory enough, the contribution of these people helps in creating employment opportunities for the poor and impoverished sections of the economy. Presently, it employs around 12 million people in USA, China and Brazil even though the total number of such people is evaluated to be over 20 millions.

C. **Collection and transportation:** The step involves assembling the solid residue and recyclables in the collection vehicles to the site to empty it. The emptying site can be a dispensation centre, a landfill dumping site or a transfer spot.

D. **Segregation, dispensing and transformation:** Segregation of wastes is generally done in the recovery sites, transfer stations and dumping places. Waste can be separated by size, manual separation, and separation of ferrous from non–ferrous metals. Organic content can be transformed biologically or thermally; often aerobic composting and incineration are used as biological and thermal control methods respectively. Transformation of waste is carried out to reduce the volume, mass, dimension and range of toxicity either with or without resource recovery through thermal, biological or a combination of methods.

E. **Transfer and transport:** This comprises the relocation of waste materials from the smaller collection vehicles to the large transport system. They are then subsequently transported over longer distances to the place of disposal or processing.

F. **Disposal:** The scientific disposal of garbage can follow various technologies like incineration, landfilling, pyrolysis, composting, biomethanation, pelletization, etc. Landfill is the most widely practiced and cost-effective way of waste disposal, where collection and transportation claims most of the expense. Incineration and pyrolysis are both thermal methods which help in reducing the volume of the waste. These methods can be employed for treating large quantities of wastes in small spaces. Waste management with energy recovery has emerged as an efficient technique and so pelletization is gaining prominence. As MSW generates poor quality fuel, pretreatment seems necessary to improve its consistency, combustion and calorific value. RDF is being prepared with a blend of pre-processed MSW and coal appropriate to be used in pulverized coal and fluidized bed boilers. The process comprises sorting, crushing and blending high and low heat value organic refuse and the solidifying it to produce pellets or briquettes. Enrichment of organic content is achieved through removal of inorganic materials like metals, glass, dehydrating and densification. These pellets can then be used to generate heat or produce electricity. 100 tons of raw waste can yield 15–20 tons of pellets. Their calorific value may be around 4000 kcal/kg as compared to that of MSW of 1000 kcal/kg.

7.3.7 Methodologies in waste management

Box 7.3: Methodologies in waste management

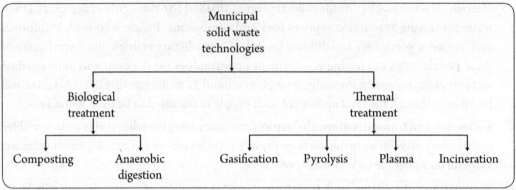

i. Thermal methods

a. Incineration:

Incineration is a thermal process where organic substances of the waste are directly combusted at elevated temperatures, usually above 800°C, in oxygen-rich environment along with the production of ash, flue gas (CO_2 and H_2O) and heat. Combustion temperature of conventional incinerators can be 760°C in the furnace and over 870°C in the secondary gas chamber. Such high temperatures prevent any bad odour from incomplete combustion.

Net energy yield depends upon the preconditions needed for complete combustion and includes density and composition of waste, its size and shape, moisture content and inert materials, design of the combustion system, attainable ignition temperature, adequate mixture proportion, constant removal of gases and combustion residues, turbulent gas flow, maintaining suitable temperature, adequate fuel material and means of oxidation, etc. Addition of road sweepings in the waste bins may affect the efficiency of incineration.

The basic types of incinerator includes mass burn, modular combustion units, refuse drive fuel (RDF)-based power plants.

An incinerator plant comprises of:

- a weighing system for controlling and recording incoming loads of wastes;
- an existing temporary storage site or waste reception site to ensure little unloading time, achieving homogeneity of wastes and smooth feeding of waste;
- feeding system adjusted to the rate of feeding velocity;
- combustion hearths with burners operating with secondary fuel to render minimum desirable temperature, appropriate combustion time and turbulence condition;
- boiler system in which the energy of the fuel material can be used for production of heat, steam, etc;
- a structure for residue removal; and
- emission control system – per ton of waste, a quantity of 4,000–5,000 m³ of air emissions are produced.

The original bulk of wastes can be reduced by approximately 90 per cent of its volume and 75 per cent of its weight. The recovered energy can be utilized for heating, production of steam or electricity production. For instance incineration of nearly 1,200 metric tons of MSW daily will approximately generate 800 megawatt-hour (MWh) of electrical energy per day. The advantage of incineration is that it can be used for the treatment of hazardous hospital wastes and unsafe waste substances where both microorganisms and toxic chemicals will be degraded at such elevated temperatures.

Ash mostly comprises the inorganic constituents and may assume the shape of solid lumps; flue gas comprise considerable amount of carbon dioxide, water, particulates, heavy metals, Polychlorinated Dibenzodioxins (PCDDs), Polychlorinated Dibenzofurans (PCDFs), SO_2, methane (CH_4) and hydrogen chlorides. The sum total of ash produced by MSW incineration varies from 15–20 per cent by weight of the original quantity of waste and the fly ash produced is about 10–20 per cent of the total ash.

Heavy metals like lead, zinc, cadmium, copper with small amount of organic compounds like dioxin and furans are present in the fly ash.

Odour pollution can be an additional problem but can be met with high temperature.

Advantages:

- Appropriate for wastes of high heat value and germ containing wastes.
- Low requirement of land.
- Relatively free from noise and odour.
- Can be located within the city.

Disadvantages:

- Least suitable for aqueous or high moisture content or low calorific value or chlorinated wastes.
- Emission of particulates, SOx, NOx, HCl and dioxins.
- Skilled personnel required.
- High moisture content affects the net energy recovery.
- High capital costs.

Figure 7.13: Incineration

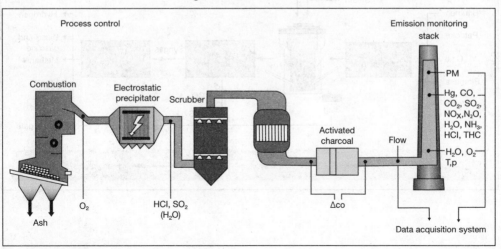

b. Gasification

Gasification refers to a thermal process that converts coke, biomass, sludge, and household waste to syngas under partial oxidation at temperatures higher than 1000°C. This can then be used to produce power, chemicals, fertilizers and hydrogen, steam and transportation fuels. Syngas primarily comprises combustible and non-combustible mixture of hydrogen and carbon monoxide and carbon dioxide.

Advantages:

■ Reduced emissions of sulphur and nitrogen oxides as a result of clean up operation of syngas.

■ Sulphur is transformed into hydrogen sulphides and nitrogen into diatomic nitrogen and ammonia; both of which are eliminated in downstream process.

■ No formation of dioxin and furan; any such things formed are destroyed at high temperature.

■ No formation of free chlorine due to lack of oxygen.

■ Resulting clean syngas can be used to generate steam or electricity.

■ The syngas obtained, after being cooled, can be used in Internal Combustion (IC) engines.

Disadvantages:

■ Includes use of large equipment and high capital investment.

■ Unsuitable for wastes with excessive moisture.

■ Transportation and burning of pyrolysis fuel problematic due to its high viscosity.

Figure 7.14a: Gasification

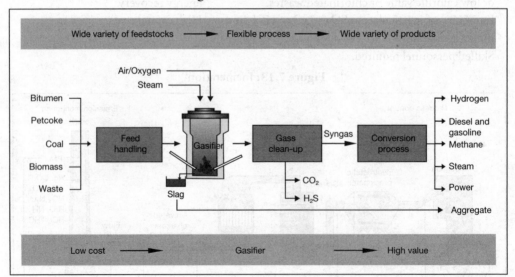

c. Pyrolysis

Pyrolysis is the thermal degradation of waste also known as destructive distillation or carbonization, involving a temperature of 450 – 750°C or even higher (900°C) in absence of oxygen or vacuum. This helps drive off the volatile organic substances and produce a syngas comprising H_2, CO, CO_2, CH_4, etc. The output is a mix of CO, methane, ethane, hydrogen, carbon dioxide, water, nitrogen, pyroligenous liquid and charcoal.

Advantages:

- Some of the ash obtained from pyrolysis may be used for making brick materials.
- A proportion of VOCs can be condensed to produce oil, wax and tar.
- The temperature for breakdown is lower than incineration temperature.
- Less air emissions due to decomposition taking place in reducing atmosphere.
- Ash content is much higher than that of incineration.
- Metals do not get oxidized and hence have high commercial value.
- No ash is produced from syngas and hence cleanup operation is simple with high initial reduction of waste volume as compared to usual pyrolysis.

Disadvantage:

- It requires pretreatment like cutting and separation of waste that increases the cost of installation and operation.
- Facilities for cleaning of gases and wastewater require high expenditure.
- The products cannot be disposed without further treatment.

Figure 7.14b: Gasification

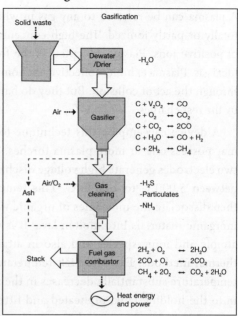

Figure 7.15: Pyrolysis

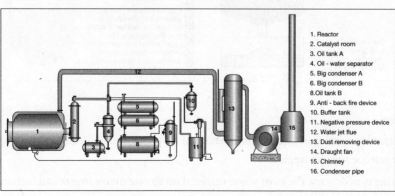

1. Reactor
2. Catalyst room
3. Oil tank A
4. Oil - water separator
5. Big condenser A
6. Big condenser B
8. Oil tank B
9. Anti - back fire device
10. Buffer tank
11. Negative pressure device
12. Water jet flue
13. Dust removing device
14. Draught fan
15. Chimney
16. Condenser pipe

d. Plasma arc gasifier

A plasma can be referred to any gas in which at least a fraction of molecules and atoms are totally or partly ionized. The high concentrations of free electrons are balanced by presence of positive ions. Plasma, accepted as the fourth state of matter, behaves differently from the ideal gas. Plasma exhibits collective phenomena and doesn't interact with surrounding particles through the act of collision. But they do have the influence of electromagnetic fields produced by the rest charges.

A device can employ this technique for effective waste management. A plasma reactor may possess one or more plasma torches. The torches under high voltage applied between two electrodes generate high voltage discharge with an exceedingly high temperature ranging between 5,000°C to 14,000°C. This temperature produces a hot plasma zone which can then dissociate the molecules of organic waste into individual elemental atoms and melt all inorganic materials into molten lava. It is very drastic due to the existence of high reactive atomic and ionic species and also in attaining high temperatures in comparison to other thermal methods. Extreme high temperatures exist in the core of the plasma whereas the temperature substantially decreases in the marginal areas. The wastes can directly be loaded in to the holding tank, preheated and fitted to a furnace. The output gas is mainly carbon monoxide and hydrogen with liquefied methanol.

Figure 7.16: Plasma arc Gasifier

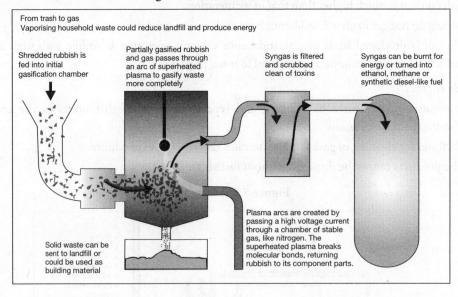

Advantages:

- Both hazardous and non-hazardous wastes can be processed. All categories of wastes except nuclear residue may be treated.

- No sorting is necessary. Packs of waste material up to one meter in size can be handled.

- No pretreatment is necessary.
- Unaffected by moisture.
- It is promising in isolated areas, for instance islands.
- It is characterized by low air emissions such as SO_2, dioxins, metals.

Disadvantages:

- To some extent, negative pressure is required by the operator. Hence, the gas has no tendency to escape. The gas has to be sucked out by compressor.

ii. Biological methods

a. Composting

Composting is a type of biological method of breakdown of organic waste such as leftover food, vegetable residue, manure, leaves and twigs, grass trimmings, paper, worms, tea leaves and coffee grounds, etc., into an enormously useful humus like material by the action of various microorganisms such as bacteria, fungi and actinomycetes in the presence of oxygen. Given enough time all these organic matters will decompose. Low temperature interferes with the progress of the process, as it lacks the required temperature to kill pathogens.

- **Backyard or onsite composting**:

  This method can be carried out by the dwellers on their own properties thus suitably used for transforming field trimmings and leftover food residue into compost. The process can be accomplished on site. Such method necessarily should not be applied for compost of animal residue or huge amount of food scraps. Residences, commercial houses, universities, schools, and hospitals can leave grass trimmings on the lawn – known as 'grass cycling' – such mass usually degrades in nature returning the nutrients back to soil. Inappropriate management of food remainders can cause bad odors and also might attract unwanted vectors and pests. So whenever the composting process requires space, NIMBY surfaces; i.e., Not In My Backyard.

- **Vermicomposting:**

  Worms can be used as a remediation of organic wastes thus regenerating the topsoil. Vermicomposting can be referred to as the process of using earthworm population to degrade organic waste into usable vermicompost or worm castings. Most of the 4,000 different species of earthworm have been identified. These are grouped into three categories on the basis of their physical and behavioural characteristics. The **anecic** worms build permanent, vertical burrows, which may be 2 cm in diameter and up to 3 meters deep. Their burrows are coated with mucus and they come up to the surface during night in search of decayed leaves. For example, nightcrawler (*Lumbricus terrestris*), *Apporectodea longa*. **Endogeic** worms build extensive horizontal burrows throughout all the layers and are geophagous. For example, *Allolobophora chlorotica*, *Apporectodea caliginosa*, *Pontoscolex corethrurus* and *Lampito mauritii*. **Epigeic** worms do not build permanent burrows and prefers to live on the topsoil. They are very active devouring on the organic matter and are commercially used in vermicomposting. For example, *Dendrobaena sp.*, *Lumbricus rubellus* and *Eisenia foetida*.

In this method, red worms or earth worms or field worms occurring in grounds are placed in containers with organic matter in order to be broken down into high–value compost called casts. Worms will consume almost anything put in a classic compost pile such as food scraps, paper, and plants. Vermicomposting can be idyllic for flat dwellers or small offices for deriving some of the advantages of composting and to minimize solid waste.

Worms are sensitive to climate fluctuations. Extreme temperature and straight solar radiations are not at all good for the earthworms. The optimal temperatures for such process ranges from 55°F to 77°F. Vermicomposting requires worms, worm bedding made up of frayed newspapers, cardboards, along with container where earthworms and organic matters are kept. The steps in the operation comprise preparation of bedding, entombing garbage, and separation of worms from their moldings.

■ **Aerated (turned) windrow composting**

In this method organic wastes are placed into extended heaping rows called 'windrows'. It is aerated by churning the pile intermittently either manually or mechanically. The perfect pile height is ranges from four to eight feet and a width of 14–16 feet allowing the pile to produce enough heat and sustain temperatures.

This method can lodge enormous quantity of wastes of various types together with plant and animal wastes. It requires large tracts of land, robust equipment, a labour supply to maintain and operate the facility, and endurance to test with various materials mixtures.

Figure 7.17: Windrow composting

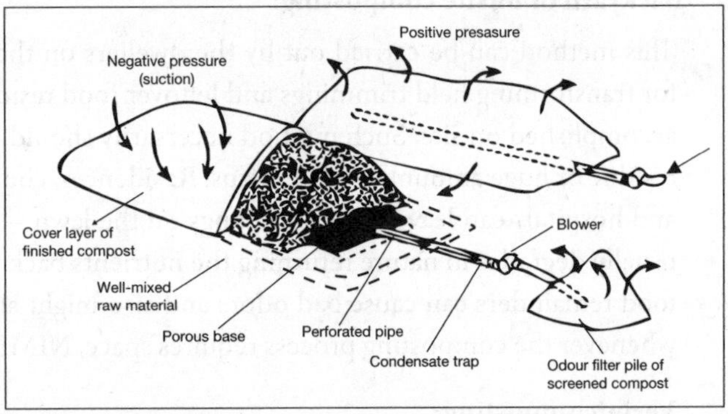

Leachate refers to the liquid released during the course of composting and thus it can pollute local ground–water and surface–water supply. Compost samples should be sent to the laboratory for verification for the presence of bacteria and heavy metal. Any release of odors should also be controlled.

■ **Aerated static pile composting**

Instead of making into rows, organic waste is churned collectively into one big mound. Layers of bulking materials (for example, wood chips, frayed newspaper) are mixed so as to allow the passage of air from the bottom to the crest of the mound. The mounds are placed over a network of pipes for the supply of air into the mixture or to draw out air out of the mixture. In some cases air blowers can be installed along with time control devices and temperature sensors.

Aerated static piles are suitable where the organic waste is homogeneous and works good for high amount of grass trimmings and compostable municipal waste. This method is not suitable for composting animal residues or grease from food processing industries.

Figure 7.18: Aerated static pile composting

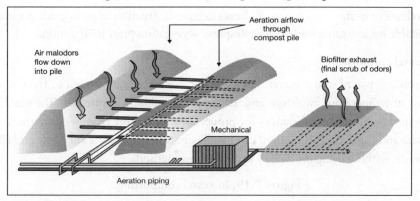

b. Anaerobic digestion:

This involves a process in which organic waste such as food residue, manure and sewage sludge are degraded in anaerobic condition. The result is generation of solid residue along with biogas; biogas is composed mainly of methane and carbon dioxide and serves as a source of energy. The process is also known as bio-methanation. The biogas production may range from 50–150 m/tonnes of wastes. Biogas is often used for making food and heating, in gas engines or steam turbines for the production of electricity. The sludge or residue can be composted and applied as a soil conditioner or as manure to the land. Thus it may help in renewable energy production, GHG gas emission and waste management.

The anaerobic digestion process is divided into three stages:

- Stage I comprises the use of fermentative bacteria (anaerobic and facultative microbes) to hydrolyze the complex carbohydrates, fats and proteins into fatty acids, alcohol, CO_2, H_2, NH_3 and sulphides.

- Stage II involves the acetogenic bacteria which devours these products to generate acetic acid (CH_3COOH), CO_2 and H_2.

- Stage III employs various types of methanogenic microbes; at first carbon dioxide is reduced to methane and in the later stage acetic acid is decarboxylated to methane and carbon dioxide.

- The factors determining the efficiency of the process are pH, temperature, loading rate, waste composition, toxic materials and mixing of waste. A good start requires a good inoculum of digested sludge and an optimum temperature of 35–40°C.

Advantages:

- Being an enclosed system there will be no escape of polluted gas in the air.

- Net production of energy.

- No GHGs.

- Since nitrogen is not lost by oxidation, the digested sludge is better with nitrogen content and a high grade soil conditioner.

- No bad odour.

- No power requirement.

Disadvantages:

- Less release of heat.
- Unsuitable for inorganic wastes.
- Less effective destruction of pathogenic organisms.
- Requires segregation prior to digestion.

c. In-vessel composting

Organic waste is placed into a barrel, storage tower, concrete-lined trough or trench, where the conditions of temperature, moisture and air supply are closely controlled. The machine has a device to turn or shake the substances for proper air supply. This type of composters can be of variable size and capacity. This method can handle enormous quantities of waste and all types of organic waste, without using much area as in windrow method.

Figure 7.19: In-vessel composting

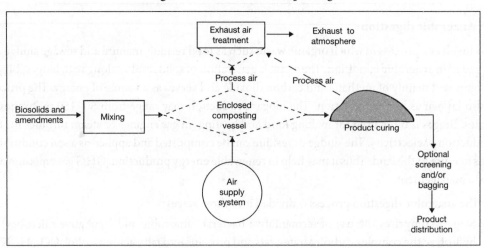

Case Study 7.1: Compost project at Nashik

The municipal corporation of Nashik receives nearly 230 MT (metric tonne) of solid wastes daily. The garbage is collected with the help of special vehicles called ghantagadi. The municipal corporation of Nashik has built of a compost plant expending ₹ 4.61 crores fully owned by them. The city waste comprises about 40 per cent of non-biodegradables which are sent for sanitary landfill located in the vicinity of the compost plant. The collected waste is transported to the compost site; the waste is mechanically separated and subjected to the process. Compost is processed by aerobic microbial composting method. The compost obtained is of fine quality and meets all the criteria beneficial to the neighbouring farmers, especially to the grape growers. The compost was sold at an affordable price of ₹ 1,700/MT since 2009 and later raised to ₹ 2,000/MT.

iii. Land filling methods

Placing waste in the Earth's crust is one of the safest practices of waste disposal. But this technique has changed with industrialization and urbanization. Uncontrolled landfill poses a severe environmental threat to mankind. To minimize such impacts regulations have been imposed in terms of site selection and operation. Environment Impact Assessment (EIA) is mandatory for siting of landfill.

a. Sanitary landfills refer to the scientific disposal of solid refuse in the form of layers, compressing it to reduce its volume and then emplacing with a layer of soil every day in a scientifically designed area. The waste is quarantined from the environment until it is rendered safe. The waste is broken down physically, chemically and biologically. The level of isolation accomplished is high in developed countries. More investment is required to improve standards but the unit cost of these improved measures decreases with increasing site size.

Prerequisites for sanitary landfill are described below:

- complete or fractional hydrogeological isolation to prevent contamination of land and ground water by the dumped wastes;

- official engineering plan, preparations and measures;

- adequately trained staff for supervision, monitoring and maintenance; and

- planned and strategic waste emplacement and covering to cut off the accessibility to pests and other animals.

Figure 7.20: Sanitary landfill

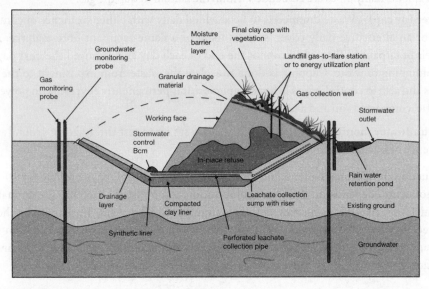

The main constituents of a typical fortified landfill are:

- **Bottom liner** is a layer made up of High Density Polyethylene (HDPE) that will separate and prevent the dumped waste from coming in touch with soil and groundwater lying beneath.

- **Cells** – refers to the approved area for waste dumping, the size of which may vary from few acres to more than 20 acres. The material is condensed and cut by compaction machinery here.

- **Leachate collection system** – The lowermost landfill surface is inclined to form a slope called sump in order to collect any leachate or trapped liquid. A sequence of pierced pipes, gravel packs and a sand layer is positioned in the lowermost region of the landfill.

■ **Storm water drainage** – This is an engineered scheme intended to regulate water overflow during rainfall or storm. Leading the water overflow through a sequence of conduits to seed ponds makes this.

■ **Methane collection system** – vertical or horizontal wells are used for the recovery of gas; the wells are drilled into the wastes where methane is generated. The most frequently used being the vertical wells that are drilled at the rate of 2 wells per acre. The wells are made up of HDPE of poly-vinyl chloride pipes. The diameter of these pipes varies from 50 to 300 mm. It is covered by a 300 mm thick envelope of 25–36 mm sized gravels. The depth of the gas well is generally at 80 per cent of the height of landfill. Wells are connected to the main collection header. The gas formed is drawn out by a blower operating under negative pressure. All the wells are not pumped together. The gas, when not used, is burnt out by a flare. Emission of landfill gas (LFG) can pose a threat to environment. Microorganisms present will decompose the wastes. With decomposition LFG is produced, which comprises 45 per cent of carbon dioxide and 55 per cent of methane (Bhide, 1998; EIIP, 1999). Its heat content is about half of the commercially marketed natural gas. LFG has a calorific value of 4500 kilocalories/m^3. Landfill system with gas recovery (LFSGR) is an inclusive approach towards managing of municipal solid waste, mitigating emission of methane and growth of nonconventional energy sources. Since methane can explode, many pipes are embedded within the landfill to trap the gas.

■ **Cover (or cap)** – Waste dumped is to be shielded daily with either six inches of compressed soil or an alternative daily cover, such as foam or a flame-retardant fibre material, or large panels of tarpaulin. This is laid over at the end of each day and removed the next day before the dumping of waste. Capping is done to isolate the waste from exposing it to the air and pests and also to control odors. After completion, it is permanently covered with polyethylene plastic, compacted soil and a layer of topsoil.

■ **Groundwater monitoring stations** – These are set up to test the adjacent groundwater for presence of leachate chemicals.

Production of LFG starts after some time of waste disposal and may continue for 10 years or more. Production depends on the waste composition and moisture content. The annual rate of gas production in major Indian cities may range from 15–25 l/kg of gas. The sites to be chosen for energy recovery by landfill should have more than 1 million tonnes of wastes. LFG can be used directly and locally, can be used to generate electricity through power grid or injected into a gas distribution grid.

Advantages:

■ Overall the cost is low. ■ Gas can be used for power or heat generation.

Disadvantages:

■ Site selection needs complete or fractional hydro-geological isolation. ■ Engineered construction. ■ Adequately trained staff for supervision, monitoring and maintenance. ■ Spontaneous ignition by means of gas flares. ■ Transportation cost involved. ■ Large land requirement. ■ Possibility of soil and ground water contamination. ■ Emission of GHGs. ■ Planned and strategic waste emplacement and covering to cut off the accessibility to pests and other animals.

7.4 Recycling of Paper

Scrap paper is frequently used for packaging. Prior to recycling, paper should be made free from metals, plastics, pins, food and other trash items. Contaminated paper unfit for recycling should be composted or burned for energy or be land filled. Different types of paper are assorted and taken to the recycling centre. The paper is wrapped into tight bales and transported to a paper mill.

The paper bales are stored in warehouses until required. The paper is shifted from the warehouses to a big vat called pulper through large conveyors, where water and chemicals are present and the paper is shredded into pieces. Heating breaks the paper fast into small strands of cellulose called fibres. Ultimately it turns into a sappy pulp. The pulp is forced through screens of various mesh sizes to remove contaminants like plastic and glue globs. Pulp is also cleaned by spinning it in large cone-shaped cylinders. Heavy contaminants like staples are thrown to the outside while lighter ones collect at the centre of the cone. The pulp then undergoes deinking to remove ink and stikies. Small ink particles are removed by washing with water while larger ones are removed with the help of air bubbles by floatation. The pulp in the floatation cell is mixed with air and surfactants. The surfactants loosen the ink and stickies and get stuck to the air bubbles and they float the top to form a froth leaving the pulp at the bottom.

The pulp is beaten during refining to make the recycled fibres swell. Chemical treatment may remove the dyes in coloured paper. It may further be bleached with H_2O_2, chlorine oxide or oxygen to make it more white or bright. Industrial paper is usually brown and need not be bleached. Recycled paper can be used or it may be blended with virgin fibre for extra strength and smoothness. Pulp is made more watery, sprayed into sheets, rolled and dried to obtain finished paper.

7.5 Recycling of Glass

The most easily type of recyclable material is glass. Glass scrap like bottle, jars, pyrex bowls, window glass etc are discarded as wastes by the consumers. They are then collected by the collectors and sorted according to colour such as green, amber and clear. This is a pre requisite since the different colours are chemically incompatible. Pyrex glass with different composition should not be mixed with container glass since even a single piece of such glass may alter the viscosity of the fluid.

The container is cleaned, rinsed, sorted for colours and types, broken into small pieces or crushed into **cullets** by a mechanical processor. Any labeling, bits of plastic, metals and caps present are removed by screening, magnetic or vacuum separation. The cullet is subsequently mixed with required quantity of silica sand, soda ash and limestone, fed in the furnace to melt it in the form of molten glass and moulded in form of new products. Glass does not degrade and so it can be subjected to continuous recycling. Such 1,000 kg of waste glass when recycled may prevent the atmospheric emission of about 315 kg of carbon dioxide.

7.6 Plastic Management and Recycling

Plastics are organic polymers derived from the petrochemicals having high molecular weight. They are usually synthetic molecules of carbon along with oxygen, sulphur or nitrogen. They can be polyethylene or polypropylene or polystyrene, etc. Plastics form an integral part of our daily lives. Globally, plastic production is more than the 150 million tons annually. As of 2008, annual plastic use in India amounts to 8 million tons which was projected to increase to 12 million tons by 2012.

It is used in containers, bags, clothing, toys, building materials, cables, films etc. Recycled plastics tend to be more harmful due to the addition of colours, additives, flame retardants, etc. Virgin plastics can be recycled only twice or thrice since the strength of the recycled plastic decreases each time due to degradation at high temperatures.

Environmental issues in relation to disposal of plastic waste:

- Release of gases during polymerization process.
- Plastic waste dumping renders the land infertile.
- Plastics on burning emit CO, Cl_2, HCl, CCl_4, CH_3CHO, C_6H_6, PCDDs, PCDFs, nitrogenous amines, nitrides, styrene, Buta-1, 3-Diene.
- Low-density polyethylene (LDPE), High-density polyethylene (HDPE) and Polypropylene (PP) use toxic metal like Pb and Cd as additives which can be leached.
- Non–recyclable plastic produces disposal problems.
- Plastics <40µ cause difficulty in compilation and recycling.
- Plastics that are overloaded add to unaesthetic look and often choke the pipelines.

The **Recycled Plastics (Manufacture and Usage) Rules**, enacted in 1999, have been amended in 2011.

- The thickness of the plastic carry bags to be used for carrying or dispensing commodities should not be less than 40 microns (µ).
- Compostable plastics may be used in the fabricating carry bags and they should comply to IS/ISO: 17088:2008
- State Pollution Control Board (SPCB) or Pollution Control Committee (PCC) shall be the standard body for registering, manufacturing and reprocessing. The municipal authorities shall be concerned with implementation of regulations pertaining to using, assembling, sorting and transport and dumping.
- Multilayered sachets for packing *gutkha* etc. shall not use any form of plastic.
- Each plastic carry should be labelled as 'recycled' in conformation to IS: 14534:1998. Each bag produced from 'compostable material' should be labeled as 'compostable' and should comply with IS/ISO: 17088:2008.
- Plastic recycling should be carried out in accordance with Indian Standard IS 14534:1998.
- Plastic materials in sachets for storage, packaging or selling *gutkha*, tobacco and *pan masala* has been forbidden.

Types of plastics:

- Recyclable plastics (thermoplastics): PET, HDPE, LDPE, PP, PVC, PS, etc.

■ Non-Recyclable Plastics (Thermoset and others): Multilayer and Laminated Plastics, Polyurethane foam (PUF), Bakelite, Polycarbonate, Melamine, Nylon etc.

The **Plastic Identification Code (PIC)** was introduced by the Society of the Plastics Industry, Inc., to pose an uniform system of identification of diverse types of polymer and thus assist the recycling houses to segregate various plastics for recovering. According to Bureau of Indian Standards Code, notified in the Plastic Waste (Management and Handling) Amendment Rules, 2011, plastics fall in the following categories:

The processing of waste plastic to a usable form is known as plastic recycling. The first step involves categorizing or sorting of plastic. The second stage is cleaning or washing to remove labels, adhesives and other impurities that might get stuck to the plastics. The clean plastic is then loaded in huge hoppers that transport the material towards the metal teeth rotating to shred or cut the plastics into small pieces. Naturally, UV B radiation can degrade the plastics. The pieces are then chemically tested and grouped according to its specifications and properties. Individual groups are then melted and extruded in form of pellets for future use with virgin plastic.

Alternatively one can use **bio-plastics**. Bio-plastics are plastics made from vegetable fat, corn starch, pea starch, etc. The commonly used plastics are derived from petroleum. Bio-plastics are intended to biodegrade in anaerobic or aerobic environments, largely depending on the methods of their preparation. Bio-plastics are usually composed of starches, cellulose, or other biopolymers. They can be commonly used in packaging materials, banqueting utensils, foodstuff packaging and insulation.

Table 7.3: Recyclable plastics as per PIC as per CPCB.

PIC	Type of plastic polymer	Properties	Applications
01 PET	Polyethelene pterephthalate (PET, PETE)	Transparency, strong, tough, blocks gas and moisture.	Soft drink, water and PET bottles.
02 PE-HD	High-density polyethylene (HDPE)	Stiff, strong, tough, resists moisture, permeable to gas.	Water pipes, milk containers, grocery bags, some shampoo and cosmetic product bottles.
03 PVC	Polyvinylchloride (PVC)	Versatile, blends effortlessly, strong, tough.	Plasticizers floorings, electrical cable insulation; vinyl records.

PIC	Type of plastic polymer	Properties	Applications
04 PE-LD	Low density polyethylene (LDPE)	Easily processed, strong, tough, flexible, can be sealed easily, blocks moisture.	Frozen food bags; films. flexible container lids.
05 PP	Polypropylene (PP)	Strong, tough, resists heat and chemicals, blocks moisture.	Reusable microwaveable ware; kitchenware; microwaveable wares; medicine wares.
06 PS	Polystyrene (PS)	Changeable, transparent	Egg cartons; foam packaging, tea cups, plates, trays and cutlery.
07 O	Others (O); may be comprised of polycarbonate or acrylonitrile butadiene styrene; plastics made from corn starch (compostable) also fall in this category; they possess the initials PLA (Polylactic acid) at the bottom.	Property depends on polymers or its combination. Polycarbonates (PC) may contain Bisphenol A (BPA), a xenoestrogen acting as endocrine disruptor.	Bakelite, compact discs; electronic apparatus housings; lenses including sunglasses, automotive headlamps, riot shields, instrument panels, dental devices, sports equipment, baby bottles.

Case Study 7.2: Plastic to oil by the Indian Oil Corporation

A method to convert plastic waste into fuel by IOC has given a new impetus to the waste management business and prospects. The next step is setting up of a ₹ 7.08 crores pilot plant to further mature the process. The process involves converting one of the world's most polluting materials into the most wanted commodity. A useful technology can lead to the setting up of a commercial plant that can change waste plastic management into a 'profit making centre', or waste to wealth.

The consequences of 11 experiments performed at the IOC, Research and Development Centre, in Faridabad, UP in 2003 have ascertained that the conversion of plastic-to-petrol process invented by the couple, Umesh and Alka Zadgaonkar does operate certainly. 1 Kg of plastic waste yields 40–60 per cent liquid petroleum. The raw materials used can be polyvinyl chlorides, PET bottles, polythene bags, broken buckets etc. The cost of production is a meager ₹ 7/liter. The result of the experiments is of important business prospects in India, which produces about 7,000 tons of waste plastic every day.

The invention was first reported in June 2003. The inventor's formula involves heating the shredded plastic waste, in oxygen-free condition, using coal and a 'secret chemical'. The products formed include about 80 per cent liquids in fuel range, 5 per cent coke and 15 per cent gases in LPG range. 1 kg of plastic and 100 gm of coal blend out 1 liter of fuel, which comprises the gasoline range. With further processing, it yields refined petrol. The inventors discovered the process and the catalyst that breaks long hydrocarbon chain molecules of plastic material into smaller segments of petroleum products.

IOC officials opine that fractionation can yield industrial furnace oil, which is ready to use, but production of gasoline requires stabilization additives and for petrol the cetane number needs to be increased. A low cetane number often causes ignition hitch, hampering the smooth running of an engine. IOC facilitated to improve the liquid quality by reducing the diene and high chlorine content.

7.7 Industrial Solid Waste Management

Environmental pollution is the foremost problem associated with rapid industrialization and urbanization that raise the living standards of people. Industrialization has caused severe problems related to environmental pollution. Hence, waste appears to be a byproduct of development and one can't let it lose as absolute waste. Raw material demands are increasing and non-renewable resources are being lost over time. Therefore, there should be efforts for reducing pollution that arises out of the discarded waste disposal by their conversion into utilizable raw materials for various uses. Disposal of industrial solid waste faces a lot of problems in absence of infrastructural amenities and also due to negligence of industries to take appropriate safety measures.

The local bodies are not accountable for the management of Industrial Solid Wastes (ISW). Industries responsible for generation of such waste are themselves responsible for such management. The industries are required to obtain prior permission from their respective SPCBs under appropriate rules. Even then it needs the combined effort of SPCB and local bodies.

In order to effectively manage ISW, the analysis of waste's nature is crucial. the industries should make necessary arrangement for collection and disposal of waste at specific disposal venue; the process should be under the supervision of SPCB or a Pollution Control Committee in case of Union Territory.

Cities and towns handling ISW confront the following problems:

- shortage of particular disposal spots;
- certain industries that produce solid wastes in small scale do not seek the consent of SPCBs and PCCs;
- the position of factories in non-conforming regions results in polluting air and water along with solid waste disposal;
- most of the industries do not have amenities to collect, treat and dispose both liquid and solid wastes; and
- lack of habitual communication between the local bodies and SPCBs/PCCs.

Thermal power plants are the leading ones to generate ISW, producing huge quantities of fly ash. To name a few, slag is generated from the blast furnace of iron and steel plants; aluminum and electroplating industries produce cyanides and heavy metals; red mud and tailings are released from Zn and Cu based plants; pesticides and pharmaceuticals produces toxic organic compounds, sugar industries generate press mud, tannery industry produces chromium sludge, pulp and paper industries produce lime and fertilizer and cement industries produces gypsum.

Fly ash is mainly produced from thermal power plants as particulate matter. They can enter the air in absence of efficient collecting device. They can enter the body simply through inhalation and cause diverse forms of respiratory ailments. These particulates may contain silica, aluminium, carbon, potassium, sodium, calcium, magnesium, titanium, sulphur, phosphorus, carbonates etc. For instance, a 30 year process, a 1,000 MW capacity station that consumes coal at a rate of 3,500 kcal/kg and ash content varying between 40 to 50 per cent would need nearly 500ha of area for fly ash dumping.

E-Waste: E-waste generally refers to the term to cover all types of wastes generated from electronic or electrical equipments (EEE). Such wastes contain precious and probable toxic materials that might be recovered by appropriate recycling or it has to be subjected to treatment towards attaining safety before it is disposed. India produces about 1.5 to 3.3 lakh tons of e-waste per year with maximum contribution from the western India. The greatest contributors are Mumbai, Delhi, Bangalore, Chennai, Kolkata, Ahmedabad, Hyderabad, Pune, Surat and Nagpur. The components are metals, motors, plastic, glass, circuits, transformers, batteries, radioactive substances, ceramics, brominated flame retardants etc. The contained metals can be lead, beryllium, cadmium, mercury, halogens, arsenic, selenium, chromium; most of which are carcinogens or endocrine disruptors. There is no specific law to handle such wastes but it is the liability of the registered manufacturers to collect, treat or recycle such hazardous wastes. The initiative by the Karnataka PCB towards e–waste management is a noteworthy step.

Categories of e-waste:

- Bulky household appliances such as washing machines, dryers, refrigerators, air conditioners, vacuum cleaners, etc.
- Minor household appliances such as coffee machines, irons, toasters, etc.
- Information and communication equipment such as personal computer's, laptops, mobile phones, land phones, fax and xerox machines, printers etc.
- Amusement and consumer durables such as video cassette recorder / digital versatile disc / compact disc players, hi–fi sets, radios.
- Electrical equipments such as sodium lamps, fluorescent lamps with the exception of incandescent bulbs and halogen bulbs.
- Machine instruments like drilling machines, electrical sawing machines, sewing machines, grass mowers etc.
- Safety tools such as close circuit television (CCTV) sets, cameras, scanners.

Healthcare instruments such as x-ray machines, ventilators, Computed Tomography (CT) scan, Positron Emission Tomography (PET) scan, Magnetic Resonance Imaging (MRI) scan etc.

Hence, e-waste comprises the **white goods** or household appliances such as washing machines, fridge, dishwashers, microwaves and the **brown goods** like TV, transistors, cameras and **grey goods** such as computers, printers, scanners and mobile phones.

E-waste (Management and Handling Rules, 2011 effective from 1 of May 2012 (Refer-http:// moef.nic.in/downloads/rules-and-regulations/1035e_eng.pdf for details)**:**

As per the Central Government notification, the primary objective is to channelize the generated e-waste of India in order to prevent environmental damage. The rules apply to the producers and consumers involved in manufacture, sale, and purchase and processing of electrical and electronic equipments as specified in the list. The rules place the responsibility of such wastes on the producers of electrical and electronic equipment by introducing the concept of 'Extended Producer Responsibility' (EPR). E-wastes are required to be disposed off separately from the usual wastes via the acknowledged collection facilities.

E-waste products should not to be mixed with usual domestic wastes, should not to be opened by the user himself and should not to be resold to any unauthorized agencies. The product should be isolated after becoming non-functional and replaced spare parts are not to be kept exposed.

E-waste may contain many recyclable materials such as: valuable metals like copper, aluminium, gold, silver, titanium, platinum, palladium, indium and others. Plastics used in the case and cover can be recycled to be used as traffic cones, car bumpers, plastic fencing. The paper used in packaging can also be use for recycling.

Industrial waste management approach is based on a two stage approach comprising:

- avoidance or prevention; and
- control of pollution that involves collection, transport and disposal.

The waste minimization approaches for waste prevention are:

- Inventory management and improved operations that trace all raw materials organizing employees' training etc.
- Modification of equipment – installing efficient equipment, modifying or redesigning equipment etc.
- Production process changes – substitution of raw materials, segregation of wastes, elimination of leakages and spills.
- Recycling and reuse for example, flyash is used in cement manufacture.

Collection and transport of wastes: It comprises manual scooping of wastes, keeping them into storage containers and then subsequently loading them into trucks. Storing ISW is the most neglected part of the operation.

Poor industrial waste disposal and health impacts : The highly toxic chemicals can affect human health in the following ways:

- skin contact – Chemicals that cause dermatitis.

- inhalation – Aerial pollutants that can directly injure the respiratory system, enter the lung and bring about systemic effects.

- intake – intake of hazardous chemicals through food and water can occur through contamination of subsoil and groundwater that results from the leachates and poorly-managed landfill places.

7.8 Biomedical Waste Management

The waste generated from hospital activities can be precarious, toxic and sometimes fatal due to their high possibility of disease communication. Its tendency to boost growth of various germs and vectors and its capability to pollute other non-hazardous/non-toxic municipal waste endangers the initiatives undertaken for general municipal waste management. The rag pickers and waste workers are the worst affected, as they scavenge various types of toxic substances as they try to salvage few things that are sellable or reusable unknowingly. This type of unlawful and unscrupulous reuse often proves to be perilous and sometimes fatal. Public are at serious health risks from diseases such as cholera, plague, tuberculosis (TB), hepatitis (HBV), Acquired Immunodeficiency Syndrome (AIDS), and diphtheria.

Since majority of the health care establishments like hospitals, clinics and nursing homes are situated within municipal areas; clearly their management has close ties with the municipalities.

Practices like stringent and cautious segregation could ease the load and cut down the cost of management of actually hazardous and contaminated bio-medical waste before it is collected, transported, treated and disposed.

Table 7.4: Categories of biomedical wastes (Schedule I)

Nos.	Category	Waste	Treatment and disposal
1.	**Category No. 1**	Human anatomical waste (human tissues and organs and body parts of human beings).	Incineration@/deep burial*.
2.	**Category No. 2**	Animal waste (animal tissues, organs, their body parts, carcasses, bleeding parts and body fluid, blood and experimental animals used in research, waste generated by veterinary hospitals, animal houses).	Incineration@/deep burial*.
3.	**Category No. 3**	Microbiological and Biotechnological waste (wastes from laboratory cultures, stocks or specimens of microorganisms, live or attenuated vaccines, human and animal cell culture used in research, infectious agents from research and industrial laboratories, waste from production of biological, toxins, dishes and devices used for transfer of cultures).	Local autoclaving / microwave/ incineration.

Nos.	Category	Waste	Treatment and disposal
4.	**Category No. 4**	Waste sharps (needles, syringes, scalpels, blades, glass etc. that may pierce and cut. This includes both used and unused sharps).	Disinfection (chemical treatment/autoclaving/ microwaving and multilation/ shredding).
5.	**Category No. 5** drugs disposed in secured landfills,	Discarded medicines and cytotoxic drugs (waste comprising outdated, contaminated and discarded medicines).	Incineration/destruction and drug disposal in secured landfills.
6.	**Category No. 6**	SW (items contaminated with blood, and body fluids including cottons, dressings, solid plaster casts, lines, beddings, waving and other material contaminated with blood).	Incineration autoclaving/ microwave
7.	**Category No.7**	SW (waste generated from disposal items other than the waste sharps; chemicals such as tubing's, catheters, intravenous sets etc.).	Disinfection by chemical treatment autoclaving/ microwaving and mutilation/ shredding
8.	**Category No. 8**	Liquid waste (waste generated from laboratoryand washing, cleaning, housekeeping and disinfecting activities).	Disinfection by chemical treatment and discharge into drains.
9.	**Category No. 9**	Incineration ash (ash from incineration of any biomedical waste).	Disposal in municipal landfill.
10.	**Category No. 10**	Chemical waste (chemicals used in production of biological, chemicals).	Chemical treatment and disinfection, as insecticides, etc. Discharge into drains for liquids and secured landfill for solids.

Notes: @@ Chemical treatment using at least 1 per cent hypochlorite solution or any other equivalent chemical reagent. It must be ensured that chemical treatment ensures disinfection.

Mutilation/ shredding must be such so as to prevent unauthorized reuse.

@There will be no chemical pre-treatment before incineration. Chlorinated plastics shall not be incinerated.

*Deep burial shall be an option available only in towns with population less than five lakhs and in rural areas.

Table 7.5: Colour coding for disposing biomedical wastes (Schedule II)

Colour codes	Waste categories	Options for treatment as per law
Yellow	Plastic bag Category 1, Category 2, Category 3 and Category 6.	Plastic bag Category 1, Category 2, Category 3 and Category 6.
Red	Disinfected container/plastic bag Category 3, Category 6 and Category 7.	Autoclaving/Microwaving/ Chemical Treatment
Blue/White translucent	Plastic bag/puncture proof Category 4, Category 7 container	Autoclaving/Microwaving/ Chemical Treatment and Destruction/Shredding
Black	Plastic bag Category 5 and Category 9 and Category 10(solid)	Disposal in secured landfill

Notes:

1. Colour coding of waste categories with multiple treatment options as defined in Schedule I, shall be selected depending on treatment option chosen, which shall be as specified in Schedule I.

2. Waste collection bags for waste types needing incineration shall not be made of chlorinated plastics.

3. Category 8 and 10 (liquid) do not require containers/ bags.

4. Category 3 if disinfected locally need not be put in containers/ bags.

Summary

▪ Waste refers to eliminated or discarded substances that are no longer useful or required after the completion of a process. Waste management refers to the collection, transport, processing/ treatment or disposal and monitoring of waste materials generally undertaken to reduce their impact on health, environment or aesthetics. Waste management is based on 3 R strategies: reduce, reuse and recycle.

▪ Wastewater is that liquid waste adversely affected in terms of quality by human interference. Wastewater treatment is a process by which the physical (large, coarse, suspended solids), chemical (dissolved minerals, gases, chemicals and heavy metals) and biological impurities (disease-causing pathogens) present in the wastewater/sewage/effluent can be reduced to an acceptable level by a series of treatment process. A wastewater treatment plant is also called STP in the municipal circle and ETP in the industrial circle. The wastewater treatment system is composed of a combination of unit operation and involves contaminant, chemicals and microorganism removal by physical forces and unit processes. Wastewater treatment involves primary, secondary and tertiary treatment. Wastewater may prove to be a valuable resource in the present world but the harmful impacts of wastewater cannot be ignored.

▪ Solid waste refers to all types of garbage, refuse or sludge from water treatment or water supply plant as a result of household, industrial, commercial, mining or agricultural activities. On the basis of its source, solid waste can be municipal solid waste, industrial or hazardous wastes and biomedical waste or hospital wastes. E-waste generally covers all types of wastes generated from EEE.

▪ The amount of total waste generated is more than 4 billion tons annually where nearly half is MSW. Global recycling markets contribute to huge quantities of materials recovery coupled with an increase in waste trafficking activities. Scrap paper, glass, metals, non-metals and plastics are often recycled. There is an urgent need to fight waste trafficking in order to establish a sound global system of waste management.

▪ Solid waste management process comprises thermal methods like incineration, gasification, pyrolysis, plasma arc gasifier; biological methods like composting anaerobic digestion or land filling methods.

▪ Solid waste management involves the solicitation of norm of ISWM to accomplish the objectives of waste minimization and operative and effective management of waste. A balance is needed that will reduce the overall environmental effects of the waste management system within a tolerable level of cost. Effective solid management systems are desired to safeguard better human health and safety and to prevent the spread of disease. Additionally, solid waste management ought to be both environmentally and economically supportable.

Exercise

MCQs

Encircle the right option:

1. Match the following types of wastes with their suitable method of disposal: Ans:
   - i.   Discarded medicine   a.   Recycling   A. ic, iia, iiid, ivb
   - ii. Scrap glass   b.   Incineration   B. ia, iid, iiib, ivc
   - iii. Vegetable residue   c.   Secured landfill   C. id, iic, iiib, iva
   - iv. Pathological waste   d.   Composting   D. ic, iia, iiib, ivd

2. During recycling of glass, _____ is not mixed with the container glasses
   - A. Optical glass   B. Pyrex glass   C. Soda glass   D. Lead crystal glass

3. Match the following types of container used for storage and transport to the category of biomedical waste: Ans:
   - i.   Yellow   a.   Categories 4 and 7   A. ic, iia, iiid, ivb
   - ii. Red   b.   Categories 5 and 9   B. ia, iid, iiib, ivc
   - iii. Blue   c.   Categories 3, 6 and 7   C. id, iic, iiia, ivb
   - iv. Black   d.   Categories 1, 2, 3 and 6   D. ic, iia, iiib, ivd

4. Hospital wastes management is required to abide by the provisions of:
   - A. Municipal Solid Waste Rules, 2000   B. Biomedical Waste Rules, 1998
   - C. Hazardous Wastes Rules, 1989   D. Both B and C

5. Recycling of glass requires its conversion into:
   - A. Shard   B. Frit   C. Cullet   D. none of these

6. Drinking water bottles are usually made up of:
   - A. PVC   B. LDPE   C. PS   D. PET

7. Which of these chemicals may be used for increasing the whiteness of recycled chapter?
   - A. Hydrogen peroxide   B. Nitric acid   C. Borate   D. Carbon dioxide

8. The ideal pile height in Windrow composting should be between:
   - A. 4–8 ft   B. 15–18 ft   C. 1–2 ft   D. 2–3 ft

9. Which of the following categories of worms is/are preferred in commercial vermicomposting?
   - A. Anecic   B. Endogeic   C. Epigeic   D. All of these

10. _____ plastics should not be incinerated:
    - A. Chlorinated plastics   B. Methylated plastics
    - C. Carbonated plastics   D. All of these

Fill in the blanks.

1. Unit operation refers to waste removal by_____ process.

2. Pyrolysis refers to the burning of waste above _____.
3. Tertiary treatment refers to the process of _____.
4. The 3–R approach of waste management refers to reduce, reuse and _____.
5. Human tissue falls under Category _____ of Biomedical waste.

State whether the statements are true or false.

1. Compost is non–biodegradable. (T/F)
2. Incineration is an eco–friendly process. (T/F)
3. Ferrous sulphate can be used as a coagulant in secondary treatment of water. (T/F)
4. Anaerobic digestion plants operate under mesophilic conditions mostly. (T/F)
5. Bakelite is a type of recyclable plastic. (T/F)

Short questions.

1. Define waste.
2. What is waste management?
3. Enlist the various types of waste water treatment.
4. What are the various sources of waste in water?
5. State the objectives of waste water treatment.
6. Why is primary treatment of waste water required?
7. Define sludge.
8. What is coagulation? Name a few coagulants.
9. What are sprinkling filters?
10. What are the components of activated sludge process?

Essay type questions.

1. Make a comparison between primary and secondary treatment of waste water.
2. Compare trickling filter with activated sludge system.
3. What are the various disinfection processes? Which method will you prefer for the same. Justify the answer.
4. Discuss the role of informal sector in waste management.
5. Do you correlate the increase in solid waste with globalization and urbanization? Justify.

Disaster and Disaster Management ____8

Learning objectives

- To develop a comprehensive understanding of the concept of vulnerability, risks, hazards and disaster as well as scope of environmental studies.
- To know about various types of disasters in general.
- To develop an understanding of disaster management, its structure and organization.
- To elicit public awareness and collective response to protect ourselves from various types of disasters.
- To know about prevention, mitigation and response in cases of floods, cyclones and earthquakes.
- To analyse the causes, impact and mitigation measures through case studies.

8.1 Introduction and Definition

Disaster is an incident or series of events that gives rise to casualties and loss or destruction of surroundings, buildings, properties, infrastructural facilities and all other vital services or survival means to such an extent that is not within normal means and competence of affected people to deal with. Disaster can be defined as *'catastrophic situation in which the normal pattern of life or ecosystem has been disrupted and extraordinary emergency interventions are required to save and preserve lives and or the environment'*.

The **United Nations** defines disaster as *'the occurrence of sudden or major misfortune which disrupts the basic fabric and normal functioning of the society or community'*.

As per the **Disaster Management Act, 2005**, disaster is defined as:

'a catastrophe, mishap, calamity or grave occurrence in any area, arising from natural or manmade causes, or by accident or negligence which results in substantial loss of life or human suffering or damage to, and destruction of, property, or damage to, or degradation of, environment, and is of such a nature or magnitude as to be beyond the coping capacity of the community of the affected area'.

8.1.1 Ingredients of disaster

- An event comprising trauma for a population/environment.
- A vulnerable position/area that bears the load of the traumatizing incident.

- The breakdown of local and neighbouring resources to cope with the problems created by the phenomenon.

8.1.2 Characteristics of disaster

- Inevitability or certainty.
- Ability to be forbidden.
- Pace of onset.
- Span of forewarning.
- Extent of impact.
- Scope and intensity of impact.

8.1.3 Factors affecting disaster

- **Host factors**

  i. age of the individuals; ii. status of immunization; iii. degree of mobility; and iv. emotional stability.

- **Environmental factors**

  i. physical factors; ii. chemical factors; iii. social factors; and iv. psychological factors.

8.1.4 Types of disaster

A. Natural or geogenic disasters

A natural disaster is the consequence of natural hazards on humans and the environment. Human vulnerability, weakness and lack of suitable emergency control leads to economic and environmental impact including that on the human beings. The ensuing loss depends on the capability of the population to survive or resist the disaster; their resilience. In other words, 'disasters occur when hazards meet vulnerability'. A natural hazard will in no way occur in areas without vulnerability.

Natural disasters can be meteorological (floods, droughts, earthquakes cyclones, snow fall, hail storm, avalanches, tornadoes, hurricanes, etc.) or biological (epidemics). According to International Federation of Red Cross and Red Crescent (IFRCRC), 2001, natural disasters are generally categorized into geo-physical disasters (landslides, earthquakes, tsunamis, volcanoes, etc.) and hydro-meteorological disasters (floods, droughts, hurricanes, tornados, cyclones, typhoons, etc.)

Each year such incidents result in the death of thousands and thousands of people along with destruction of property. On the other hand, natural disasters may also hit unpopulated areas and hence do not culminate into disasters. But rapid uncontrollable growth of human population and increased population density in risky environments have amplified both the frequency and severity of natural disasters. Practically, developing nations are more disaster prone and suffer more chronically from natural disasters because of tropical climatic conditions, unstable topography and increased deforestation, unplanned population growth, unplanned townships and settlements, non-engineered constructions, delayed communication systems and poor or no budgetary allocation. Asia is first in the list in terms of casualties due to natural disasters.

B. Man-made or anthropogenic disasters

The range of man-made disasters broadly cover dispute and rampage, civil discord, wars and industrial disasters. These disasters are mostly an outcome of technology and mental disagreement. Examples are innumerable from stampedes, road accidents, railway accidents, airliner accidents, oil spill, nuclear disasters to chemical and other industrial disasters. These unnatural disasters foul the environment, damages public and personal property and kills a large number of people. Take for example, the 11 September 2001 attack at the World Trade Center in New York, US.

Environmental emergencies: These are technological or industrial accidents (mining, chemical, biological, nuclear) with hazardous materials and generally occur at the site of production, usage or during transit like gas leakage, risk posed by LMOs, radiation. Factory fires and forest fires are also included in this group and are either caused naturally or by human negligence.

Complex emergencies: These include plundering or robbery, breakdown of authority, a premeditated sudden attack on installations during conflict situations, war or riots.

Traffic emergencies: Such emergencies occur during air, rail, road or ship related accidents.

Pandemic emergencies: These include social and financial loss by abrupt inception of a contagious disease affecting health, disrupting essential services, business, etc.

8.1.5 Components at risk

The following components are at risk:

- people;
- rural housing stockpile;
- electricity poles and cables, telephone lines, vegetation including crops and trees;
- individual or personal property;
- infrastructural support.

- livestock or farm animals;
- vulnerable houses;
- boats, looms, working equipments and tools;
- electric, water and food supply; and

8.1.6 Effects of disaster

Effects of disaster includes:

- loss of life;
- spread of epidemics, water borne, food borne and air borne diseases;
- vandalization;
- disruption to essential services – shortage of food, medicines;
- disruption of lifestyle;
- psychological trauma.

- injury and disability;
- damage to national infrastructure and governmental systems.
- damage to and destruction of property;
- manufacturing loss and National economic loss;
- loss of livelihood; and

8.2 Disaster Management

Basically the process of disaster management involves dynamism. Its mission includes the duty of making a plan, systematizing the entire process, employing and training crews, leading, administering and supervising. It involves multiple organizations that should act together in order to avert (prevent), alleviate (mitigate), prepare for (arrange), respond (react) and recover (recuperate) from the consequences of a disaster. Thus, it includes prompt response, regaining or retrieval, avoidance, diminution, alertness; the cycle gets repeated on and on. Disaster management is a complex humanitarian process involving international, national and local organizations each with a distinct role to play. A coordinated effort is required in order to respond to a disaster.

It is defined as 'an applied science which seeks, by the systematic observation and analysis of disasters, to improve measures relating to prevention, mitigation, preparedness, emergency response and recovery'. Disaster management can be referred to as the organizing and managing of the resources and responsibilities in order to handle all emergency situations with a humanitarian approach; especially, its prevention, mitigation, preparedness, response and recovery so as to reduce the impact of disasters.

Any sort of disaster is likely to disrupt the essential services, healthcare facilities, supply of electricity and water, sewerage systems, removal of wastes and garbage, transport and communication systems. Such interruption gravely affects the health, the socio-economic set up of the local people and countries. Even if the immediate impacts have been mitigated, it usually leaves a predominant long lasting effect on people. Ill planned assistance and rescue activities negatively affect both the victims of disaster and the donors and aid organizations; hence it is foremost that the physical therapists should also participate in collective and recognized efforts rather than individual programmes.

8.2.1 General objectives of a disaster management plan

- To guarantee the safety, security and stability to secure business sites and facilities; to safeguard and make available critical resources; to ensure supply of equipments; to make sure security and resurgence of accounts from expected disasters and to protect all vital information and records.

- To reduce the risk of disasters caused by human error and to be well equipped in order to pull through from a major natural disaster.

- To recover lost or damaged records or information ensuring the organization's capacity to carry on its operation and service even in the post-disaster period (sustainability).

8.2.2 Principles of disaster management

- Disaster management denotes accountability at all levels of government.
- Disaster management arrangements must identify the contribution and prospective role of non-governmental agencies.
- Disaster management should utilize existing resources for daily purpose.
- The concerned organizations should work and endeavor as a supplement to their central business.
- Persons are to be responsible for their own safety.
- Planning of disaster management should be focused on large scale episodes.

8.2.3 Organizations involved in disaster management

Disaster management arrangements must recognize the involvement and prospective roles of non-governmental organizations and coordinate with government organizations, involving communities with clear lines of authority and unity of command.

A. The National Disaster Management Authority (NDMA)
B. The Indian Military and para-military forces
C. Utilization of total governmental structures/resources i.e., at national, state and local level.

   i.   People: individuals, households, volunteers
   ii. Gram panchayat: sarpanch, panchayat secretary, panchayat members
   iii. Village elders: irrespective of caste, community and religion
   iv. Leaders, teachers, doctors, engineers
   V.   Retired army and police personnel

D. NGO: The international agencies that provide humanitarian assistance are the UN agencies like

   i.   Office for the coordination of Humanitarian Affair (OCHA)
   ii.   World Health Organization (WHO)
   iii. United Nations International Children's Emergency Fund (UNICEF)
   iv.   World Food Programme (WFP)
   v.   Food & Agricultural Organization (FAO)
   vi.   International Organization for Migration (IOM)
   vii. Office of United Nations High Commissioner for Human Rights (OHCHR)
   viii. United Nations Development Programme (UNDP)
   ix.   United Nations High Commission for Refugees (UNHCR)

   Other organizations like:

   i.   Co-Operative American Relief Everywhere (CARE)
   ii. International Committee of Red Cross (ICRC)

E. Local clubs and social institutions

F. Special system requirements:

   i.   Emergency Operation Centre (EOC)/Control Centre
   ii. Coordinating Authority
   iii. Information management and Communications
   iv. Warning Systems: **Indian Meteorological Department (IMD)**, under the **Ministry of Earth Sciences (MoES)**, plays a very significant task in weather forecasting and early warning of disasters. It has six regional meteorological centers at Mumbai, Chennai, New Delhi, Kolkata, Nagpur and Guwahati with various types of operational units from forecasting offices and flood meteorology, hydrometeorology, seismology, cyclone warning centers, etc. They also have several observation posts along the Indian coast.

    v.   Search and Rescue Team

    vi.   Survey and Damage Assessment Team

    vii. First Aid and Triage Group and Mobile Medical and Health Team

    viii. Evacuation Team

    ix.   Animal Husbandry/Veterinary Management Team

    x.   Emergency Welfare Team

    xi.   Emergency Shelter

    xii. Emergency Logistics

    xiii. Survey and Assessments, specialists from field of disaster studies and research (geologists, meteorologists, etc.)

8.2.4 Disaster management cycle

8.2.4.1 Pre-disaster phase

Before designing for prevention and mitigation, risk assessment is necessary. The hazard assessment will target and convey precise details of disaster about individual positions. Susceptibility to a particular hazard will take account of:

- important products, services, records and operations;
- hazardous materials;
- impending effects of damage on stakeholders;
- expected financial costs;
- personnel and duration available for making arrangements; and
- status of insurance cover.

    The permutation of hazard and vulnerability appraisal will lead towards formulating total risk assessment.

A. Prevention and mitigation

    i.   Prevention includes all such procedures that are premeditated to delay or hold back the incidence of a catastrophic event and/or arrest the happening of such incidence in the future. Actually these actions are planned and undertaken to afford permanent fortification against disaster. It should be remembered, not all sorts of disasters can be intercepted, such as natural disasters. The possibility of loss of lives and injuries can be mitigated

    ii. On the other hand, mitigation assumes the shape of precise, explicit programmes with an intension to lessen the impacts of a catastrophic event on the community or the entire country. Excellent evacuation strategies, good environmental planning and design standards can mitigate the possibility of the loss of lives and injuries. Say for instance, the evolution and practice of building codes (to lessen the destruction and damage due to earthquakes and storms) can be placed in to the category of mitigation. On the basis of the type of disaster, mitigation measures protect the vulnerable population and vulnerable structures. Minimization of medical casualty by improvement of the structural qualities of schools, colleges, houses and other construction; ensuring the safety and security of health services to public health facilities such as supplying water, effective sewerage system to trim down the cost of rehabilitation and reconstruction, all act to compliment disaster preparedness and the disaster response activities.

B. Preparedness

This stage comprises all such steps that enable the government authorities, the organizations, community groups and individuals to react **fast** and **efficiently** to disaster circumstances. These actions or response are intended to minimize the loss of lives and damage of any kind. For example, relocation of people and shifting of movable property from a vulnerable location and by catering well timed effectual rescue, relief and rehabilitation operation. Preparedness is the principal method of reducing the effect of disasters. In physical therapy practice and management, community based preparedness should be highly prioritized.

The measures for disaster preparedness comprise:

i. The foundation of a latest and justifiable counter disaster plan, a plan that should be able to differentiate between incidents and disasters. The operational arrangements are supplementary to and do not substitute the incidental managements. Any planning in disaster management must take into account the physical environment and the pattern of population. Planning involves a lucid and logical move in dealing with disasters. It should be able to provide common reference for all the departments and authorities with roles they are to play. It should be able to furnish with information for setting up a multifunctional organizational set up. The plan should rest on foundation for synchronized action. The plan provides allotment of responsibilities. The plan should have provision for reviewing and evaluating current and future disaster management requirements. It should focus for training related to disasters.

ii. Special rations for crisis action

iii. Necessary forecasting systems

iv. Emergency information and correspondence

v. Community edification and awareness

vi. Instruction and training programmes comprising drills and tests. The training includes:

- recognition of training requirements;
- extent of training schedules;
- strategy and guidelines for training;
- execution of instructions; and
- disaster drill, which is a type of workout in which the individuals create or replicate the disaster situations and scenario in order to have the chance of practicing their reactions.

In this case training should be designed to be tuned to support the tasks required to be performed after a disaster.

Disaster preparedness targets to minimize the undesirable effects of a hazard or danger with the aid of effectual preventive or deterrent measures. It includes the following steps.

- Evaluate from past experiences about risk location of disaster prone areas.

- Ensure timely, appropriate and efficient organization and delivery of emergency response following the impact of a disaster.

- Analyse susceptibility or weakness and map for inclusion of resources.

- Evaluate or appraise the requisites for strengthening and implement.

- Arrange the appropriate funding for preparedness.

- Develop public education programme.

- Organize communication, information and warning system, national and international relations, coordinate with media, create lead time by interpreting warnings, ensure peoples' assistance and teamwork through politicians, elderly persons, volunteers and NGOs.

- Ensure co-ordination and response mechanisms plan to incorporate the mobility of resources within the time frame, keeping stock of foods, drug and other essential commodities.

- Target to lessen the potentially vicious disasters, provide suitable and timely relief or assistance to the victims and rapid and resilient recovery.

8.2.4.2 Disaster phase

Case Study 8.1: Cyclone Phailin

Indian Ocean has warm waters and is probably a cyclone hot spot. It was the deep depression over the north Andaman Sea that culminated into a severe cyclonic storm. The Cyclone hit the Bay of Bengal on 12 October, 2013 on the Indian east coast, specially Andhra Pradesh and Orissa, followed by heavy downpour. It made a hit on Gopalpur with winds more than 200 kmph speed. IMD confirmed it to be as strong as Category 5 hurricane.

Soon after the 'Red Message' from IMD, Delhi, control rooms were being set up in various parts of the eastern coast with updated mobile numbers and leaves being cancelled for all to be on their toes. Enough food was stocked by the concerned states. The early warning systems were worth admiring. Messages were being continuously passed through mobiles, TV and radio and this salvaged lakhs of lives. The first warning or 'Cyclone Watch' was conveyed 72–96 hours in advance of the land fall. The speed, the direction and the target area of hit was closely followed. Information was assembled from the satellite imageries using Doppler radars. The second warning, that is, 'Cyclone Alert' was given 36–48 hours prior to the hit followed by the third 'Cyclone Warning' just 24 hours before land fall, when evacuation was ordered. Rapid evacuation measures were being deployed both by national and state authorities. The series of concrete constructions provided vital shelter to the refuge. The last 'Post land fall Alert' was done stating its direction, speed and amount of precipitation.

The NDMA deployed many teams comprising 2,000 National Disaster Response Force (NDRF) personnel in all vulnerable places; 29 in Orissa, 19 teams in Andhra Pradesh and 7 in West Bengal. It also advised the states to keep their State Disaster Response Force (SDRF) ready and the central medical teams were ready. Nearly 5 lakh people have been evacuated from the danger zones to safe places. Helicopters, fixed winged aircrafts, and army were shifted to the areas likely to be affected. Indian navy and coast guard helped the ships and the fishermen from the clutch of Phailin. The National Highways Authority of India (NHAI) and the Deptartment of Communication were directed to restore communications as soon as possible. Indian Railways were conveyed not to run trains in the affected area.

The loss of lives was 15 as compared to 12 million people that lived in the route of cyclone. Around 5,000 km² of paddy cropland were destroyed with probably a loss amounting to $320, the recovery of which will need some time. What remains praiseworthy is the proactiveness of the NDMA and the concerted effort of authorities at all levels.

C. Response

The response actions refers to those which are employed right away before and after the disaster impact like epidemiologic surveillance, nutrition and disease control by vaccination, etc. The characteristic measures include:

- execution of plans;
- activation of the counter-disaster system;
- search and rescue operations;
- provision of emergency food, shelter, medical assistance etc;
- survey and assessment; and
- evacuation measures.

The Government of India (GoI) Departments for Disaster Response Ministries/ Departments comprises Rural Development, Drinking Water Supply, Power, Telecom, Health, Urban Development, Food and Public Distribution, Shipping, Surface Transport, Railways, Civil Aviation, Women and Child Development, Water Resources, Animal Husbandry, IMD and Ministry of Defence - Armed Forces, Ministry of Home Affairs – Central Para Military Forces and Ministry of External Affairs – International Response.

D. Relief

Relief is a coordinated multi-agency response in time to reduce the impact of a disaster and its long-term consequences. Such measures embrace recue, repositioning, supplying food, water and medicines, intercept the spread of disease and disability, restoring crucial services like telecommunications and conveyance, rehabilitating in temporary shelters and immediate health care services.

8.2.4.3 Post-disaster phase

E. Recovery

In this step, the societies or communities or nations are provided with help so as to revert to normalcy or optimum level of operation following the disaster. Still, subsequent to the crisis needs being met, the affected people and the communities are still in danger. Three important activities that come under recovery are:

- restoration of the existing situation so as to prevent it from turning even worse;
- reconstruction and rebuilding infrastructure; and
- rehabilitation of water supply, food safety, basic sanitation, personal hygiene and vector control.

All these measures should merge with the developmental activities, such as building of human resource for health and formulating strategies and practices to evade similar situations in the coming future. By all means disaster management should be coupled with sustainable development, especially in relation to vulnerable people like disabled, elderly, children and marginalized groups.

Figure 8.1: Disaster management cycle

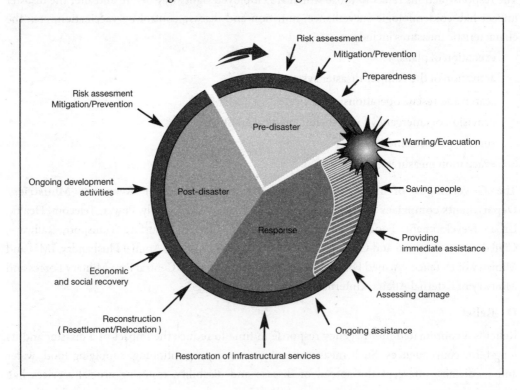

8.2.5 Evolution of disaster management

8.2.5.1 *Disaster management at international level*

According to United Nations (2002), United Nations International Strategy for Disaster Reduction, 2003, disaster management was defined as *'An integrated, multi-hazard, inclusive approach to address vulnerability, risk assessment and disaster management, including prevention, mitigation, preparedness, response and recovery, is an essential element of a safer world in the twenty-first century'.*

The 1994 conference popularly known as **World Conference on Natural Disaster Reduction** held at Yokohama in Japan affirms:

- increase in the occurence of natural disasters;
- disaster prevention, mitigation, preparedness and relief to be the key elements in sustainable development and to be integrated with developmental strategies; these steps are superior over disaster response;
- need for bilateral, multilateral assistance as well as financial resources;
- application of appropriate information, knowledge and technologies;
- involvement and active participation of community for perceiving risks; and
- acceptance of the Yokohama Strategy and a Plan of Action for a Safer World with the following doctrines:
    - i.   prevention and preparedness to be integral part of developmental policy;
    - ii.   early warning to be key in prevention and preparedness;
    - iii. development and consolidation of capacity;
    - iv.   involvement of people at all levels; proper education and training;
    - v.   reducing vulnerability through proper design and development; and
    - vi.   disaster reduction policies to require risk assessment.

World Conference on Disaster Reduction, 2005, Kobe, Hyogo, Japan

A review of the Yokohama strategy was carried out that identified few gaps and challenges in the following areas:

- organizational, legislation and policy framework;
- identification of risks, its evaluation, monitoring and early forecasting;
- education and information management;
- mitigation of the fundamental risk factors; and
- preparedness along with response and recovery.

After review of the Yokohama strategy, gaps were identified and subsequently the **Hyogo Framework for Action (HFA) for the decade 2005–15,** was adopted with five priorities for action:

- ensure that lessening of disaster risk is of national and local priority with strong institutional basis for implementation;

- identification, assessment and monitoring of disaster risks and enhancing early warning systems;

- using knowledge, innovation and education in order to create a culture of safety and flexibility at all levels;

- mitigating the necessary risk factors; and

- consolidation of disaster preparedness for effectual response at all levels.

8.2.6 Disaster management in India

As disaster management evolved in India it showed a shift from imprudent, action based arrangement towards realistic, proactive, organized arrangement; from single power domain to multiple stakeholder based set up; from an approach essentially relief based to a practical, integrated comprehensive move towards many dimension for the purpose of reducing the danger.

8.2.6.1 Disaster management in the period of British rule and after independence

At the time of British governance in order to combat emergency at the time of disasters departments of relief were set up. Such a reactive approach based on activity based set up was operational only in the post disaster scenarios that included programmes like designing relief codes and initializing food for work.

Post-independence, the responsibility of disaster management was retained with the State Relief Commissioners. They carried out the task under the Central Relief Commissioner with a restricted job such as to allocate relief materials and money in the affected parts.

The subject of 'Disaster Management' did not find any place in the Seventh Schedule of the Indian constitution. A review through the five year plans revealed that in those days disasters were only referred to droughts and floods. Quite obviously, Drought Prone Area Programme (DPAP), Desert Development Programme (DDP), National Watershed Development Project for Rainfed Areas (NWDPRA) and Integrated Water Development Project (IWDP) served as paradigm programmes.

8.2.6.2 Emergence of institutional arrangement in India

After the UN General Assembly and the announcement of 1990s as the decade of 'International Decade for Natural Disaster Reduction', a permanent institutionalized set up for the management of disaster was started under the Ministry of Agriculture. The country had gone through a sequence of catastrophe, such as, earthquake of Latur in 1993, landslide in Malpa in 1994, super cyclone in Orissa in 1999 followed by earthquake in Gujarat in 2001. Such a series of events led to the formation of a high power committee. The committee was chaired by J. C. Pant, the Secretary, Ministry of Agriculture. The committee was bestowed with the responsibility for formulating a systematic, comprehensive and holistic approach towards disasters. The committee submitted its report on 2001 which strongly recommended spending at least 10 per cent of the funds that should be earmarked and allocated at national, state and district levels for prevention, mitigation, reduction and preparedness of disasters.

In 2001, Ministry of Home Affairs (MHA) took over the charge of the department from Ministry of Agriculture after notified by the Cabinet Secretariats. This marks the evolution of the hierarchy for disaster management in India. The Bhuj earthquake left a massive impact on the Indian government. It was the Government of Gujarat that enacted the Gujarat Disaster Management Act in 2001 for the first time, prior to which there were no such acts at any level to complement disaster management.

It was in the Tenth Five Year Plan (2002–07) that for the first time an entire chapter 'Disaster Management: The Development Perspective' was included. The Tenth Plan emphasized that sustainable development is impossible without integration of disaster mitigation in the development process. The Eleventh Five Year Plan thus aims in strengthening this process by giving momentum to the projects and programmes that nurture the culture of safety. Some catastrophes have led to paradigm shifts in the disaster management policy and created new organizations. A path breaking strategy was the enactment of the Disaster Management Act, 2005. Many institutions have come up with a focused mandate on disaster management. The footsteps of Gujarat State Disaster Management Authority (GSDMA), Orissa State Disaster Management Authority (OSDMA) and Disaster Management Institute (Ahmedabad) are considered as strategies in the proper direction.

Concerted effort can be noticed on part of state towards disaster management strategies in the Rural Development Schemes also. The synchronization between the Ministry of Rural Development and the MHA is one such example. The present nodal ministry is the Ministry of Home Affairs. It collaborate both relief and response measures as well as the natural disaster management in general. It modifies the guidelines of the various schemes such as Indira Awas Yojna (IAY) and Sampoorn Gameen Rogar Yojna (SGRY) to ensure that the houses built under IOY and schools/community buildings built under SGRY are earthquake/cyclone/flood resistant.

Figure 8.2: Hierarchy in the disaster management, India

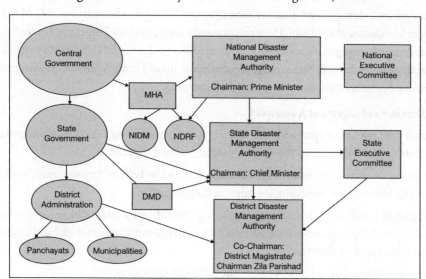

8.2.6.3 Disaster Management Act, 2005

The enactment of this act was to provide for effective and efficient supervision of disasters and for matters connected herewith or incidental thereto. The act affords an organized operational mechanism for designing and supervising the execution of disaster management. The Disaster Management Act makes sure that ample amount of measures are taken by different governmental wings to avert and alleviate disasters as well as a fast response to any sort of disaster situation. The act features the following:

- The NDMA with Prime Minister as the chairman, State Disaster Management Authority (SDMA) with Chief Ministers as the chairman, District Disaster Management Authority (DDMA) with collectors/district magistrates/deputy commissioners as the chairman.
- Additionally, both national and state level executive committees need to be set up under this act.
- For the purpose of capacity building, the National Institute of Disaster Management (NIDM) and for the purpose of immediate response, the NDRF was established under the auspices of this act.
- The act also commands all the concerned ministries and departments to formulate their strategies in compliance with the national strategies.
- It provides for making the financial provisions like financial resources for response, NDMF as well as other analogous means at lower levels of administration such as state and district.
- The Act has provisions for the precise role of the local bodies.

With further seventy-third and seventy-fourth constitutional amendments and surfacing of local self government like urban and rural as primary tiers of governance, the function of local dominion becomes most evident. The Disaster Management Act of 2005 did foresee the particular responsibilities to be shouldered by the local dominions in management of catastrophe.

Organization and structure of disaster management: The Joint Secretary (DM) of MHA leads the Disaster Management Division. He receives assistance from three Directors, Under Secretaries, Senior Economic Investigator consultants, Section Officers, Technical Officer and various support staffs. The uppermost level of this arrangement is made up of Home Minister, Minister of State in charge, Home Secretary, Secretary (Border Management).

8.2.6.4 Disaster management framework

- Disaster management began addressing issues like early warning, forecasting and setting up of monitoring stations to deal with various weather related hazards.
- Emergence of an arrangement of information flow in the type of forewarnings, vigilance and up to date information regarding the approaching hazard.
- Setting up of a multi-stakeholder high level conglomeration by integrating people from various ministries and departments whilst declaring some of the ministries as nodal authorities to deal with certain types of disasters.
- Emergence of a multilevel link between the different ministries.

Environment relief fund: On 4 November 2008, under Public Liability Insurance Act (PLIA), 1991 and in order to implement the command as per Section 7(A), the Environment Relief Fund Scheme was set up by the central government of India.

Figure 8.3: Structure of disaster management in India

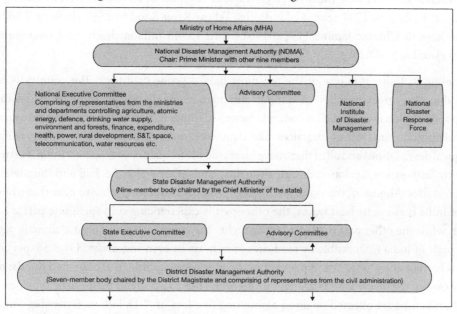

India's association with the world: Besides setting up institutions in the country, India has also been associated with various international organizations.

A. United Nations Development Programme

Since the 1993 earthquake UNDP had been tied up with India. Subsequently, UNDP was involved in capacity development programme in 1996. UNDP assisted community based disaster risk management programme in 20 blocks of Orissa after the super cyclone in Orissa in 1999. Under this programme, in a joint venture with the MHA, UNDP furnished a monetary support of US $ 41,000,000 to the GoI and 17 state governments in order to reduce the threats basically at the society level in 176 susceptible districts for multiple hazards.

B. Indo-Swiss agreement

On 10 November 2003, the GOI signed an agreement with the Government of Switzerland for spreading out cooperation in order to prevent and prepare for handling disasters.

C. Indo-Russian agreement in the field of emergency management

On 21 December 2010, India signed a bilateral agreement with the Russia Federation during the visit of the President during Indo-Russia Summit.

8.3 Disaster Profile of India

Before taking a journey through Indian disasters it is good to remember that globally there were more than 200 disasters in the second half of twentieth century alone claiming nearly 1.4 million lives. Notable disasters were the Shaanxi earthquake of China in 1556 with over 8,30,000 deaths, cyclone in Kolkata in 1737 with 3 lakh deaths, Yellow River flood in 1887 claiming 9 lakh to 2 million lives to Chinese famine (1958–1961) with 15–43 million deaths and Cyclone Bhola (1970) claiming 5 lakh lives.

Coming to India, it is one of the ten most hazard prone countries. The country is highly vulnerable to a great number of natural and man-made disasters owing to its geological, topographical, climatic and socioe-conomic factors (demographic pressure, poverty, urbanization, industrialization, unscientific practices like dam construction, deforestation, overgrazing and ecological degradation) and all of these have increased the frequency of disasters. India is extremely prone to earthquakes, cyclones, deluge, droughts, landslides, avalanche, hail and thunderstorms and forest fires. Almost all the states are disaster prone. It is not surprising to hear that while one part of India is swept by heat waves, the other part is experiencing cold spell; one part is having deluge, while the other part is affected by drought. The complicated and erratic climatic patterns make parts of India inaccessible by roadways or railways or even waterways. Over 58 per cent of India is earthquake prone, over 8 per cent of the area is susceptible to erosion and floods; nearly 68 per cent of the arable land is drought prone while the coasts are in the peril of cyclones. Each year 8 million ha are affected by flood amounting to a loss of ₹ 18 billion. Extensive damage is caused not only to the crops but also to the houses and other utilities. The frequencies of disasters are remarkable, humanity has largely adapted to these catastrophic events, but the social and economic costs are mounting every moment.

India exhibits distinct regions on the basis of vulnerability – the Himalayan region, the alluvial plain region, the hilly peninsula, deserts and the coasts. The Himalayan region is susceptible to earthquakes, landslides, avalanches; the plains (Indus, Ganga and Brahmaputra) suffer regular flood especially in UP and Bihar; the arid part (Rajasthan, Gujarat and parts of Maharashtra) is affected by famine and drought and the coastal regions are very prone to storms and cyclones.

The country experience two seasons of precipitation; three fourth of the annual spell happens during June–September. Rainfall varies greatly in different places. Average precipitation may vary from 13 cm or less in west Rajasthan, 250 cms in Western Ghats and Khasi Hills to 1,141 cm in Mausinram of Meghalaya per year.

Deficiency of rainfall causes drought. It was more frequent in the 1960s with 27 major drought years from 1871–2009. Drought is found to bear a close tie with El Nino Southern Oscillations (ENSO) patterns resulting in decline of ground water and increased salinization.

Floods mostly occur during monsoon season and are linked to tropical storms and depressions. They occur almost in all the river basins when the rivers flow with inadequate drainage capacity.

Landslides, typhoons, cyclones and cloudburst also causes flood. 23 states are flood prone covering nearly one-eighth of India's area or 40 million ha of land. In 1954, the National Flood Control Programme was launched after which considerable progress was made in flood control. From 1953–2005, over 84,000 lives were lost.

Table 8.1a: Significant droughts in India

Drought year	Regions affected	People affected
1972	Most of India	Approximately 200 million
1987	Haryana	Approximately 300 million

Table 8.1b: Significant floods in India

Flood year	Regions affected	People affected
2005	Maharashtra (994 mm rain)	1,094 deaths and several injured and missing
2008	North Bihar (Kosi)	527 deaths along with cattle, houses damaged
2010	Leh	

India with a coastline of nearly 7,516 km is exposed to storms and cyclones. The states of Andhra Pradesh, Orissa, West Bengal, Gujarat, Maharashtra, Karnataka, Kerala, Tamil Nadu, Puducherry, and Goa alongside with Andaman and Nicobar Islands and Lakshadweep are particularly vulnerable to tropical storms. On a regular basis 5–6 cyclones strike each year. The wind speed varies between 65 km/hr to 164 km/hr.

Table 8.1c: Significant cyclones in India

Cyclone year	Regions affected	People affected
1977	Andhra Pradesh	10,000 people and 40,000 cattle dead
1999	Orissa (Super cyclone) up to 250km/hr	10,000 people died
2008	Tamil Nadu	204 died with damage amounting to 800 million Us dollars.
2008	Myanmar (Nargis)	1,40,000

The Indian subcontinent, at the boundary of two continental plates is susceptible to quakes. As per BIS, India is divided into four seismic zones – II, III, IV and V according to India's susceptibility to earthquake of given intensity in Modified Mercalli Scale. Northeast India, northern part of Bihar, HP, Uttarakhand, Jammu and Kashmir, Gujarat, Andaman and Nicobar Islands lies in Zone V represents the highest risk and is referred to as Very High Damage Risk Zone while zone II is the Low Damage Risk Zone. Earthquake beneath the sea may cause tsunamis. Tsunamis may have wavelengths of hundreds of kilometre, amplitude of 1 km and a speed of 800kms/hr. The earthquake should be of magnitude 7 or more to hit tsunamis in Indian Ocean. The most tentative places are Andaman – Sumatra or Makran.

Landslides mostly affect the Himalayas and Western Ghats. The Himalayan region of Sikkim and Garhwal experience 2 landslides/km². Himalaya's have a history of landslides owing to its unique features.

When temperatures rises much above the normal temperatures, heat waves result. Casualties have increased due to abnormally high temperatures in states like Orissa, Bihar, UP, Maharashtra, etc.; maximum temperature went up to 49°C at Gannavaram in Vijayawada in May 2002. Similarly extremely low temperatures with cold winds give rise to cold waves, prevalent in the Northern provinces. The highest number of cold wave casualties are from UP and Bihar.

The southern part of India and Kolkata gets thunderstorms; dust storms are quite frequent in the north and northwestern regions while hailstorms are common in Assam, Uttarakhand, Jharkhand, Maharashtra.

Man-made disasters may be chemical disasters, stampede in socio-cultural situations, road accidents, rail and air accidents, mining disasters (Chasnallah Colliery), epidemic outspread, riots, terrorism and crime. These are quite frequent in India.

8.3.1 Flood management

Risk assessment in India reveals around 40 million ha of the country to be flood prone. About 75 lakh hectares of land is affected and more than 1,500 die due to flood each year severely damaging the crops, houses and public utilities worth ₹1,805 crores. It is worth to mention that in 1977 more than 11,000 lives were lost. Torrential downpour and overflowing of the river banks, excessive irrigation, inadequate drainage, unscientific encroachment of waterways, cyclones and cloudbursts, all contributes to flood. Urban flooding in the cities of Mumbai, Kolkata, Chennai, Ahmedebad, Bangalore, is now a recurring phenomenon these days owing to population pressure, lack of regulations, encroachment, poor or no drainage.

- The NDMA has shifted the focus of flood management to preventive, mitigation oriented and preparedness by regular monitoring, modernization of early forecasting support systems, incorporating flood resistant design and construction, improved awareness of the stakeholders, education, training and capacity building, thorough research and development, implementation of a sustainable, time bound, techno-economically viable, environmental friendly measures along with consolidation of emergency response.

- Identification of the flood prone areas at national, state and district levels are to be carried out by Central Water Commission (CWC), Ganga Flood Control Commission (GFCC) or

Brahmaputra boards and other state boards in association with National Remote Sensing Agency (NRSA) and Survey of India (SOI); joint formulation of flood forecasting by CWC, IMD or NRSA or the states.

- The topic of flood management is to be incorporated at various levels of education from schools, colleges, technical institutions, defence, etc, by the MHA, MHRD and Ministry of Defence (MoD) in consultation with councils or secondary and higher secondary education, University Grants Commission (UGC), All India Council of Technical Education (AICTE) etc.

- The roles of various groups like police force, civil defense, fire services, NGOs, women groups, youth organizations, and corporate houses should be institutionalized apart from NDRF and SDRF. Massive awareness and education programmes including preparation of guidelines and manuals for general people in several languages.

- The by-laws are to be amended keeping pace with time and need of the people, to comply with the flood resistant designs and structures. Integrated Water Resources Management (IWRM) for intra-state rivers, water shed management, Catchment Area Treatment (CAT), afforestation and building of reservoirs for storage should be developed. If need be for reservoirs, negotiations with Nepal and Bhutan should also be persuaded.

Case Study 8.2: Flash floods in Leh (Ladakh), 2010

It was quite surprising for general masses to hear about flood in the mountainous and arid regions of Leh. On 6 August 2010, midnight, massive flash floods triggered and swept the Leh region of Ladakh. The flash flood resulted due to unprecedented cloudbursts. The downpour was about 14 inches in two hours. Leh and Kargil are two districts of Ladakh; with Leh comprising the Leh town and 113 villages of which one is uninhabited. Leh lies on the plateau at an elevation of 3,500 km above mean sea level. Rainfall is low. The flash floods carried muddy waters from the higher regions and followed three diverse routes. It buried the houses and washed out almost everything in its course leaving large chunks of boulders and deposits of silt. 34 villages were affected by the event and the village of Leh town, Choglumsur and Saboo were the worst hit of all. The death toll was 234 with 424 seriously injured. 800 people were reported missing including few foreigners. There were cases of death in Rajouri, Reasi, Kathua, Poonch, Doda, Varmul, Udhampur, Kargil, Budgam and Shopian. Varmul and Shopian, famous for apple orchards, suffered huge loss both in horticulture and agricultural sector. There was scarcity of food and water. Stagnancy in water increased the risk of silt contamination and with corpses buried in the debris. There was increased chance of water borne diseases. Major transport routes were blocked like the Leh-Srinagar and Leh-Manali highways. Leh Civil Hospital (SNM) and the Army Hospital was rendered dysfunctional on account of destruction; so was the mobile network and other forms of communication.

The GOI responded immediately. Rescue and relief operations were primarily by the Indian Army, Jammu and Kashmir State Government, Central Reserve Police Force (CRPF) and Indo-Tibetan Border Police (ITBP). The center released ₹ 5 crores for the erecting of pre-fabricated structures. The injured and dead people were sent to Army Hospital. The runway of Leh airport was cleared soon after and tertiary level healthcare was provided at the Army Command Hospital in Chandigarh. Army engineers started restoring and reconstruction of the collapsed bridges immediately.

An effective, emergency response involves search and rescue operations to minimize the loss of lives and properties, providing first aid and medical assistance along with providing immediate relief to the flood affected people.

8.3.2 Landslide management

Landslides are downward movement of rocks, debris or land under the gravitational force. It occurs in the mountainous terrains which may be due to rainfall, cloudbursts, earthquakes, floods, cyclones or unscientific human activities. Over 8,500 individuals have died between 1990–99 because of landslides and avalanches along with damage to livestock, infrastructures, building, roads, railway tracts, water supply, power supply, ropeways, mines, tunnels, holy places, pilgrimage routes and other resources. It costs about ₹ 300 crores per year.

The largest landslide is the Ambutia landslide in Darjeeling. Landslides are quite frequent in the NH1A and NH1B in Rishikesh-Badrinath route, Uttarakhand, Sikkim and Darjeeling, Dimapur--Imphal, and Shillong-Shilchar highways. Well-known landslide events are Kapkot (2010), Leh debris flow (2010), Konkan (2005), Nilgiri(2009) and Malpa (1998), etc.

Mapping of landslide vulnerability, geo-technical investigations, scientific research, abiding of land use regulations, should be done and building codes are to be abided by. The vulnerability can be reduced by early warning, proper land use planning, proper engineering techniques, preparedness and rapid response and recovery after landslides as well as awareness and education including preparing guidelines and manuals for general people in several languages.

8.3.3 Cyclone management

Cyclones or tropical cyclones refer to the winds with a minimum speed or 34 knots or 62 km/hour. The winds blow around the low pressure in anti-clockwise direction in Northern Hemisphere and clockwise in the Southern Hemisphere with little wind free from cloud and rain in the centre or eye of the cyclone. It is usually characterized by destructive winds, storm surges and torrential rainfall. More cyclones are prevalent in the Bay of Bengal. The incidence of cyclones in Arabian Sea and Bay of Bengal is in the ratio 1:4. Cyclones with wind speed between 89–118 kmph are grouped as Severe Cyclones that with 119–221 kmph are Very Severe Cyclones while that above 222 kmph are Super Cyclones as per the IMD. Cyclones are known as typhoons in North West Pacific, hurricanes in the North Atlantic, Willy-Willies in North Western Australia and Tropical Cyclones in the North and South India Ocean. The process of naming cyclones is carried under the aegis of the WMO. It started in 2000 for the cyclones in Indian Ocean where eight countries,

India, Bangladesh, Pakistan, Maldives, Sri Lanka, Thailand, Myanmar, Maldives and Oman have contributed few names, which are to be assigned sequentially. Say for instance, 'Nilam' was given by Pakistan while 'Phailin' was given by Thailand.

The first initiative in India was the **Cyclone Emergency Reconstruction Project (CERP)** assisted by World Bank initiated by Andhra Pradesh following the severe cyclone in May 1990. A **National Cyclone Risk Mitigation Project (NCRMP)** was formulated by the MHA, GOI and subsequently transferred to NDMA in 2006 with four components.

A. Improving early warning broadcasting system

- Establishment of Early Warning System (EWS) comprising, observation, predictions, warning and advice.
- Commissioning of Aircraft Probing of Cyclone (APC). Facility with both manned and unmanned aerial vehicles (UAV) to fill up the critical gaps over Arabian Sea and Bay of Bengal.
- Commissioning of National Disaster Communication Infrastructure (NDCI) at NDMA, SDMA and DDMAs of the 84 coastal districts that are cyclone prone, such as high end computing, 3D Virtual Reality Visual Studio, Central Comprehensive Data Bank, etc.
- Warning distribution through Direct-To-Home (DTH) transmission system in the remote areas.

B. Cyclone risk mitigation investment

- ensuring safe lifeline infrastructure in the coastal region;
- construction of tough multipurpose cyclone shelters, cattle mounds;
- ensuring cyclone resistant design structures;
- constructions of all weather roads linking all shoreline habitations;
- designing of drainage systems to carry additional flow;
- mapping coastal wetlands and mangroves;
- water quality monitoring;
- developing Integrated Coastal Zone Management (ICZM);
- restoring and spreading bio-shields and mangroves; and
- generation of personal and community awareness, preparing guidelines and manuals for general people in several languages.

C. Technical assistance for hazard risk management and capacity building

- It covers all aspects from prevention to recovery. The main response is undertaken by the NDRF and SDRF along with district administrations, civil defence, NGOs and locals.
- It involves training knowledge management, technology framework, etc.

D. Project management and support from institution

- This covers development of cyclone impact assessment, rescue and evacuation plans, medical preparedness, identifying vulnerable populations, conducting meetings, documentation etc.

- National Cyclone Disaster Management Institute (NCDMI) will set up a platform to synergize various stakeholders starting from the central government to the local level, from academicians and scientists to lay people to offer the best disaster risk reduction options.

Case Study 8.3: Hurricane Katrina, 2005

Hurricane Katrina, 2005 started as a very low pressure weather system but strengthened to become a tropical storm and eventually a hurricane as it moved west and neared the Florida coast on the evening of 25 August. Hurricane Katrina impacted about 90,000 square miles. It was a Category 5 hurricane, with winds up to 175 miles per hour. The gale surge from Katrina was 6 m high. The death toll was nearly 1,836 mostly from Louisiana and Mississippi. Around 80 per cent of New Orleans was under water, up to 20 feet deep in places. It caused about $81 billion property damages; the total economic impact in Louisiana and Mississippi may be more than $150 billion, earning the title of costliest hurricane ever in US history.

Case Study 8.4: Cyclone Aila

Cyclone Aila was a strong tropical cyclone of the North Indian Ocean in 2009 causing extensive damage in India and Bangladesh. More than 300 people have been killed, nearly 8,000 or more are missing and around one million people became homeless. The damage from Aila totalled $ 552.6 million.

8.3.4 Earthquake management

International experiences have revealed that greatest benefit from earthquake management can be obtained by consolidating the pre-earthquake mitigation and preparedness efforts. It is realized that there is a greater than before need for a much organized, holistic and comprehensive endeavour to embark upon the grave areas of apprehension responsible for feeble seismic protections in India. A continual improvement in structural and non-structural measures in reducing of seismic risk can bring seismic safety.

The Himalayan region was regarded as quake prone to a high magnitude above 8 on Richter scale. Notable earthquakes such as Shillong in 1897 with M 8.7, Kangra in 1905 with M 8.0, Bihar-Nepal in 1934 with M 8.3 and Assam-Tibet in 1950 with M 8.6 strongly support the fact. This opinion was ousted after the earthquakes in Koyna in 1967, Latur in 1993 and Bhuj in 2001. A pressing requirement for the revision of the Indian seismic zone map was felt.

With assistance from the DST and other specialized institutes like National Geophysical Research Institute (NGRI) in Hyderabad, Wadia Institute of Himalayan Geology (WIHG) in

Dehradun, Indian Institute of Geo Magnetism (IIG) in Mumbai and Centre for Earth Science Studies (CESS) in Thiruvananthapuram, the Global Positioning System (GPS) measurement throughout India was started by the Survey of India (SOI) in order to supervise tectonic movement.

The GoI had established the MoES by combining Earth Commission, IMD and other scientific and technical institutions. The nodal or crucial ministry to address the problems regarding supervision of seismic activity and networking of forewarning is the MoES. It coordinates its activities with IMD, Earthquake Risk Evaluation Center (EREC), Bureau of Indian Standards (BIS) and the GSI along with other expertise. NDMA have outlined certain guidelines for the individuals to follow.

There are six pillars or guidelines that have been outlined for earthquake management. They are:

A. **Safeguard the integration of strategies in the building of new construction that can withstand quakes**

   - Most of the constructions are non-engineered and do not obey earthquake-resistant construction principles. Indigenous earthquake-resistant structures, such as, bhongas in Kutch, dhaijidiwari in J&K, and ekra in Assam are now being replaced by modern cement concretes without complying building codes and byelaws.

   - To ensure seismic safety, the central ministries and state departments will implement and enforce constructions with seismically safe design standards as specified in the byelaws that will have to go through a process of scrutiny by competent authorities through a general review and verification before building approval and technical audit process developed by MHA.

B. **Expedite selective intensification and tectonic retrofitting of the existing prioritized construction in regions susceptible to earthquake.**

   - Approximately there are 12 crores buildings in seismic zone III– V. It is not feasible to retrofit all the existing buildings either.

   - A compilation of the vulnerable building inventory is to be done to formulate a seismic risk profile. Either Rapid Visual Screening (RVS) or Detailed Vulnerability Assessment (DVA) can be used for assessment of vulnerability.

   - Selections of vital lifeline constructions and highly prioritized edifice for retrofitting should be based on the degree of seismic hazard, population size and potential for loss of life, condition and importance of the building.

   - The strengthening and retrofitting will be undertaken under Urban Earthquake Vulnerability Reduction Project (UEVRP) in phases in collaboration with United Nations Development Programme or through National Earthquake Mitigation Project.

C. **Development of compliance system through suitable regulation and enforcement**

   - The essential techno-legal and techno-financial mechanisms will be established by the state governments after consulting with the State Earthquake Management Committees (SEMCs) and Hazard Safety Cells (HSCs) and ensure all stakeholders to abide by it.

- The design codes are developed by BIS, Indian Roads Congress (IRC), Ministry of Shipping, Road Transport and Highways (MoSRTH), Research Designs and Standards Organization (RDSO), Ministry of Railways (MoR) and Atomic Energy Regulatory Board (AERB), Department of Atomic Energy (DAE).

- The codes should be updated time to time and all professionals will have to be certified through licensing.

- Self certification for all edifices forms an essential element of the procedure, it is also essential that critical structures will have to be reviewed by competent external agencies.

D. Develop awareness and preparedness of all the shareholders

- Awareness involves sensitization and educating all stakeholders; preparedness involves formulation of emergency plans for family and community, conduction of mock drills, training for administering first aids, preparation of handbook and manuals in many languages for general mass.

E. Announcement of suitable capacity development intervention for effective earthquake management including education, training, research and development and documentation

- Educating the mass about earthquake mitigation, preparedness and response strategies. Disaster related curricula have been included in the secondary education of Central Board of Secondary Education (CBSE).

- State governments should incorporate disaster related curriculum at higher secondary and undergraduate studies in all institutions.

- Medical colleges to have the subject of disaster medicine to handle emergencies, etc.

- The subject, Earthquake Engineering, was initiated in 1960 in School of Research and Training in Earthquake Engineering (SRTEE), IIT Roorkee and thereafter Central Building Research Institute (CBRI) and various other institutes.

- Device installations were done by the IMD for the recording of strong motion data; EREC was established to add to national efforts; IIT, Roorkee have designed Structural Response Recorders (SRRs).

F. Reinforcement of emergency response competence

- Over 80 per cent of the searches cum rescue operations from collapsed structures are done within localities before government team intervenes, as observation reveals.

- So along with local participation, NDRF battalions, SDRF, Police Force, Fire services, Home Guards, Civil Defence, NGOs, National Cadet Corps (NCC), National Service Scheme (NSS) and Nehru Yuva Kendra Sangathan (NYKS) should all be adequately trained to the search, rescue and relief operations.

- The corporate sector as a part of Corporate Social Responsibility (CSR) can also provide services to hospitals, communication, power, relief transport, equipments and logistics supply.

Case Study 8.5: Gujarat earthquake in 2001

Gujarat is a western province of India. The indo-Australian and Eurasian continental plates are moving at a pace of 2cm/year beneath India. Fold Mountains like the Himalayas are the product of such movements at the compression boundary. Such movements create stress within the rocks and earthquake may happen with release of such shocks. The epicentre of Gujarat earthquake was about 20 km northeast of Bhuj in Kutch district of Gujarat. The quake on 26 January 2001, was generated 23 km below the surface, measuring 7.9 on the Richter scale. The intense shaking shook almost two-third of India and reached Pakistan and Nepal. The Indian Meteorology Department has recorded almost 500 aftershocks above 3.0 magnitude that continued during the first quarter of the year. Kutch is in the seismic zone V and the only zone outside the seismic belt.

21 districts out of 25 was affected including 18 towns, 182 talukas and 7,904. Bhuj, Anjar, Bavhau and Rapar and 450 villages were totally devastated. Ahmedabad, Rajkot, Jamnagar, Surendranagar, Gandhiram, Morvi and Patan were highly affected.

As per government records, nearly 20,000 people died with over 1,60,000 injured. 3,50,000 houses were destroyed and more than 6 lakh people were homeless. Over 20,000 cattle were killed. The total affected was 15.9 million people with an estimated damage varying between 1.3 to 5 billion US dollars.

Soon after the disaster the National Disaster control room was activated with mobilization of para-military and army personnel. The Gujarat government announced four packages to about 1 billion US dollar for reconstruction and economic rehabilitation of 3 lakh families. 2 lakh blankets and over 20,000 tents and tarpaulins were dispatched. Medical facilities were made available and petroleum supply was resumed from Jamnagar by the railways. For communication, hot lines were activated between Ahmedabad and New Delhi and satellite phones were installed.

World Bank announced financial assistance of 300 million and Asian development bank announced 500 million US dollars. The Housing and Urban Development Corporation and National Housing Bank (NHB) announced a financial help of 400 million US dollars. Groups like Reliance, VSNL, L&T, Tata Steel, FICCI, Coca Cola, Associated Chambers of Commerce and Industry of India (ASSOCHAM), ESSAR, etc decided to take part in the relief and rehabilitation programmes. EU, United States Agency for International Development (USAID), Canadian International Development Agency (CIDA) and other international agencies also came forward for help.

Following drought there were reports of raised water table to surface level. A number of new springs with fresh and salt water appeared. The once dry rivers started flowing again. Gujarat earthquake was very significant from the view point of disaster mitigation in India and evoked an urgent need of proactive disaster management in India. Apart from extensive destruction to property it left thousands of people injured, bruised and handicapped tormenting people physically, psychologically and economically.

Case Study 8.6: 2004 Asian tsunamis

Commonly known as Indian Ocean or Indonesian tsunamis it is also referred to as Christmas tsunamis on account of its incidence on 26 December 2004. The hypocenter was located about 160 km in the Indian Ocean, north of Simuelue Island to the west coast of Sumatra. The tsunamis lasted for 10 minutes and measured 9.9.3 on Richter scale. The epicentre was centered in Indian Ocean, near the west coast of Sumatra of the Indonesian island. Due to the lateral movement of the plates (the Indian plate slid under Burma plate), a fault of 250 km occurred, the sea floor rose by about 20 m, opened a rift of 10 m depth, displacing about 30 cu. km of water. The entire planet seemed to vibrate a few centimetres so much so that the tremor was felt even in Alaska. The resultant were the harbour waves more than 30 m high killing masses in Sri Lanka India, Thailand, Indonesia, Somalia, Myanmar, Malaysia, Maldives, to name a few. Nearly 2,30,000 to 3,10,000 people died, over 45,000 went missing, over 1 lakh injured, over 1.6 million displaced and many bodies swept in the sea; 9,000 deaths among them were of tourists. Environmental assets such as coral reefs, sand dunes, mangroves, ground water, land, vegetation and animals incurred sever damage, there were reports of salt infiltration and spread of epidemics like cholera, typhoid, dysentery, hepatitis, etc.

Summary

- Disaster is a catastrophic situation in which the usual pattern of life or ecosystem has been disrupted and extraordinary emergency interventions are required to save and preserve lives and or the environment. It involves an event comprising trauma for a population/environment, a vulnerable position/area that will stand the load of the traumatizing incident and the breakdown of local and neighbouring resources to cope with the problems created by the phenomenon. Disasters are characterized by inevitability, pace of onset, extent of impact and intensity of impact.

- Disaster can be a natural disaster such as, meteorological or biological. It can also be man-made disasters like those airing from stampedes, blaze, road accidents, railway accidents, airliner accidents, oil spill, nuclear disasters to chemical and other industrial disasters.

- The components at risk in disasters are the people, livestock, vulnerable houses, crops, trees, telephone lines, electric poles, boats, looms, working implement, electric wires and cables, water and food supplies.

- The impacts of disasters comprises loss of life, injury and disability, spread of epidemics, water borne, food borne and air borne diseases, psychological trauma, disruption to essential services – shortage of food, medicines, vandalization, national economic loss, loss of livelihood and damage to national infrastructure and governmental systems.

- Disaster management is basically a dynamic process comprising the task of planning, systematizing, recruiting and training crews, leading, administering and supervising. It involves multiple organizations, which must work concertedly in order to avert (prevent), alleviate (mitigate), prepare for (arrange), respond to (react) and recover (recuperate) from the effects of disaster. Disaster management is a complex humanitarian process involving international, national and local organizations and individuals each with a distinct role to play. Disaster management is the accountability at all levels of government and needs contribution and prospective role of non-governmental agencies as well.

■ Disaster management cycle involves the pre-disaster phase (prevention and mitigation and preparedness), the disaster phase (response and relief) and the post-disaster phase (recovery and rehabilitation).

■ Disaster management in India has evolved from an activity-based reactive set up to a proactive, practical institutionalized organization; from single faculty domain to a multi-stakeholder arrangement; and from a relief-based approach to a multidimensional proactive holistic approach for reducing risk. The Disaster Management Act, 2005 provides for the effective disaster management and for matters connected herewith or incidental thereto.

■ Globally there were more than 200 disasters in the second half of twentieth century which alone claimed nearly 1.4 million lives. Coming to India, it is one of the ten most hazard prone country with almost all the states being disaster prone. The country is highly vulnerable to a great number of natural and man-made disasters owing to its geological, topographical, climatic and socio-economic factors.

Exercise

MCQs

Encircle the right option:

1. The place of origin of earthquake is:
   A. Fault   B. Epicenter   C. Focus   D. Eye

2. The study of earthquake is:
   A. Seismology   B. Siesmology   C. Etomology   D. Phorosics

3. During tsunamis, the largest wave is the _____ wave.
   A. First   B. Second
   C. Unpredictable, could be any of them   D. Third

4. Tsunamis are destructive due to their:
   A. height of the wave   B. Uncertainty
   C. Momentum and long wave length   D. Temperature

5. Drought is:
   A. A year without rainfall   B. A week without rain
   C. Water scarcity over a long period   D. A season without rain

6. Flood is:
   - A. Excessive rain
   - B. Water overflows to submerge land
   - C. Excess water
   - D. Excessive water depth

7. The naming of tropical cyclones is useful:
   - A. For reporting
   - B. For the forecasters
   - C. For research
   - D. For tracking, forecasting and reporting

8. Pre-disaster management includes:
   - A. Prevention
   - B. Mitigation
   - C. Response
   - D. A and B

9. The charge of disaster management just after independence was with:
   - A. Ministry of Water Resource
   - B. Ministry of Home Affairs
   - C. Relief Commissioners
   - D. Ministry of Human Resource

10. The Disaster Management Act was enacted in:
    - A. 2005
    - B. 2002
    - C. 2006
    - D. 2000

Fill in the blanks.

1. Disaster Management Act was enacted in _____.
2. The uranium bomb dropped at Hiroshima is known as _____.
3. World Conference on Natural Disaster Reduction, 1994, took place in _____ , Japan.
4. As per BIS, India is divided into _____ seismic zones.
5. The minimum wind speed requisite to qualify as cyclones should be _____.

State whether the statements are true or false.

1. Disasters are characterized by inevitability. (T/F)
2. Riots are a type of natural disaster. (T/F)
3. Phaillin was a Category 5 cyclone. (T/F)
4. Disaster management implies relief and rehabilitation. (T/F)
5. Disaster management is a matter under the Ministry of Defence.(T/F)

Short questions.

1. What is a disaster?
2. How is disaster different from hazards?
3. State the characteristics of a disaster.
4. What are the factors influencing a disaster?
5. Classify disasters and cite suitable examples.
6. Enlist the elements at risk during disaster.
7. Enumerate the general impacts of any disaster.

8. How many seismic zones is India divided? Which are the most earthquake prone areas in India?

9. What are tsunamis?

10. Define disaster management. Why do we need disaster management plan?

Essay type questions.

1. What are the important provisions of Disaster Management act, 2005?

2. State the major disasters that have hit India in the recent past?

3. What is the extent of flood risk in India? What are the important mitigation measures adopted to control flood situation in our country?

4. 'We must, … shift from a culture of reaction to a culture of prevention. The humanitarian community does a remarkable job in responding to disasters. But the most important task in the medium and long term is to strengthen and broaden programmes which reduce the number and cost of disasters in the first place. … Prevention is not only more humane than cure; it is also much cheaper'. … Briefly describe the evolution of the present disaster management strategy in India.

5. What is a landslide? Which regions of India are vulnerable to landslide?

Social Issues and Environment ——————— 9

Learning objectives

- *To understand the concept of sustainable development.*
- *To be informed about urban problems.*
- *To identify the need for water conservation and its various practices.*
- *To identify the aspects of resettlement and rehabilitation due to development.*
- *To develop a concept of environmental ethics.*
- *To get acquainted with different global environmental issues towards seeking solution.*
- *To know about the various existing environmental laws and their implications.*
- *To incite awareness in the general people.*
- *To know the effective role of citizens and groups towards environmental protection.*

9.1. Unsustainable to Sustainable Development

Industrial Revolution unleashed in its wake, massive development, especially in Europe and America. Fueled by fossil fuels, these economies witnessed industrial activities in scales hitherto unimaginable with newer and bigger orbits of growth being achieved as a matter of natural consequence. However, in this self-perpetrating rush for growth and development, two vital issues were conveniently ignored.

- The Earth's resources are finite and therefore they will run out sooner or later. With exponential growth in their consumption and usage the prospect of their getting extinguished can only be hastened. Peak oil, is a prime example.

- There are adverse effects of burning fossil fuels. The rate at which fossil fuels are burnt releasing carbon dioxide in the atmosphere cannot be matched by all the plants on Earth that convert such CO_2, resulting in increasing concentration of the gas with its attendant problems. Again, the pollutants that such industrial activities give rise to as byproducts are seriously damaging the fragile eco system. The result is what may be loosely termed as climate change and global warming.

By the time mankind really took cognizance of the facts, the damage had already been done with the wheels set in motion that will have far reaching (and needless to say, terribly adverse) consequences for the world we live in.

The rich, industrial nations that have by now have become used to a lifestyle of conspicuous consumption with inordinately huge carbon footprints, were and are, still not ready to sacrifice their desire for industrial development to offset the corrosive results of their ecological sins. To appear responsible, they sought out the common high ground, that of sustainability – a lifestyle that urges one to consume in a manner that ensures one leaves enough for the future generations. It is about the optimal usage of resources by inducting efficient technologies that reduce wastage and increase per unit efficiencies on the one hand, while consciously taking steps to reduce the adverse effects of pollution on the other.

By classical definition given by World Commission on Environment and Development (Brundtland Commission), 1987, sustainable development is defined as that which '*meets the needs of the present without compromising the ability of future generations to meet their own needs*'. Since then the definition is constantly being revised, extended and refined but it still lacks uniform interpretation. But it is clearly understood that this intergenerational equity is impossible to achieve in the current scenario of societal inequity; especially if the economic activities of a number of groups go on to jeopardize the welfare of other groups or of those people that subsist on other parts of the globe.

Sustainable development otherwise could presumably be called 'equitable and balanced'. It means for development to persist for the foreseeable future, it should balance the interest of the various communities, within the same generation and among generations in three most important areas – economic, social and environmental aspects. Thus, sustainable development explains equality of opportunities for the well being of people in general and about extensiveness of goals. With so many varied objectives, striking a balance while formulating environmental policies, becomes a challenge for any nation. No fixed method can be applied to carry out evaluations and comparisons. The government must plan and decide at frequent intervals that they reflect the welfare of major sections in the society. These decisions ought to be accepted in the most democratic and participatory basis.

The idea is not to put a cap on development. While the industrially developed countries in the west have already reached optimum levels, there are a number of nations, BRICS (Brazil, Russia, India, China and South Africa) for example, that are today waiting to keep their tryst with economic development. Thus, it will neither be possible for the world at large to sacrifice development on the altar of climate change, nor will it be just and equitable. Therefore, the logical way out will be to embrace sustainability – to walk the path of development, but like responsible world citizens aware of the pitfalls on the way and concerned, not only about conserving but also about consuming only what is essential so that one may leave a cleaner, greener and more prosperous world for the children of tomorrow. The present day ideals ought to be knowledgeable enough to reflect the benefit of future generations so as to guarantee that the upcoming generations accede to all the essential situations that will supply for their own well-being.

Figure 9.1: Objectives of sustainable development

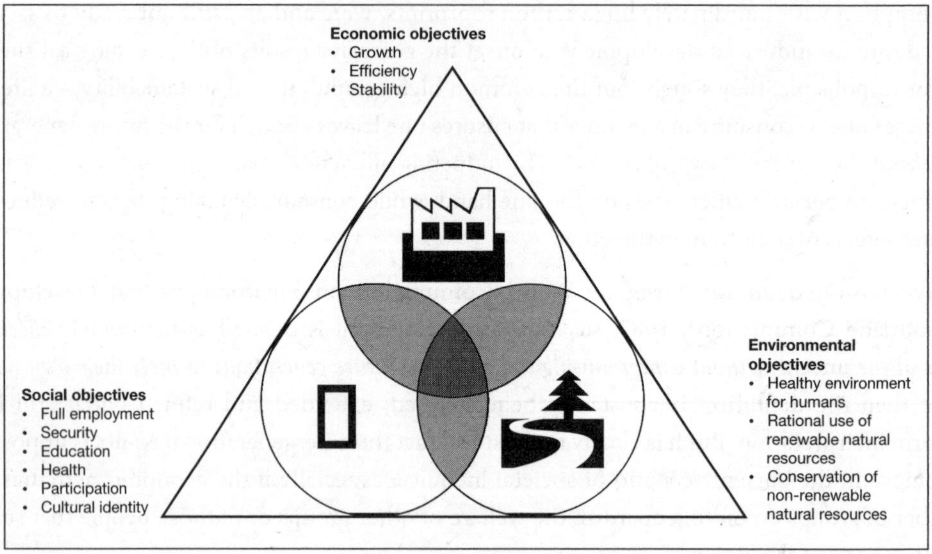

9.2 Urban Problems Related to Energy

The world is increasingly becoming urban with more and more people migrating to the cities in search of food, work and shelter. While some are being forced to relocate as opportunities dry up in their traditional lands, others are travelling in search of a better life while still others are forced by climatic conditions that make their homes inhabitable.

As urban communities consume more energy per capita than their rural counterparts, such an influx is sure to have an adverse effect on the demand for energy. With prevailing systems that are heavily dependent on fossil fuels as the primary source of energy, such growing demands are creating energy bottlenecks which threaten to blow out of proportions if not addressed immediately. The global demand for energy far outscores the supply for the same and with increased industrialization and urbanization the gap is becoming wider by the hour. Add to it the fact that energy production cannot be scaled up the way previous generations had been doing, considering the fast depleting resource base along with concerns of sustainability and climate change. The per unit cost of generation will also become a limiting factor as the cost of energy, following natural laws of demand and supply, go up to unimaginable levels. In urban areas, energy is required for household lighting, street lighting, transport lighting, industrial lighting, energy for electrical and electronic gadgets, energy for waste treatments and energy for pollution control technologies.

The social and economic fallout of such an energy crisis is certain to be catastrophic and the debate about the steps needed to address and mitigate such a crisis have been raging for long. The answer is a massive drive to harness non-conventional energies. However, these sources generally have high cost of technology as effective entry barriers and may not be as abundantly acceptable as fossil fuels are. The result too is predictable. While people are aware of the impending doom, they can do precious little than inch towards the inevitable.

9.3 Water Management

The belief that water is abundant led people to utilize plenty of water for drinking, food preparation and washing clothes; much more water is used for manufacturing articles such as foodstuff, paper, yarn clothes, etc.

'The virtual-water content of a product (a commodity, good or service) is the volume of freshwater used to produce the product, calculated at the site where the product was actually manufactured (production-site definition). The virtual-water content of a product is the volume of water that would have been necessary to manufacture the product at the place where the product is consumed (consumption-site definition)'.

It denotes the summation of the water used at different stages of the production process. The human impacts on freshwater systems are related to the patterns of human consumption. Problems of water scarcity, pollution and calamity is better comprehended and addressed by taking into consideration both the production and the supply chains as a whole. Water issues are closely knotted with global market. Many nations have started externalizing their water footprint. Rather they import water intensive products from other nations. This leads to exerting pressure on the water wealth in the exporting countries. Judicious and wise water governance and conservation are the need of the hour. Governments, consumers, businesses and civil society communities all must take active part in order to achieve a better management of water resources.

Water footprint thus serves as an indicator to express direct and indirect utilization of water by the consumers or producers. The water footprint of a person, company or nation can be defined as *'the total volume of freshwater that is used to produce the commodities, goods and services consumed by the person, company or nation'*. The concept of the water footprint is somewhat similar to that of ecological footprint, but it focuses on the use of water. The global water footprint for the period 1996–2005 was 9087 Gm^3/year (74 per cent green, 11 per cent blue and 15 per cent grey) of which agricultural production contributes 92 per cent, 1 Gm^3 is 1 cubic gigameter equal to 1 billion cubic meters. Imagine, 15,000 liters of water (93 per cent green, 4 per cent blue, 3 per cent grey water footprint) is needed to produce just 1 kilogram of beef. The water footprint of a 150 gram Netherlands' soy burger is about 160 liters and that of a beef burger requires about 1,000 liters. Japan has around 77 per cent of its total water footprint outside the country. About 20 per cent of the US water footprint is external.

Blue water refers to the freshwater; surface or ground water, i.e., water in the lakes, rivers and aquifers.

Green water refers to the precipitated water on the land or that stored in the soil or vegetation; it does not include the precipitation that runs off the groundwater.

Grey water or sullage refers to the wastewater generated from domestic activities such as bathing, bathtub or shower, washing clothes and washing machine, dishwashing, sinks, etc. It derives its name from its cloudy appearance and can be recycled onsite for landscape irrigation and constructed wetlands. Grey water differs from water from the toilets which is designated sewage or black water to indicate it contains human waste. Dishwasher and washing machines that run only for full loads can cut household water consumption by an average of 1,000 gallons per month.

Black water or sewage is the wastewater from toilets, garbage grinders, feces and other human body fluids, considered hazardous. The sewage is different from grey water as it contains bacteria, pathogens and food particles, which can rot and decompose. The filtering system for black water treatment is generally located outside the home. Black water recycling systems are expensive – both installation and maintenance. It is also more difficult and costly to treat. Water recycled from black water should never be used as potable water or on food crops because they may not be fully disinfected. Recycled black water can be used to water the lawns and non-food crops/gardens. Plants grown with recycled black water hardly need any fertilizer.

9.3.1 Water conservation

Human life is threatened by unequal rainfall. Such uneven distribution of rainfall hampers their livelihood and economic wellbeing. This ever increasing scarcity of water is due to exploding population especially in the developing countries, increase in demand for foods, sanitation, urban inflow and improved standards of living. Three-fourth of the earth is covered by water, yet the world is thirsty and needs to depend on the skewed freshwater supply for its sustenance. World Bank report states that about two billion people are subjected to inadequate sanitation facilities and about a billion do not have access to clean water. The years from 2005–2015 has been declared as the 'water for life' decade by the UN, while the cities continue to dump and dispose sewage into the water supply. Such unhealthy water has become the reason for all health maladies infested with a number of pathogens and parasites. Water scarcity in water stressed countries is likely to jump six fold in the coming 30 years, which raises concerns about all the activities associated with water, like agriculture, industry and household use. Inequity in water sharing is another problem. Researches show that developed countries with raised standards of living demand more and more water than a developing country. The problem might be solved by assigning a monetary value to freshwater. The value of water is often undermined as it is still a free commodity. Several countries like Israel, Japan and Saudi Arabia have installed desalination plants. Large scale installation of such plants may generate freshwater but is likely to create other issues. It is estimated that some 11,000 desalination plants exist in about 120 countries all around the world. The population has doubled in the last few years coupled with economic growth and urbanization. Relocation in newer areas requires more food, more shelter, more energy, more clothing; more daily needs all thrusting additional pressure on the freshwater supply. Almost 41 per cent of the world population resides in the river basins that are under severe water stress. Climate change seems to result in erratic climatic behaviour – drought in some parts, flood in the others; melting and receding glaciers etc. Increased energy production, increased industrial waste, increased waste water, pesticides and fertilizers all aggravate this problem to a considerable extent. Moreover, these toxic residues may leach and contaminate the underground aquifers. Much of the water is also wasted due to leaky irrigation practices to clean water.

Clean water is essential to life. As stated, about 1.1 billion people do not have access to water. Two-third of the world population may face acute water shortage by 2025. This means there will not be enough water to drink, irrigate, or wash, resulting in outbreak of water borne diseases. Half of the wetlands have been destroyed. Wetlands provide an array of ecosystem services, for example they are a habitat for fishes, birds, amphibians, reptiles, mammals and other invertebrates and serve as nurturing beds for various other species. They act as effective filtration system, protect people from storm surges, flood etc. They also provide for eco-tourism spots. Natural landscapes are often wiped out. Land degradation, forest depletion, erosion, desertification and other ecological catastrophes are obvious consequences of water scarcity. Notable example is the Aral Sea that has an area of almost the size of Lake Michigan.

When water is scarce and the supply is limited, effective and efficient water management seems inevitable at global, national, local and personal levels. Human beings were least bothered to reserve water due to its abundance. But now, every drop counts. Our irresponsible attitude has resulted in the deterioration of both water quality and quantity. There is a dire necessity to go for proper management and conservation of water. Projects like 'water foot printing' and 'water stewardship' are undertaken under United Nations and World Economic Forum. Water risks, water use and their impacts, climate change adaptations, are been studied and promoted by WWF in collaboration with industries.

Several international treaties are being negotiated to protect different types of water bodies. The water management and distribution by government has disrupted the system and mentality of community participation.

For efficient water conservation and management, the following points are to be kept in mind:

- to promote public awareness about water necessity, indispensability and conservation of water;
- active participation of local people in all water management activities, like rainwater harvesting, drip irrigation, etc;
- not to use treated water for gardening, washing toilets and so on;
- prevention of drying up of underground aquifers and promoting recharge of ground water;.
- not to discharge pollutants and contaminate water; and
- to use water shed, which is a basin of a tributary, for integrated development.

Traditional water harvesting in India: Rainwater is a copious source of water, but, it is neither uniform in distribution nor ensured everywhere. There are seasonal variations with river overflow during monsoon. If such water can be channelized and stored, it can be used in necessity. India has a tradition of water harvesting since the Vedic times. Rainwater may be captured from the place where it showers; flood water may be captured from the sides or embankments while runoff can be collected from its path.

Such harvesting serves to recharge ground water, increases the surrounding vegetation, reduces soil erosion, decreases silting and controls flood while neighbouring wells become full of water.

Table 9.1: Few traditional water harvesting systems in India

Sl. No.	Name of traditional waster harvesting systems	State/region they are found in	Brief description
1	Zing	Ladakh	tanks meant for collection of water from melted ice
2	Khatri	Himachal Pradesh	chambers made of hard rock for collection of water
3	Zabo	Nagaland	impoundments of runoff
4	Bamboo drip irrigation	Meghalaya	water from the streams in the hills is flown into the plains through bamboo pipes
5	Dungs	West Bengal	small canals connecting the paddy fields and the stream
6	Baolis	Delhi	step well
7	Kunds	West Rajasthan	underground store
8	Baoris	Rajasthan	community wells
9	Yeri	Tamil Nadu	tanks
10	Korambu	Kerala	temporary wall of grass and mud laid across the channels to raise the level of water

9.3.2 Rainwater harvesting

Rainwater Harvesting can be defined as the 'conscious collection and storage of rainwater to cater to demands of water, for drinking, domestic purpose and irrigation'. Conventionally there was a system of collecting rainwater and storing it for future use in India. This system was the result of a time-tested wisdom. Traditionally water was harvested in *jheels*, *bawaries*, step wells, lakes, tanks etc.

Groundwater is naturally recharged through percolation. But due to rapid urbanization and reduced exposed surface of soil there is a reduction in rainwater percolation, thereby reducing ground water store. Harvesting of rainwater is a practice of increasing the infiltration of natural rainwater under the ground artificially. There are two common methods of rainwater harvesting –

- surface run off harvesting; and
- rooftop harvesting of rainwater.

In surface run off, the overflow can be caught in the cities and towns and then used for recharging aquifers. In rooftop harvesting, rainwater is caught from the roof, stored and passed through an artificial recharge system. If properly implemented, such practice assists in raising the level of groundwater level of that region.

The surface receiving rainfall directly is known as the catchment which can be the courtyard, terrace, covered (tiles or cemented) or uncovered. The roof can be flat or sloped.

Rainwater collected should be carried down with the help of UV resistant water pipes to the storage/harvesting system. Gutters are used to catch water from the slanting roof and drained down through pipes. The mouth of each drain should be fitted with a sieve or a wire mesh to restrict unwanted substances.

First flush is a device to wash out the first spells of rain so that the probable contaminants in the rainwater and any deposition on the roof top do not enter. There should be provision for placing separators at the outlet of each drainpipe.

Appropriate means of sieve should be adopted in order to avoid rainwater contaminating ground water. Filters are used to effectively remove turbidity and microorganisms. There are diverse types of filters in practice. The sand gravel filter, charcoal filter, sponge filter and PVC-pipe filter are widely used.

In storage and direct use, the rainwater collected from the rooftop is channeled to a storage tank, designed according to the necessity of water, quantity of precipitation and availability of catchment. If possible, each and every tank should possess excess water over flow system. This water can be used for washing, gardening and bathrooms.

The excess water is to be driven and diverted to recharge system by various methods to ensure percolation of water rather than draining from the surface. The most frequently used recharging methods are percolation tanks, dug wells recharging, recharge trenches, recharge shafts, bore well recharging and recharge pits.

Case Study 9.1: Rainwater harvesting (RWH) in Frankfurt Airport, Germany

On an average the rainfall in Germany ranges between 563–855 mm. It had cost about US $63,000 in 1963. It was installed with and expectation of handling 13 million people/year and saves about 1,00,000 cu m of water annually. The roof with an area of 26,800 sq. m provides as a catchment. The water is transported to six storage tanks of 100 cu m capacity in the basement. The water is used for washing clothes, toilet flushing, gardening and cleaning of air conditioners, etc.

9.3.3 Watershed management

Watershed is an area from where the run off flows to a common point. Every river, stream or the tributaries is coupled with a watershed. Smaller watersheds together comprise larger watersheds. Watersheds can be of many types such as macro-watershed, sub-watershed, milli-watershed, micro-watershed or mini-watershed depending upon its size, drainage, land use pattern, etc.

The primary purpose of watershed management is to promote water and soil conservation:

- to control run off;

- to utilize the run off for useful purpose;
- to improve the land of watershed;
- to check soil erosion and sediment deposition on watershed;
- to increase rainwater infiltration;
- to increase water holding capacity;
- to regulate the flood peak in the downstream regions;
- to be able to supply clean and sufficient amount of drinking water; and
- to improve the income by simultaneous regeneration of natural resources like increase in the yield of timbers, fodder and wildlife resource.

It acts as a socio-political-ecological entity and is critical in resolving food, social and economical security and offers life support services to the rural people.

Various watershed programmes were developed such as Drought Prone Area Programme (DPAP), Desert Development Programme (DDP), River Valley Project (RVP) and National Watershed Project for rain-fed areas (NWDPRA) and Integrated Wastelands Development Programme (IWDP). Subsequently all have been integrated in to a programme called Integrated Watershed Management Programme (IWMP) during mid 1980s and early 1990s. The watershed guidelines were revised in 2001 as Hariyali guidelines with a view to simplify and involving people's participation along with community empowerment. The programme was launched in 2009–2010.

The different watershed management practices are:

- Vegetative measures like strip cropping, pasture cropping, grassland farming and woodlands.
- Engineering measures like contour bunding, terracing, earthen embankment, check dams, farm ponds, rock dam, permanent grass, making diversions, stone barriers, silt tanks distensions etc.

Watershed management is the study of the characteristics of a watershed with the idea of planning and setting up of efficient systems with a view towards sustainably and optimally use the water resources of the area for the greatest common good. The concept is to utilize and enhance the water within the watershed boundary for the plants, animals and the humans within the space. Water supply, water quality, drainage, rainwater harvesting, storm water run off, ground water levels, water rights are critical components that are paid heed to the managers of watersheds.

One of the biggest problems of watershed management is ground water contamination or the pollution of water, which can either be due to industrial effluents or the unscientific use of pesticides and chemical fertilizers in agriculture. The most important question that watershed managers have to answer is whether to consider water as an economic good, with its attendant costs and benefits or as a social good, which is for all to access and consume, paid for by the society. The answer of this question leads to whether the water that is made available to the population within the domain as a commercial good, paid for the individual depending on his usage or is given away by the authorities for all to enjoy as other ubiquities like sunlight and air are?

9.4 Resettlement and Rehabilitation of People

People migrate or are forced to displace out of their habitats owing to natural and manmade disasters such as cyclone, earthquakes and tsunamis. Even developmental projects like construction of roads, dams, canals, flyovers, mining displace people from their homes. Thousands to millions of people become homeless and are forced to leave the land for safety and resettle in other areas. *'Displaced persons are those who are forced to move out of their habitat, whether it is individually and formally owned, or a traditional, customarily, and collectively owned areas'.* Many, such as those who depend on forests, are not given admittance to their livelihood if the area is notified as a sanctuary or a national park. Such people are called Project-Affected Persons (PAP).

Resettlement or physical relocation of the displaced persons is a onetime event. Whereas, rehabilitation is a long term process of rebuilding individual's physical and economic means of survival, their social and cultural connections along with psychological acceptance of the altered situation. It involves both Displaced Persons (DPs) and the PAPs. It should begin long before the actual process of physical displacement starts.

9.4.1 Basis of displacement

- Primarily natural disasters like earthquakes, tsunamis, cyclones etc.
- Anthropogenic accidents like industrial disasters, oil spills, toxic chemical contamination, nuclear accidents.
- Development and construction projects like urbanization, dam, mining, roads, flyovers etc.
- For better employment opportunities and livelihood.
- Oil and gas exploration activities, laying of pipelines.
- Restriction in the protected area networks.
- Conflict induced displacement:
    i. political causes and secessionist movements like the Naga Movement and Assam Movement and killing of Kashmiri pundits;
    ii. autonomy movements like Bodoland, Punjab, Gorkhaland etc.;
    iii. caste and communal displacements like in cases of Bombay and Gujarat riots; and
    iv. environment and development induced displacement due to building of dams, roads, mining, nuclear power plants, urbanizations, etc.

9.4.2 Objectives of resettlement

- People should be allowed to lead their own way of living with proper livelihood.
- They should be provided with a means to develop and promote their art and culture.
- They should have the option of shifting out and mixing with others to live social and community based life.

- They should be provided with all basic amenities of life and infrastructural facilities.
- They should be given education, training and employment opportunities.
- They should be given full rights and responsibilities of a citizen.
- People should be involved in decision making.

9.4.3 Issues related to resettlement

Displacement usually hits the poor and rural people who do not figure in the priority list of any political parties. The compensation for such displacement is often not paid or delayed or the amount is too low in terms of money. The new area of resettlement lacks basic infrastructure and amenities. The temporary camps become their permanent shelter. Displaced people often engaged with traditional livelihood do not possess any other skills and are forced to pick up alternative occupations. There is a trend of decline in the general health of such people owing to food insecurity, poor sanitation and heath facilities. They cannot carry forward their traditional and cultural practices and lose connection with nature. Their families break up. Their lives become purposeless and they are often unable to bear the shock of such a trauma. They often get engaged into various addictions and immoral and antisocial activities.

For many years the developmental projects have encompassed misappropriation and have forcefully dislodged people by not giving any safety and security to them. The only existing law has been the Land Acquisition Act (LAA), 1894 that states how land can be acquired with payment of compensation. It does not contain provisions of resettlement and rehabilitation. According to Fernandes (1998), nearly 21.3 million people were displaced or project affected in between 1951–90. Fernandes (2007) stated that the total number of DP and PAP between 1947 and 2000 was over 60 millions. The effective means of resettlement and rehabilitation requires reliable statistics and database on DPs and PAPs. Internally Displaced Persons (IDPs) for a variety of reasons are not categorized as in-house refugees since they have not traversed national boundary. In 1985, the Government of India (GOI) started drafting a policy after the National Commission for Scheduled Castes and Scheduled Tribes stated that 40 per cent of these DPs and PAPs were tribal people. In 2004, the Ministry of Rural Development, GOI published a draft of the Resettlement and Rehabilitation Policy (R&R Policy) which had been revised in 2007. The important provisions are:

- covers all cases of involuntary displacement;
- compulsory consultations with the *gram sabhas* and public hearing;
- support for development of skill and preference to be given in project jobs (one person per family);
- compensation in terms of land for land;
- rehabilitation grants to be given;
- housing benefits and basic infrastructural facilities;
- options for shares in the companies that are implementing the projects;
- monthly pensions to disabled, orphans, widows, etc; and
- ombudsman for grievance redressal.

9.4.4 Dam-induced displacement in India

In the past five decades, over 50 million people have been uprooted from their homes, lands, forests and streams. They sacrificed for the sake of national interest and now they are a mere witness to their losses. Even then it remains a non-issue to the government, politicians and bureaucrats. As per world banks review an average of 13,000 people gets displaced for each dam constructed. Walter Fernandes predicts the displacement numbers to be more than 40 million against the government underestimates. Only 730 out of the displaced 2,108 families have been relocated.

- Against the research estimates of 1.8 lakhs, the official estimates for displacement in Hirakud project was only 1.1 lakh. Farakka project claims no affected ones while World Bank gives the number 63,325.
- Number of displaced families for Sardar Sarovar rose from 6,000 in 1979, 12,000 in 1987, 27,000 in 1991 to 40,000 in 1992.
- NBA estimates it to be 85,000 families or 50,000 lakh people.
- Nearly 70,000 people of 101 villages were to be displaced in case of Bergi Dam (Jabalpur) but after the reservoir was filled in,162 villages got submerged.
- More than 2 lakh people were displaced in Rihand Dam project in 1964.
- Resettlement consumes time as in case of Tungabhadra Dam in AP which took five years while Machkund took more than a decade.
- In case of Pong Dam of Himachal Pradesh, 16,000 families out of 30,000 were found to be eligible for compensation of which ultimately 3,756 were relocated hundreds of miles away to a completely different ecological, cultural and linguistic atmosphere in Rajasthan.
- Soliga tribals in Karnataka suffered multiple displacement, first by the Kabini Dam in 1970s and then by the Rajiv Gandhi National Park.

The process of displacement is dehumanizing, de-empowering and excruciating so that dams are no longer the temples of modern India but rather have turned into burial grounds.

9.4.5 Displacement in India for wildlife conservation

The displacement and relocation of humans from Protected Areas (PA) culminates in a host of impacts and such initiatives were being practiced in many countries for wildlife conservation and management. In most of these cases, the affected people are the indigenous tribals. Such displacement aims to decrease the human pressures and disturbance in the conserved areas. In Corbett tiger reserve, the tiger population increased by 52 per cent and in Kanha National Park Barasingha increased to more than 400 after relocation. There have been reduced incidences of forest fires and man-wildlife conflicts.

- When Kanha National Park was declared a Tiger Reserve (TR), nearly 650 families from 24 villages were relocated voluntarily outside the boundaries of tiger reserve. There are 19 villages in the core area and 169 villages left in the reserve to be relocated as per MoEF, 2005.
- In Gir National Park, 60 Maldhari villages of 580 families were displaced in 14 settlement villages in the buffer zone. In this case, it was disastrous for the Maldharis to shift from pastoral

activity to farming without acquiring any skills for such transition. 54 villages are still there in the park with 65 people in the core zone.

- Three villages with 417 families were shifted from core areas to multiple use areas (MUA) in Bandipur National Park with remaining 54 hamlets in the core and some 200 villages neighbouring the park.

- In Palamau sanctuary and Betla National Park, the forest was a resource to 200 indigenous people. There are still three hamlets in the core area and 16 within MUA and habitation management zone.

- 24 families of Surma village were forcefully relocated outside the TR of the Dudhwa National Park in UP. One village still lies in the core and about 37 villages in the reserve.

- There are no hamlets in the core area although 20 hamlets are in the buffer area and 142 villages are located on the fringe in case of Valmiki Sanctuary in Bihar.

- 1,390 Van Gujjars were proposed to be displaced outside Rajaji National Park in Uttarakhand in Pathri and Gaindikhata. Such forceful resettlement faced opposition and led to serious impacts like beating, illegal fines, threats, coercion, etc, by the forest departments. National Human Rights Commission orders were violated and the rights of Van Gujjars were denied. Relocation in the second phase with the help of 'Friends of Doon' was better. A Taungya village is also relocated though some of them still remain.

- In Tadoba National Park, out of six selected villages, two have been displaced outside the TR in Kaiselghat near Mul. Their lives were dependent on non-timber produce which is absent in the new area. Significant forests have been cleared for such purpose.

Where displacement brings social and economic trauma in case of unwilling and forced displacement, from a non-monetized to money dominated economy, from indigenous culture to imposed cultures. In most cases, the land given is a degraded forest not suitable for cultivation. There is destruction of the natural ecosystems arising from clearance for faming, housing, roads or other needs. In Satpura, over 30,000 trees were cut down to build a rehabilitation site that involves loss of wildlife too. The biotic pressure in the new area increases. People from Rajaji National Park faced problems of water scarcity. Sometimes the relocation packages do not meet the needs of the people as in case of Pench National Park in MP.

9.4.6 Mining induced displacement in India

Nearly about 2.5–2.6 million people have been displaced from 1950–90 due to mining and such Mining Induced Displacement and Resettlement (MIDR) imposes social problems that challenge the human rights. Not even a quarter have been rehabilitated. The most notable determinants in India for MIDR are coal, copper, bauxite and uranium mining. These are mainly in the states of Orissa, Andhra Pradesh, Jharkhand and West Bengal. Major Indian coal companies are reported to have displaced more than 32,700 families from 1982–86. The setting up of National Aluminium Company Limited (NALCO) Refinery Plant in Damanjodi displaced nearly 600 families. Mining is an unstable source of income. On an average the life time of open pit exploitation is 10–40 years. People suffer from loss of access to previously owned land, receive little compensation, along with joblessness, homelessness, unemployment, food and water insecurity, health risks, lack of education, social marginalization, deforestation and nature devastation – all leading to disastrous effects. Mining practices based on sustainable development not only contribute to profit maximization but also to the prosperity of localities, their health and conservation of environment.

9.4.7 Climate change and climate refugees

Melting polar caps (due to global warming) are releasing water into the oceans, leading to rising sea levels. This rising sea promises to inundate vast tracks of land around the world. Let us concentrate on the Sunderban areas of India and Bangladesh which are today threatened with sea water submersion.

Thickly populated and impoverished, the rising sea levels threaten to render the entire populations of this area homeless as their lands get submerged. Already reeling under poverty, over population and food shortage, Bangladesh will not be in a position to relocate, rehabilitate and resettle these people – the climate refugees – within her boundaries giving rise to a terrible humanitarian problem, which will raise a number of pertinent questions that will need to be answered, not only by Bangladesh but the world community at large.

The climate refugees and their status: The Geneva Convention that laid down the meaning of the term did not extend its meaning to embrace climatic causes rendering people homeless. Thus, today, while political persecution and enemy action may accord one the right to be treated as a refugee, one forced to become homeless by the rising seas cannot be considered so. Consequently, these people, homeless, helpless and hapless will have no protection under the global laws. A free radical of a refugee with no defined rights is a liability.

Their destination: As the primary cause leading to the rising sea level can be traced back to the burning of the fossil fuels for development, the lion's share of the blame should logically go to the industrially developed nations. These are the nations that should be morally responsible for the plight of these people and therefore should help in their resettlement and rehabilitation. But the sad fact remains that these nations, by coincidence or sinister design, have the most stringent immigration laws. The writing on the wall is clear – those who contributed the most to global warming that led to the loss of habitat, have their doors firmly shut on the face of the displaced and homeless masses.

Financial arrangement: Removing an entire population, re-skilling, resettling and rehabilitating them will call for a huge sum of money. As of now, there is no concrete proposal as to where these sums will come from. Who should foot the bill for this colossal effort is something that is not even in the agenda of discussion among the global leaders, leave alone the working out of a mechanism so that the United Nations (or some independent body) by the imposition of a Sin Tax (Carbon Tax) raise the money, prepare a war chest and use the proceeds to ensure that the plight of these people are addressed, that too with the least possible hassle and pain.

Our concern: The answer is an overwhelming no. Primarily because people in a third world country are normally considered to be sub human, whose plights do not concern the citizens of the first world. The fact that climate is a great leveler and that the wheels that have been set in motion by the wanton consumerism and pursuit of instant gratification may claim its first victims in the mangrove forests of the Sunderbans may not concern the perpetrators. That the same climate will one day bare its fangs closer home is still something that has not dawned in the west. The sooner the world realizes that climate change and global warming are not issues that can be contained by artificial and man-made boundaries and that they threaten all, equally the better.

Possible solution: We have brought the world to such a pass and it is on mankind at large, to frame the response. The problem is global. The problem is universal. The treat equally spread, irrespective of region or boundaries. The response too, has to be uniform – a worldwide effort. It is not about Bangladeshis, it is about humans and the faster we realize this and take steps to mitigate the situation the better.

Case Study 9.2: People of Maldives and Tuvalu

Small Island Developing States (SIDS) such as Maldives and Tuvalu are already in danger of extinction. Lying about 2,000 mile to the east side of Carteret islands and nearly in midway between Australia and Hawii, the population of Tuvalu was less than 12,000 and is habitually portrayed as the inhabitants in southwestern Pacific whose country is doomed to disappear and devoured by the mounting and tumultuous oceans of that region.

The island group is just about 10 cms above the sea level. If temperature increases, the sea level will rise and there will be nothing there in just 50 years time. The situation is indeed dreadful. The pace at which the underground drinking water is turning salty, it is a peril to the coconut and taro farming. The total area of nine islands together is 26 km²; the fourth smallest nation of the world with some islands not habitable already. There are very few industries, no military forces, few vehicles and just 8 km of paved roadways.

Small island states around the world are particularly vulnerable to sea level rise because in many cases (the Bahamas, Kiribati, the Maldives and the Marshall Islands) much of their land is less than 3 or 4 m above present sea level. One 1999 analysis estimated that, by 2080, flood risk for people living in small island states will be 200 times greater than if there had been no global warming. Other island states tend to have high levels of development and high density population around their coasts. Half the population of the Caribbean, for example, lives within 1.5 km of the shoreline.

9.5 Environmental Ethics

Morality is an expression of human culture and it is the moral behaviour that develops moral virtues. A number of philosophies came up that deals with man's relationship with the environment. In anthropocentrism, humans are of prime importance and value. Humans hold the crown position of creation and are of immense importance. All other life forms are subordinates and important only to the extent to which they affect human beings. In this ethic, nature is only considered as its protection or degradation can benefit or harm human beings. It is, simply put, human chauvinism that holds humans to be the measure of all things. Western thinking is largely anthropocentric. Books like *Confessions of an Eco-warrior* by Dave Foreman and *Green Rage* by Christopher Manes have postulated anthropocentrism. This gave birth to shallow ecology which considers human beings above and outside nature and is the source of all value. This ecology fails to vision the intrinsic value of all and their value to each other.

In *Deep Ecology*, the living environment is taken as a whole where all have the same right to live and flourish. The term deep ecology was coined by Arne Naess, a Norwegian philosopher later epitomized by Gifford Pinchot who viewed earth as a set of natural resources that is required to be managed for the present and future generations of the human race. It emphasizes on the intrinsic equality of species including humanity. It asks deeper questions such as why and how. In the concept logic and facts are unable to answer the ethical questions; for this, an ecological wisdom that focuses on deep questions, deep thinking and deep commitment is needed. It is much more than science; it involves spirituality and respect for all beings. Much of these ideas can be related to ancient religion like Buddhism and Jainism. A natural follow up is eco-centrism that considers ecosphere rather than any individual organism to be the source of life and sustains life. It believes that biotic community as a whole maintains a homeostasis in all ecological processes and ecological structure. This ethic conceived by Aldo Leopold advocates that all species are a product of the evolutionary process and are intricately related in their life processes. The philosophy is also outlined in 'Gaia Hypothesis' by James Lovelock where earth in its entirety is treated as living and single organism functioning through a vast array of relationships.

9.5.1 Resource consumption patterns and the need for their equitable utilization

A resource is anything that can be used by humans. Resources that occur freely in nature are natural resources, while resources developed by man for his own consumption are human resources, like schools and hospitals. Resources can be further classified as renewable – those which can be replenished and non-renewable, those which once used cannot be reused like fossil fuels.

When one speaks of resource consumption patterns and the need for equitable distributions, one is speaking primarily of non-renewable resources. As such resources are not distributed equally; the question of their use comes into sharper focus. It is natural to presume that a country that is rich in natural resources will be economically developed while a country that is not endowed naturally will lag behind.

Japan, which does not have any mineral base to talk about, is an example of the opposite, which has walked the development path by up-scaling its human resources, while the Democratic Republic of Congo is an example of underdevelopment despite being well endowed with natural resources.

As development is directly related to the consumption of resources, the countries that have reached development having consumed the resources of their colonies are today in a position to continue on the growth path by accessing the resources of less developed countries or by securing their supply lines from resource rich countries like Australia where the demand for the consumption/conversion of such resources is relatively lower.

If a holistic view is taken, then it will be amply clear that the resource consumption by the countries to the North of the divide is much higher than those in the South. What is more alarming is the fact that consumption in the north is much more frivolous than in the south with scarce resources being diverted to quench needs for instant gratification and conspicuous consumption than for meeting the natural needs for growth and development. This consumerist use of scarce resources is giving rise to concerns about ethicality, which when viewed in the light of Malthus' prophecies take menacing proportions.

To put it bluntly, countries in the North are ecological debtors (whose ecological footprints are bigger than their bio-capacities – ability of the biosphere to carry the people and consumption patterns), while those in the South are generally ecological creditors, their lack of development leaving resources underutilized.

And it is exactly here that the questions of equitable utilization come in. Not only will resource utilization have to be sustainable in the future for a just and equitable world order to emerge, the consumption patterns will have to change drastically to ensure that the scarce resources are first used to meet the basic needs of the world's poor before they are deployed to feed the fancies of the rich.

Case Study 9.3: Pani Panchayat scheme in Pune

In Maharashtra, the councils established for the conservation, management and distribution of water resources on equitable basis are known as 'Pani Panchayats'. Such councils have been established by Mr Vilasrao Salunkhe and Gram Gaurav Pratisthan Trust. The councils address all such issues concerned with water scarcity in the drought prone areas of Maharashtra. Mr Vilasrao Salunkhe selected the village of Naigaon of Purandhar block in Pune district for embracing and implementing of soil and water conservation programmes that would contribute towards the rural development; Purandhar block in Maharashtra is very vulnerable to drought. The area experiences a rainfall of 250 mm to 500 mm, but, above all the pattern of rainfall is uncertain and erratic. On an average the area experiences drought almost once in every three years.

The technological and social innovations in this council helps refurbish and restore the degraded watersheds. This in turn increases the water harvests and guarantees each family an equal share of water that is harvested thus meeting the goals of equity. Every member of the family, on an average, is allowed water for cultivation at the rate of half an acre per head, subjected to a maximum of 2.5 acres per family. In other words, 1,000 m^3 of water per person per year to a maximum of 5,000 m^3 per family per year is supplied.

Mr Salunkhe began the programme by trying to persuade the neighbouring locality regarding the utmost importance of water and soil management. He was rather encouraged taking the responsibility to indicate the impacts of watershed development programme towards achieving sustainable development. He started his mission of dry farming that was based on watershed management approach in just 14 acres of leased land taken by him. Out of the total area, 2.4 hectares were utilized for afforestation, 9.6 hectares were used under protective irrigation and the rest was used under percolation tanks, wells, field bunds, tracts and other infrastructures. In the beginning, the farmers expressed little interest. They were amazed to see the production of 200 quintals of food grains whereas 40 acres of the land they owe could hardly produce 10 quintals. Progressively they began to grow interest.

The number of lift irrigation schemes has gone over 100 in about 10 years and most of them are functioning well. The village people could harvest two crops per year with a provision for irrigating eight months. Another implication of Pani Panchayat has been the generation of employment opportunities. This has stimulated reverse movement. The rationale behind such accomplishment lies in the harmony amongst the village people and their stringent obedience to the rules and regulations formulated by Pani Panchayat, a collectively nominated village body.

9.5.2 Equity – disparity between northern and southern countries

The disparity in resource consumption takes two extreme stands, depending upon from where the problem is viewed. From the north, it looks like a sea of overpopulated swarms ever hungry and capable of eating away all the resources of the world. Check population growth or else, Malthus will unleash war and pestilence, being their common refrain to the brethren in South.

From the South however, it is simply a matter of conspicuous consumption – a case of the rich diverting the scarce resources of the world for frivolous pursuits. Resources that should be utilized to create wealth for the masses and are capable of eradicating hunger and poverty, they feel, should be rightly deployed as opposed to current usage patterns that are not only uneconomical, but unethical to boot.

The truth, naturally struggles somewhere in between, pulled as it is by these two diverse forces. The fact however, is that the more we are able to shrink this gap, the so-called North-South divide, and the faster we do it will not only be good for the people but also the world at large.

The division between the North and South is basically on the basis of socio-economic and political differences. The North comprises the West, first world and the richer and developed nations in contrast to the South having the poorer and less developed countries. All the G8 countries belong to the North Block and comprise North America, Western Europe, Australia and Japan. As nations become more and more economically developed, they gain influence and so become a part of North, irrespective of their geographical location. The South covers Africa, Asia and Latin America. 95 per cent of the people here lack adequate food and shelter. They are rich in resources but lack proper education and knowledge to utilize them. They do not have appropriate technology and are politically unstable. A good measure to find out which block a country is located in is the Human Development Index (HDI). Countries with value nearer to 1.0, indicates greater level of the country's development. A visual representation of the North–South Divide is the Brandt Line, proposed in 1980 by Willy Brandt, the West German Chancellor. It moves around the world at approximately 30°N latitude but dips down to include Japan, Australia and New Zealand into the North.

9.5.3 Urban – rural equity issues

Rural lifestyles are more sustainable – cleaner, greener and healthier. However, the lure of the cities, with their tinsel tomes and seductive opportunities have been a draw since the dawn of mankind with a steady trickle of migrating people becoming townsfolk. With the rural economies coming under increasing pressure, the urban-rural divide is becoming wider with the traditional trickle of migrants becoming a steady flow. These migrating hordes are not only putting the civic amenities of the cities under severe pressure but are also creating in their wake pressures on the environment. Also, life in the city is generally less sustainable than in the villages. More and more electricity is being consumed; water has to be provided for drinking and sanitation as the other basic amenities. All this ultimately lead to the burning of fossil fuels and it has been seen that marginal increases in the urban population leads to a disproportionately bigger carbon footprint.

9.5.4 Need for gender equity

Even before we strive for gender equality, understanding gender dynamics of resource management should be the starting point of mankind's response to resource degradation. It is man who harnesses the forces of nature with women traditionally providing him with the support systems.

The developing world's 1 billion plus the poor people of the rural areas make up the world's largest group users of natural resources. The starting point for reversing the process of environmental degradation is to understand their roles and responsibilities – including the gender dynamics of natural resources utilization and management. Women manage natural resources all the time in their roles as farmers and household felicitators. Typically, they are in charge for growing crops and often possess unique knowledge of local cropping patterns. To meet family needs, rural womenfolk walk long distances to collect fuel wood and water. Despite their dependence on natural resources, historically, women have lesser access to and control over resources than men. Usually it is the men who set land, water, flora and fauna to commercial exploitation, which is more valued over and over again than women's domestic uses – just like the lionesses of the pride doing the actual hunting while the lions have the manes.

Gender inequality is most apparent in access to land and property. It is customary to prohibit women from owning land in many countries and cultures. Mostly, women have only usage rights, that too, intermediated by men, and those privileges are extremely uncertain to put things mildly.

Without secure land rights, farmers (increasingly women) have limited right to credit and little motivation to invest in superior management and conservation practices. Both women and men are more liable to make environmentally safe and sound land management decisions when they have protected ownership and know they can benefit by accessing credit facilities at cheap rates.

Improved water management, especially irrigation and fertilizers, is precarious to higher crop yield and resource conservation. Women farmers have limited access to network of irrigation. Membership of water users' groups is often connected to land ownership which naturally goes against them. Women's with constrained water rights are often forced to use subsistence agricultural practices that may result in soil erosion, a major basis of instability in watersheds, apart from huge opportunity losses.

Over generations, small-scale farmers have created a wide variety of crop species and several animal breeds. Commercialization of agriculture, determined partially by global trade in high-yielding varieties of crops and animals, is responsible for a speedy decline in biodiversity, which puts to risk not only local production but also global food security, apart from pushing many a marginal farmer over the threshold.

To safeguard their natural resources, rural folk (both men and women) must be authorized to take part in decision making that affect their requirements and vulnerabilities. The policymakers will be able to formulate more efficient and helpful interventions for conservation and sustainable use of resources, by properly addressing the gender perspectives of natural resource management, which can and should be put on use for the greatest common good.

9.5.5 Preserving resources for future generations

The very core of man's response to climate change and global warming is sustainability. Sustainability, put simply, is a practice that leads to rational use of resources in the present so that enough resources are left for the use of the future generations. It is opposed to wanton consumerism and stresses on optimal and rational consumption in the present.

9.5.6 Rights of animals

Being the most evolved of all the creatures on Earth, and sitting bang at the top of the food chain, it is for the humans to be compassionate about the rights of the animals. But what exactly are animal rights? At its simplest, it holds the ideal that all non-human animals are free to live their own lives. Animals must have a right to their fundamental requirements and welfare so that they are not tormented and their rights ought to be considered at a similar level as of mankind. The advocates or supporters of animal rights resist the mission of moral values and elementary protections on the basis of species relationship only (a thought that gained currency in the 1970's as speciesism, when the term was coined by Richard D. Ryder). They agree for the most part that animals should no longer be looked upon as property, or used as foodstuff, clothing, research subjects, entertainment (as in a circus), or beasts of burden. Every animal is a cohabitant on Earth and its rights to food, life and procreation should be honoured. All animals should be left to live the life their instinct leads them to live and should not be subjected to human intervention, leave alone use them as staples of human consumption, be that for physical needs like food and clothing or of entertainment.

A charitable organization:

People for the Ethical Treatment of Animals (PETA), was established in the year 1980 by

Ingrid Newkirk and Alex Pacheo, based in Norfolk, Virginia in the US is perhaps one of the most important and single largest world's animal rights association. The fundamental belief, 'animals are not ours to eat, wear, experiment on, or use for entertainment'.

There are four main concerned areas of work: resisting factory farming practices, farming for fur, animal testing in labs and employment of animals in the field of entertainment. Its battle includes fishing, pest killing, chaining of dogs, cock fight, dog fight, bull fight, etc. It also works in the field of public awareness and education, investigative work, animal rescuing, legislative works and media coverage which in many ways have bettered the lives of animals, saved countless lives of animals and encourages vegetarianism.

9.5.7 Ethical basis of environment education and awareness

Environmental education comprises an ethic that ensures improved behaviour and attitudes among the communities towards the environment. It enables us to achieve social and economic goals based on ecologically sound principles to strive equilibrium between resource management and nature.

EE is very much within the concept of sustainable development, i.e., *'meeting the needs of the present without compromising the ability of the future generations to meet their own needs'* (Groharlem Brundtland, 1987). The needs can be social, political and economic. The needs of future generation can be met by minimizing the use of non-renewable resources and reducing the generation of wastes and also keeping in mind to keep the limit of waste within absorptive capacity of local and global sinks. Fien in 1993 has defined the notion of environmental education for the purpose of sustainability as

'The development of human capacity and creativity to participate in determining the future, encourage technical progress as well as fostering the cultural conditions favouring social and economic change to improve the quality of life and more equitable economic growth while living within the carrying capacity of supporting ecosystems to maintain life indefinitely'.

Improving the environment needs a change in human behaviour and attitude. A new approach to education and awareness goes further than conveying of knowledge and consciousness raising activities and also focuses on the perspective in which the individual finds himself, the concept that was upheld during the United Nations Conference on Environment and Development, in Rio de Janeiro in 1992.

The basic principles of EE are

- to promote environmental education for all communities;
- to elicit analytical thinking, variation and newness in environmental education;
- to build up a feel of global nationality;
- to focus on inter disciplinary and holistic approach in the relationship amidst nature and society;
- to guarantee parity and reverence in the field of human rights;
- to promote cultural interchange;

- to address critical global issues within their social and historical context;

- to involve all stake holders in decision making;

- to have knowledge of the past in as well as native cultures so as to endorse variedness in the field of culture, languages and ecology;

- to make empowerment and advocate grass-root democracy; to endorse opportunities and involvement which assures societies control over their own fate;

- to recognize various dimensions in knowledge;

- to design environment education course in such a manner so as to facilitate individuals to deal with their contention in justifiable and humanitarian way;

- to hold conversation and collaboration between individuals and institutions for the innovation of novel lifestyles meeting every ones fundamental requirements, despite of ethnicity, sex, age, religion, bodily or intellectual differences;

- to utilize mass media in order to furnish to the curiosity and significance of all sectors of society through the exchange of information, morals and wisdom;

- to integrate understanding, proficiency, principles, approaches and actions, that will renovate every prospect into an enlightening incident for sustainable civilizations; and

- to grow moral consciousness, by respecting all life forms and imposing substantial limits on human exploitation of other forms of lives.

9.5.8 Conservation ethics and traditional value systems of India

For 5,000 years, the Indian civilization has evolved a highly developed way of life that taught successive generations how to live in sync with nature. The traditional value systems taught people the lessons of taking from nature only that part that was needed and to ensure that by conscious acts they ensured replacement and replenishment of the resources used.

However, under British rule, when the focus had shifted to exploiting India's resources much of this traditional knowledge was discarded, often banished by arguments in favour of modern science. Since independence, when the focus of the nation shifted to the utilization of the resources in an often unequal fight against poverty and under development, the needs of the hour ensured that these traditional value systems and knowledge banks were largely ignored.

Today, as we stand face to face with a prospect of India keeping her tryst with destiny, in a world order that is facing the resource crunch, we are becoming increasingly aware of our legacy. On the one hand is the agenda of feeding the teeming millions and helping them rise above the poverty line, while on the other is the acute need to optimally use our resources so that we may morph into a strong and vibrant economy with a solid industrial base. These two extremes have to be addressed using sustainable means so that we may achieve our development without causing harm to the fragile ecosystem in which we operate.

And it is here that the traditional knowledge about the forces of nature and the means to harness them in the most sustainable and eco-friendly way comes. Yes, we will have to use the raw power of science but with the ethically correct mindset that is enriched by traditional ways so that we may win, win and win again.

9.6 Global Environmental Issues

9.6.1 Climate change and global warming

'Global warming' and 'climate' change are terms that are used almost interchangeably. However, careful scrutiny will reveal that one is the cause, while the other is the effect.

Global warming is the result of increase in the Earth's temperature. Some of the sun's rays bounce back from the atmosphere, whereas some are trapped by the greenhouse gases like carbon dioxide and methane which keeps the Earth warm. This process is known as the green house gas effect and contrary to popular beliefs it is as old as life and is actually beneficial. These GHGs are both naturally created and anthropogenic and comprise the following: carbon dioxide, sulphur hexafluoride (SF_6), hydrofluorocarbons (HFCs), methane, nitrous oxide, ozone, perfluorocarbons (PFCs) and water vapour.

What is harmful is the fact that due to various reasons like the indiscriminate burning of fossil fuels, the green house gases in the atmosphere are currently at a historic high which is naturally trapping more heat than ever before. This in turn is leading to the Earth gradually heating up, at alarming rates according to some scientists. The higher atmospheric GHG concentration is also adversely affecting the natural balance which has evolved over millions of years and what effect they will have is still unknown to a very great extent.

GHGs, their effect and global warming are, contrary to popular belief, not terms that have come into being in the last decade or so. In fact, the effect of GHG was known ever since Joseph Fourier postulated in 1824 and stated that Earth would have been much colder had it been without atmosphere, it is the greenhouse effect that makes the Earth habitable. It was, again way back in 1895 when the possibility of enhancement of GH effect on mankind by release of carbon dioxide was put forward by a Swedish chemist, Svante Arrhenius.

Table 9.2a: List of countries by 2011 emissions estimates

Country	GHG emissions in million tons CO_2	Percentage of global total	Emission per capita or emission/per person tonnes of CO_2/year
World	**42669.72**	**100**	**4.49**
China	9679.3	22.7	5.83
United States	6668.79	15.6	17.67
India	2432.18	5.7	1.38

Country	GHG emissions in million tons CO$_2$	Percentage of global total	Emission per capita or emission/per person tonnes of CO$_2$/year
Russia	2291.57	5.4	11.23
Japan	1257.10	2.9	8.64
Germany	903.98	2.1	9.30
Canada	710.72	1.7	16.15
South Korea	661.69	1.6	10.89

The emission of CO$_2$ from burning of fossil fuel and industries showed a trend of retarding in which the discharge increased only by 1.1 per cent. This is less than half the average increase of 2.9 per cent per year that was observed in the last decade. China tops the list in terms of total emission followed by USA and India in the third place. But if per capita emission is concerned, the world's smallest countries and islands emits the most like Gibraltar (British overseas territory) with 151.96 tons and Virgin islands, US with 113.71 tons etc. The ranks per capita GHGs emissions are given below in Table 9.2b.

The result of this global heating of the earth is manifesting itself in various ways – the polar ice caps are melting, leading to a rise in the sea levels. As the Earth continues in its eternal spinning motion, the increase in temperature helps to pick up more and more moisture from the oceanic surface, allowing them to settle as rain elsewhere. This dance of creation gone berserk is changing the rhythm of climate as we have come to know it. This in turn has given birth to the monster called 'climate change'. To put things simply; climate change is the change in the climate that is caused, either directly or indirectly by human activities. An estimated 57,73,000 cu. miles of water is locked up in the cryosphere. If all ice were to melt, the seas would rise by about 230 feet, as per the National Snow and Ice Data Center. At the current rate of withdrawal, it is estimated that all the glaciers in Glacier National Park will be gone by 2070. The earth would be left in imbalance if all the ice caps would melt. Such sources being freshwater resource will desalinate the ocean after they melt and mix with sea water. The desalinization of the Gulf will 'screw up' ocean currents that regulate temperatures. The temperature rises and changing landscapes will endanger several species of life forms in the Artic which will enable only the most adapted to survive. Some polar bears are reported to be drowning as they have to swim longer distances to reach ice floes. Without proper action, the polar bear are likely to become extinct in Alaska by 2050. Ice caps reflect back sunlight into the space, further cooling Earth. In absence of icecaps the only reflector to be left are the oceans. Scientific prediction states that a 1.5 °C global temperature rise may witness the disappearance of 25 per cent of the Earth's animals and a 3°C rise may see almost 30 per cent of biota to disappear. According to IPCC, a warming of 1°C above 1990 levels would result in the bleaching of all coral reefs and transformations of 10 per cent of global ecosystems, whereas a rise in 2°C above 1990 levels will kill the coral entirely.

Table 9.2b: List of few countries by per capita emissions according to WRI, 2000

Ranks in terms of GHGs emission	Per capita GHGs (tons CO_2 equivalent)	Countries	CO_2 (tons)	Ranks in terms of CO_2 emissions
1	67.9	Qatar	60	1
2	36.1	UAE	25.2	3
3	31.6	Kuwait	26.8	2
4	25.6	Australia	17.3	7
5	24.8	Bahrain	20.6	4
6	24.5	USA	20.4	5
7	22.1	Canada	17.1	8
8	21.7	Brunei	13.7	10
9	21	Luxembourg	19.2	6
10	19.3	Trinidad and Tobago	16.7	9

Since the thermometer was invented, we have already witnessed eleven of the twelve hottest years between 1995 and 2006. This decade has been the warmest since 1880. According to the National Oceanic and Atmospheric Administration (NOAA), the year 2010 was coupled with 2005 as the warmest on record (See Table 9.5). With the Earth becoming hotter the menace of climate change too will occur – what is alarming is the fact that such changes will occur faster than many species can adopt to the change with disastrous consequences. According to the US Environmental Protection Agency (EPA), temperatures could rise an average of 2 to 11.5°F (1.1 to 6.4°C) by the end of the twenty-first century. Rising sea levels are already threatening to inundate vast tracks of coastal areas with many islands with their unique eco systems facing obliteration. The weather will only become harsher and move towards extremes – the winters will become colder, the summers warmer and the monsoons more erratic. While some parts of the world will face severe draught and desertification, others will be inundated, needless to say causing extreme and untold misery.

As countries in the north warm, vectors and carriers migrate north, carrying pathogens with them. There will be more genetics variations and mutations in microorganisms. This may lead to outbreak of malaria, dengue fever, Hanta virus and several air and water borne diseases that may assume the magnitude of epidemics. Global warming is likely to become a threat factor for heat strokes, cardiac and pulmonary problems. Heart patients are particularly susceptible as their cardiovascular organs need to exert more for cooling the body in situations of hot weather.

As the ocean temperature rises, the probability of frequent and stronger hurricanes also increases as people have experienced in the year 2004 and 2005. Hurricanes cause damage of billions of dollars and more are expected in the Caribbean, the United States and Burma. More devastating floods are likely to occur like those in Pakistan (in 2010), Brazil and Australia (in 2011). A rise of just 1 metre (approx 3 feet) could lead to the submergence of substantial portions of the eastern coastline in USA; one sixth of Bangladesh could permanently vanish with an increase of 1.5 meter (5 feet). This would imply relocation of power stations, refineries, hospitals, homes etc.

There are some areas that will experience more precipitation on account of global warming, whereas other areas are likely to undergo severe droughts and heat waves. Africa will face the most awful consequences with more ruthless droughts. Droughts are also anticipated in Europe, Australia, Southern Europe and parts of China and India. The 2003 summer heat wave killed 14,800 people in France and in July 2006, severe heat waves in North America resulted in at least 225 deaths.

Hot and dry weather led to a record of about 100,000 wildfires in 2006 with nearly 10 million acres burned.

Water has already become a rare commodity and according to scientists all over the world, global warming will worsen such conditions leading to conflicts and water war situation.

The fact is that the developing countries will be affected the most because they do not have the amenities to combat like that of developed countries; the colossal scarcity and areas with severe weather conditions that lies in these developing nations. Small island countries by now are witnessing the rise of sea level.

The measure of GHGs: *'Global-warming potential (GWP) is a relative measure of how much heat a greenhouse gas traps in the atmosphere'*. It compares the amount of heat absorbed by a given mass of the gas to the amount of heat trapped by the same mass of carbon dioxide. A GWP is usually calculated over 20, 100 or 500 years' time interval. GWP is stated as a factor of carbon dioxide. The GWP of CO_2 is considered to be 1. The GWP of a GHG depends generally on the absorption of infra-red radiation and its atmospheric lifetime. Thus a GHG with a large infrared absorption and a longer atmospheric lifetime will have high GWP.

Box 9.1: Nations to be worst affected from climate change

A **Global Climate Risk Index** published by German Watch, towards the end of 2011, has enlisted countries to be worst hit by climate change. They are -

1. Bangladesh	2. Myanmar	3. Honduras	4. Nicaragua	5. Haiti
6. Vietnam	7. Dominican Republic	8. Pakistan	9. Korea, DPR	10. Philippines

Table 9.3: Global warming potential of a few GHGs as per IPCC

Formula	Species	Lifetime (years)	GWP		
			In 20 years	**In 100 years**	**In 500 years**
CO_2	CO_2	Variable §	1	1	1
CH_4	Methane*	12 plus minus 3	56	21	6.5
N_2O	Nitrous oxide	120	280	310	170
CHF_3	HFC–23	264	9,100	11,700	9,800

Formula	Species	Lifetime (years)	GWP		
			In 20 years	In 100 years	In 500 years
CH_2F_2	HFC-32	5.6	2100	650	200
$C_2H_2F_4$	HFC-134	10.6	2,900	1,000	310
CH_2FCF_3	HFC-134a	14.6	3,400	1,300	420
$C_2H_4F_2$	HFC-152a	1.5	460	140	42
SF_6	SF_6	3200	16,300	23,900	34,900

Measures to be initiated by the local government and individuals to curb GHGs emission:

- replacing the existing lights by low wattage lamps;
- reducing the number of lights and hour of operation;
- switching to PV powered street lighting; replacing incandescent bulbs by LEDs;
- promoting fuel efficient vehicles;
- regular checking and monitoring; switching to BOVs;
- increase the use of bicycles and walk;
- shifting to the use of bio-fuel; improving route efficiency;
- for public, use of mass transit system; car pool;
- infrastructural improvements;
- reduction of public transit fares;
- increase in PV or electric charging stations;
- increase toll taxes to reduce the use of private vehicles;
- to conduct energy audit for municipal water pumping;
- retrofit of energy efficient facilities;
- new construction of energy efficient sewage and waste water treatment systems;
- to promote increase of recycling of paper, cans, glass containers cartridges, etc;
- recover food wastes for composting;
- to prefer purchase of recyclable items;
- establishment of centres for reusing salvageable foods;
- distribution of compost bins, and to arrange for community based recycling drop – off sites;
- set up of waste to power generation facilities;
- improving the light efficiency of buildings by using CFL, LED, sensors etc;
- using energy efficient motors, equipments, air conditioners, refrigerators;
- having rooftop gardens and greening the neighbourhood;

- building specific fuel switches for a shift from electricity to natural gas;
- using roof for setting up of PV and solar thermal systems;
- arranging for efficient heating and ventilating systems;
- to implement cogeneration and heat recovery;
- retrofit of appliances in parks, stadiums, markets;
- suitable insulation of buildings;
- regular energy audits for identification of the problem areas;
- to implement energy conservation in industries;
- pre-requisite of energy efficient measures for granting of permits to new industries;
- prudent use of renewable along with heat recovery and cogeneration systems; and
- purchasing of energy efficient products and appliances.

In Dec 1997, at Kyoto in Japan, the representatives from most of the countries worldwide met to negotiate a binding agreement to set targets or deadlines to reduce GHGs emission. The Kyoto Protocol wanted to devote developed countries to individual targets of emission for the period of 2008 to 2012. The aim was to reduce the overall emission by 5.2 per cent from the base level of 1990 to be achieved by 2012. To put Kyoto into effect at least 55 nations should ratify and a total of these countries' emission to be at least 55 per cent of the total emission. This clause was fulfilled in 2005 with Russia's ratification and hence the delay in implementation. George W. Bush, President of the biggest greenhouse gas producing country announced on 2 April 2001 that the US has no intention to comply with the treaty.

9.6.2 Acid rain

Acid rain is downpour comprising droplets of water which are remarkably acidic. It is caused by atmospheric pollution, particularly by enormous amounts of sulfur dioxides and nitrogen dioxides emitted by vehicles and industries. Acid rain causes a change in the pH balance of water and cause damage whenever it occurs. Sometimes acid rain is also referred to as acid deposition as it incorporates acid precipitations of snow.

There may be both dry and wet form of acid deposition. Wet deposition in the form of rain can remove the acid particles in air and deposit them on the Earth's surface. On the other side, dry deposition refers to the pollutant and gas particles that glue and attach to the ground with the help of dust and smoke in absence of rainfall. It poses more hazard as the precipitations ultimately overflow these contaminants to rivers, streams and lakes helping the menace to spread geographically.

The pH level of water droplets determines the extent of acidity in acid rain. A pH scale can usefully measure the amount of acid in water. The pH scale has a range from 0 to 14 with 7 as the neutral. Lower pH less than 7 indicates acidic condition while a higher pH value more than 7 indicates alkalinity. Normally rainwater is slightly acidic with a pH value lying between 5.6 and 5.8 due to the presence of carbonic acid caused by CO_2 in water. Acid deposition is all that is lower to this scale. It is also imperative to remember that the pH scale is logarithmic and each digit on the scale represents a 10 times change.

Cause of acid rain: Acid deposition may take place through volcanic eruption though burning of fossil fuel seems to be the principle cause of emission of sulfur dioxide and nitrogen oxides. Upon release of SO_2 and NO_2 in the air, they get to react with moisture, oxygen and other gases already existing to form sulfuric acid, ammonium nitrate and nitric acid. These particles then diffuse over extensive regions due to wind currents and plunge on the earth as acid precipitation.

Reactions: Sulphur dioxide is oxidized by the following reaction and then dissolves in water vapour.

$$SO_2 + O \rightarrow + SO_3$$
$$SO_3 + H_2O \rightarrow H_2SO_4$$

Nitrogen dioxide reacts with water to form nitric acid:

$$NO_2 + OH \cdot \rightarrow HNO_3$$

These gases are normally the byproducts of thermal power generation, vehicular exhaust and coal combustion. During the post Industrial Revolution period such air had in-flown in excess amounts, first discovered by Robert Angus Smith, a Swedish scientist in the year 1852. The phenomena hardly received any public attention till the term 'acid rain' was coined in 1972. Acid rain caught attention with the publication of issues like Hubbard Brook Experimental Forest in New Hampshire (late 1970s) in New York Times.

Effects of acid rain: Acid rain hit the aquatic environment quite obviously as it showers upon them directly. Both dry and wet deposition falls on the forest, field and roadways and ultimately it is carried to the rivers, streams and lacustrine systems. After the pouring on to the large water bodies they undergo dilution over time. Such acidic deposition causes the liberation of aluminium and magnesium. This will lower the pH even more in some areas. Below the pH of 4.8, the plants and animals face the risk of death. Fishes, frogs and insects may die if the pH falls below 4.5. At a pH of 5.5 all the benthonic bacterial decomposers starts dying. Freshwater shrimps die when the pH becomes 6.0. Estimates say that more or less 50,000 lakes in USA and Canada have a pH far below normal.

Trees can lose their leaves, on exposure to acid deposition while barks can be damaged and result in stunted growth due to lack of body functions. The trees become susceptible to disease and insect attacks. Acid rain showering on forest's soil affects the soil nutrients, kills micro-organisms in the soil, cause calcium and other mineral deficiencies and disrupts the ecosystem balance. Trees at high altitudes are also susceptible as the moisture in the acidic clouds blankets them. Acid rain reacts with aluminum readily in the soil converting it to aluminum sulfate or aluminum nitrate thus preventing the trees from absorption.

The damage to forests by acidic deposition is globally realized all over the world with notable cases all over in Eastern Europe where industrialization is running amuck and already claimed its toll. Estimate says that half of the forests are damaged in Germany and Poland, while 30 per cent have been affected in Switzerland.

The impact of acid rain on human health is yet to be correlated, but some particles formed from sulfate and nitrate ions can affect respiration. Fine particles easily enter indoors. Many scientific studies have identified a possible relationship between elevated levels of fine particles and increased chances of heart and lung disorders, such as asthma and bronchitis.

Acidic deposition exerts tremendous impacts on the architectural designs and marvels, mostly owing to its ability to wear away specific materials. As the acid starts falling on the houses specially those that are made of limestone, it starts its reaction with the minerals in the stones causing it to corrode, crumble, dissolve and get washed away – giving rise to the ailment called stone cancer. Acid deposition can also eat away present day buildings, vehicles, railway tracks, aircrafts, steel bridges and pipes both above and below the ground. Evidenced throughout the world some notable examples are the Acropolis in Greece, Renaissance buildings in Italy, Taj Mahal in India, Yucatan peninsula in Mexico, Cologne Cathedral, Washington Monument, Statue of Liberty, Notre Dame, the Colosseum and Westminster Abbey etc. Temples, murals and ancient inscriptions which were preserved for centuries are currently displaying austere signs of corrosion.

A number of steps are taken to ensure the reduction in the emissions of sulphur and nitrogen. Most of the governments are now putting pressure on the industries to clean the smoke stacks using scrubbers before their release and fitting catalytic convertors in automobile to reduce the emissions. Furthermore, subsidies are being given to alternative eco-friendly resources along with their increasing importance to refurbish the environment that had been damaged by the acid rain.

9.6.3 Ozone layer depletion

The ozone layer is a region of naturally occurring ozone gas of about 15 to 30 km above the Earth's surface in the stratosphere. Only 10 or less of every million molecules of air are ozone. This stratospheric ozone layer protects the Earth from the UV rays of the sun for more than billions of years. If the ozone layer is worn out by human activity, the consequences on the planet could be disastrous and appalling, an understanding that is just about dawning on mankind. Ozone is a bluish gas; an ozone molecule comprising three atoms of oxygen. The type of oxygen that we breathe in, comprises two oxygen atoms, O_2. When found in troposphere, ozone is considered as dangerous and unsafe, and is one of the substances responsible for producing the greenhouse effect.

The Antarctic ozone hole discovery in 1985 was led by British scientists Joesph Farman, Brian Gardiner, and Jonathan Shanklin of the British Antarctic Survey.

Considerable loss of ozone layer in the lower stratosphere was noticed over the Antarctic region during monitoring in 1970s by a research group from the British Antarctic Survey (BAS). When the initial measurements were recorded in 1985, the drop in ozone thickness was presumed to be due to instrumental errors. Replacement of the instruments revealed the same results and the phenomena of ozone depletion was accepted as a fact.

The thickness of ozone is measured in Dobson Unit. 'One Dobson unit is the number of ozone molecules needed to make an ozone layer of 0.01 mm thickness at 0°C and 1 atmospheric pressure'. Generally, the air has an ozone measurement of 300 Dobson Units, equivalent to a layer of 3 mm ozone thickness. If the level of ozone falls below 220 Dobson Units, the depleted area is referred to as 'ozone hole'. The ozone hole has steadily increased both in dimensions to about 27 million sq. km and span of existence from August through early December over the last 20 years. Now global ozone is about 4 per cent below the 1964–80 average.

Antarctica was an early sufferer of ozone depletion. A massive puncture in the ozone layer exactly above Antarctica threatens not only the continent, but many other regions that could be the sufferers of Antarctica's melting ice caps which are being melted by the harmful rays of the sun sipping through hole in the ozone layer. The principal cause of this is the liberation and discharge of CFCs, used in refrigeration systems, air conditioners, aerosols, solvents and in the production of some types of packaging materials. Other compounds containing bromine and other halogen compounds, nitrogen oxides (NOx) are also responsible for ozone depletion. CFCs are a common industrial product, nitrogen oxides are mainly released as byproduct of combustion processes, such as aircraft emissions.

Figure 9.2: Impact of various fractions of UV radiation

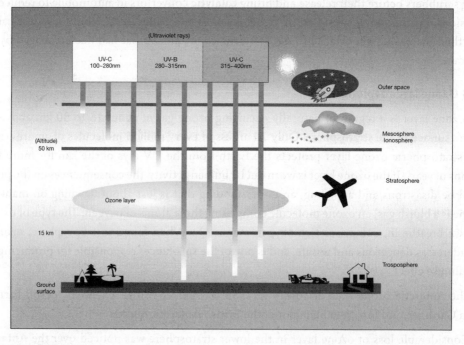

CFCs to undergo photo dissociation in presence of UV radiation, producing highly reactive chlorine free radicals.

$$CF_2Cl_2 + hv \; (< \text{about } 260 \text{ nm}) \longrightarrow Cl + CF_2Cl$$

dichloro difluoro methane

Chlorine free radicals may then react with ozone, thus destroying it from the atmosphere. Highly reactive ClO free radicals were formed.

$$O_3 + Cl \longrightarrow ClO + O_2$$

The ozone formation and decomposition continues in the stratosphere, the oxygen free radicals undergo reaction with ClO free radicals to reform chlorine free radicals:

$$ClO + O \longrightarrow Cl + O_2$$

ClO (hypochlorite) free radicals can undergo photo dissociation alternatively to reform chlorine free radicals.

$$2ClO + hv \ (< about \ 260 \ nm) \longrightarrow 2Cl + O_2$$

Chlorine free radicals reformed begins the process again, resulting in a chain reaction.

One chlorine atom can destroy about 1,00,000 ozone molecules, before it reacts with methane, nitrogen dioxide or itself.

Clouds are not generally formed in the stratosphere due to its extreme dryness.

Antarctica is completely surrounded by water and geographically isolated from air at higher latitudes during the winter season. This results into an isolated air mass whirling above Antarctica called the **south polar vortex**. With arrival of winter, a whirlpool of wind forms near the pole and separates the polar stratosphere. The temperature falls to very low levels in the stratospheric polar vortex, lower than 80 degrees, resulting in the formation of thin clouds of ice, nitric acid, and sulphuric

Figure 9.3: Ozone hole over Antarctica

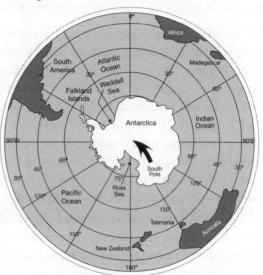

acid mixtures. At such low temperatures, the stratospheric chemicals freeze out and form **polar stratospheric clouds (PSCs)**. The chemical reactions occurring in the PSCs results in the large depletion in ozone during every austral spring over Antarctica. This is known as the Antarctic ozone hole. In spring, with the rising temperature, the ice starts melting and the ozone layer begins to mend and recuperate.

$$HCl + ClONO_2 \xrightarrow{\text{on ice}} Cl_2 \ (gas) + HNO_3 \ (ice)$$

PSCs acts as the medium on which reservoir chlorine compounds are converted into chlorine radicals and promote denoxification. Essentially there are two types of PSC: known as type 1 and type 2. PSCs I are supposed to be nitric acid and water mixtures that forms just above the frost point and can be either solid or liquid depending on the conditions. PSCs 2, less common, are formed of water-ice crystals at lower temperatures. Antarctic temperatures below −88°C (−126°F) are quite often but such low temperatures are rarer in the Arctic.

Every time 1 per cent of the ozone layer is depleted, 2 per cent more UV-B will effectively hit the surface of Earth. Decrease in stratospheric ozone will increase in higher levels of UVB radiations. Epidemiological studies show that UV-B triggers non-melanoma skin malignancy. It plays a key function in development of malignant melanoma. UVB is also said to be associated to cataract of eyes – a clouding of the eye's lens. Photokeratitis is an inflammation of the cornea and an acute syndrome that occurs after ultraviolet irradiation of the eyes. A dangerous form of photokeratitis is snow blindness.

Figure 9.4: UV Protection by ozone layer

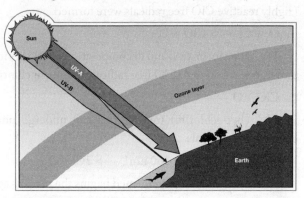

The life cycles of plants and their growth will change, disrupting the food chain and energy flow in ecosystems. Severe effects on animals are expected particularly in the early developmental stages of fishes, lobsters, shrimps, crabs, amphibians and others including stern effects in reduced reproductive capacity and defective development of larva; other effects are difficult to foretell. Seas and oceans will suffer hard. Both the locomotion and orientation of phytoplanktons are affected by Ultraviolet-B (UVB) exposure. This in turn reduces the rate of survival for these organisms. Researchers documented alterations in the reproductive rates of young fishes, shrimps, and crabs as well as frogs and salamanders that are exposed to excess ultraviolet B. Increase in UV radiation could have an effect on land and aquatic nutrient cycles bringing about changes in source and sinks of vital elements and major trace gases like carbondioxide, carbon monoxide, carbonyl sulphide and ozone.

As an international measure, the Montreal Protocol was finally approved upon on 16 September 1987 at the Headquarters of the International Civil Aviation Organization in Montreal to phase out the use of ozone depleting substances. The production and consumption of ODS in the stratosphere such as CFCs, halons, carbon tetrachloride, and methyl chloroform are to be phased out by the stipulated frame of time.

9.6.4 Nuclear accidents and nuclear holocaust

Today, nuclear energy is utilized to generate about 13 per cent of the world's electricity, with almost zero green house gas emissions. It is considered as one of the cleanest sources of energy and has the capability of meeting the growing needs of an increasingly energy hungry world order. However, the bombings of Hiroshima and Nagasaki and the gruesome images of Chernobyl (not to mention the nuclear holocaust movies) have terrorized entire generations who are still unwilling to embrace the fruits of nuclear power.

Because of its almost unending abilities – a single pellet of uranium fuel contains as much energy as 807 kilos of coal or 149 gallons of oil and 480 cubic meters of natural gas. Nuclear power plants can generate electricity incessantly for many months at a time, without interruption – nuclear power is looked at with much skepticism, general people being scared of the impending meltdown.

However, to put things in the right perspective, from the very onset, there was always a strong alert about the criticality of nuclear power and release of radioactive substance emissions. Till date in the history of civil nuclear power, Three Mile Island, Chernobyl and Fukushima were the three most notable accidents that have ever happened in over 14,000 cumulative reactor years of large scale productions in over thirty nations. Viewed in this light and considering the fact that it has no record of harmful green house gas emissions, nuclear power is not only sustainable but is also extremely safe for use.

Case Study 9.4: Fukushima Daiichii meltdown

On 11 March 2011, subsequent to a tremor of great magnitude, 15 m tsunamis totally impaired the functioning of electric supply and cooling of the 3 reactors of Fukushima Daiichi culminating into an accident. Though the reactors were seismically sound, it was the tsunami that led to their meltdown. During first three days all the three cores had melted. On International Nuclear Event Scale (INES), the accident was rated as 7 and the first few days had high radioactive discharges. Four reactors of net 2,719 Megawatt electric (MWe) net capacity were shut down.

The three reactors were quite safe with the addition of water after two weeks time. There was no provision for heat sink in order to remove the decay heat from the fuel. By the month of July the reactors were made to cool with the help of recycled water from a fresh treatment plant. The temperature of the reactors had dropped below 80°C by the end of October. In the middle of December a formal closure was announced.

Other than cooling of reactors, the basic challenge was to prevent release of radioactive materials, mainly from the contaminated water that had undergone leakage from 3 units. No reports of deaths or cases of radiation sickness from such meltdown have been declared. Nearly 1 lakh people had been evacuated and displaced from their houses to ensure safety and security. The psychological fallout of Fukushima – telecast live to the world – is something whose impact is yet to be assessed. Suffice to say, policymakers in every corner of the globe are seriously considering their options as far as nuclear plants are concerned.

9.7 Wasteland Reclamation

According to Integrated Wasteland Development Programme,

'Wasteland is a degraded land which can be brought under vegetative cover, with reasonable effort, and which is currently under utilised and land which is deteriorating for lack of appropriate water and soil management or on account of natural causes.'

Categories of wasteland:

- **Culturable wasteland** refers to the land which has a possibility for the development of vegetative cover but has not yet been used due to different factors such as erosion, water logging, salinity etc. The category lands are not cultivated for more than five years.

■ **Unculturable wasteland** refers to the land that cannot be used or developed for vegetative cover, involves very high cost if they are to be brought under cultivation, such as, the desolate rocky areas and glaciers covered with snow.

They can be landslides, industrial wastelands, ravines, sheet, rill and gully erosion, sand dunes and bar, coastal lands, shifting cultivation and lands affected by mine spills, water logged and saline lands etc.

The main reasons for wasteland formations are over cultivation, deforestation, overgrazing and improper irrigation practices. The geomorphic processes become active in the absence of land management practices. Erosion of soil layers makes the land infertile and useless. According to Wasteland Atlas of India-2005, NRSA, a total 552692.26 sq. km was designated as wasteland in India, with Jammu and Kashmir, remaining unsurveyed.

This is much enhanced by deforestation. Further, felling of trees intensify the lowering of water table and drought like conditions. The loss of fertility following erosion also results in the transformation of marginal forest lands into wastelands.

Urbanization and migration of population to towns and cities have led to the rise of many problems like more utilization of agricultural lands for housing construction, roadways, industries etc.

The balanced use of land resource is only possible by adopting an integrated land use policy. This involves prevention of land misuse and reclamation of unutilized, degraded land, fallows, etc. Reclamation of abandoned mines and brick kilns may add some more land. As much as possible agricultural lands should not be sacrificed for non-agricultural purposes.

Table 9.4: Land use pattern in India

	% of land	Activities
1	44	Agriculture
2	23	Forest cover
3	8	Housing, agro forestry, roadways, industries etc
4	4	Grazing and pastures
5	8	Miscellaneous
6	14	Barren

Wastelands are broadly categorized under two groups: barren and uncultivable waste land and cultivable wasteland. The first category includes lands which cannot be brought under cultivation and economic use except at a very high cost, whether they exist as isolated pockets or within cultivated holdings. They are mostly lands such as hilly slopes, rocky exposures, stony or leached or gully land sandy deserts.

Important measures for wasteland reclamation:

■ By using abundant water and fertilizers the land can be made cultivable.

■ Adoption of afforestation practices and agronomical practices for soil conservation. Soil testing is to be conducted and appropriate crops are to be selected and cultivated according to the local conditions.

■ Irrigation patterns needs to be changed.

- To afford safe disposal of water of the catchment areas, contour bunds are constructed. Guidance is to be provided to prevent water logging and salinity.

- Land may be used for settlement of landless agricultural workers.

9.8 Consumerism and Waste Products

Consumerism refers to the use of resources by people. Early human societies used to consume a smaller amount of resources; with the onset of the industrialization, consumerism has exhibited an exponential increase. It is a social and economic order that is based on the systematic conception and promotion of a craving desire to purchase more and more goods and services and is not always related to the needs and requirements. A product is something that is prepared, manufactured, grown or procured in great quantities so that it can be sold to the consumers. Production forms the basis for consumption. Production generally increases with its demand. This increase in production attains sustainability only when the product quality is maintained; the price is regulated and claims in advertizing are adhered to.

Our forefathers led a simple life with fewer wants. In modern era, our needs have multiplied and amplified. This is related to an increase in demands and due to change in the life style. As a result we have to confront a number of inter-related ecological and resource problems.

In developing countries, the growing population size and the degradation of renewable resources are the key factors in the total environmental impact. Excessive population pressure causes over utilization and degradation of the finite resources, and the consequence is complete poverty, starvation, malnutrition and early deaths. This is prevalent in the less developed countries (LDCs). The per capita resource use in LDC is small. On the contrary, in developing countries, rate of per capita resources use is high. The elevated level of pollution and environmental dilapidation per capita, determines the overall environmental impact. In more developed countries (MDCs), the population size is small, the resources are in abundance, people lead luxurious lifestyles and consequently per capita use of resources is very towering. More the resource consumption more is the waste generation and hence greater is the deterioration of the environment. The average US Citizen uses about 35 times more than an average Indian citizen and 100 times more than the average person in the poorest countries. The overall environmental impact of both the types of consumerism may be similar. USA is well-known for maximum consumerism.

Present day environmental management stresses on structural changes like:

- large scale technological substitutions;

- proper resource utilization;

- less pollution, adoption of waste minimization techniques;

- adopting 3 R's – reduce, reuse and recycle technologies; and

- to provide direct economic benefit and stimulate national economy.

Inefficient consumerism results in the following problems:

- uncontrolled manufacture of foods and inferior quality of goods;

- uncontrolled adulteration which may lead to health and hygiene problems;

- improper services may result in customer dissatisfaction and stress; and
- production of tons and tons of waste leading to natural resources depletion, the outcome of which is environmental imbalance.

Efficient consumerism is the most indispensable element to minimize waste and to uphold economy of a nation. This can be achieved by:

- Product standards may be verified before buying from market. Product may also be standardized by certification audit such as ISO.
- Waste minimization techniques can be adopted.
- Recycling and proper disposal of waste materials can be done.
- Enactment and implementation of strict legislation is another step.

9.9 Environmental Legislation

9.9.1 The Environment Protection Act, 1986

The Act came out as an important constitutional implication with the background of Bhopal gas tragedy. The Act was passed to 'provide for the protection and improvement of environment and for matters connected therewith'. As per the Act, environment is defined as which includes 'water, air and land and the inter-relationship which exists among and between water, air and land, and human beings, other living creatures, plants, micro-organisms and property'.

The law gives general powers to the central government. Central government can undertake any measures or steps that become necessary for protecting and improving environment. The Act provides for the standards that defines quality of air, water and land for different areas for different purposes. The highest permissible limits for various pollutants are also specified under this law.

The Act makes EIA mandatory for the specified 29 industries. Under this Act all companies must possess a type of Spill Prevention Control and Counter-measures plan. The law requires Environmental Audit since 1993, the report of which is to be submitted to the SPCB. The law provides for the declaration of CRZ.

The law prevents the reckless handling of hazardous substances except as is prescribed or stated. The act is supplemented with National Environmental Tribunals Act, 1995 and National Environmental Appellate Authority Act, 1997. Violation of act calls for penalty which may be liable to be punished by up to seven years imprisonment or a fine up to ₹ 1, 00,000 with an additional fine up to ₹ 5,000 for every day of infringement.

9.9.2 The Air (Prevention and Control of Pollution) Act, 1981

The government enacted this Act to provide for 'Prevention, control and abatement of air pollution.' According to the law an air pollutant is 'any solid, liquid or gaseous substance present in the atmosphere in such concentration as may be or may be or tend to be injurious to human beings or other living creatures or plants or property or environment'.

The Act provides for the establishment of boards and also assigns them with powers and responsibilities to look into such matter. The Act prohibits the discharge of PM, lead, sulfur dioxide, nitrogen oxides, VOC, carbon monoxide and other toxic substances into the air, beyond the prescribed emission level. The SPCB is assigned with the responsibility to monitor the pollution levels in air at specific sites. The Act requires an approval to any industrial plant prior to its operation.

The government may suggest selling of controlling equipment before granting consent to the concerned industry. The board specifies specific standards to the industries. Penalties for non-compliance are a minimum of six months custody to a maximum of seven years along with a fine of ₹ 5,000 for each day of violation.

9.9.3 The Water (Prevention and Control of Pollution) Act, 1974

This Act was enacted in order 'to provide for the prevention and control of water pollution and the maintenance or restoring wholesomeness of water'. The Act provides for the establishment of Boards to prevent and control water pollution. The law confers the boards with powers and responsibilities to look into the matters connected therewith. The Act prohibits discharge of any poisonous, noxious or polluting matter into the water bodies whether static or free flowing.

The organization has to get permission from the State Pollution Control Board prior to any category of new release into the water body and this applies for temperature discharges also. Individuals are required to comply with the effluent standards during any discharge of sewage or sullage. The law specifies the standards for small scale industries. The central and state boards are required to set up a laboratory for sampling, testing and analyzing water samples. A fee is to be charged for such services.

The penalties for non-compliance involves three months imprisonment and a fine of ₹ 10,000 or a fine up to ₹ 5,000 every day of violation or both plus the expenses incurred by the Board during inspection, sampling and analysis. If any offence has occurred with the consent of any director, officer, manager or secretary of a company and proved, he is deemed to be guilty.

9.9.4 Wildlife Protection under Wildlife Protection Act, 1972

Wildlife Protection was seen for the first time in certain notification by the state administrative authorities around 1972. The main objective was to protect hunting reserves. There was no separate legislation for the protection of wildlife. The National Forest Policy 1952, emphasized on the wildlife protection through the formulation of special laws and by setting up of sanctuaries and parks. It also established the Central Board of Wildlife under the Ministry of Food and Agriculture.

The Wildlife Protection Act, 1972 is a major step to protect wildlife with a complex administrative structure. The Act defines the following as:

- Animal – *'amphibians, birds, mammals, and reptiles, and their young, and also includes, in the cases of birds and reptiles, their eggs.'*

- Wildlife – *'includes any animal, bees, butterflies, crustacean, fish and moths; and aquatic or land vegetation which forms part of any habitat.'*
- Hunting – *'includes:*
  i. *capturing, killing, poisoning, snaring, or trapping any wild animal, and every attempt to do so*
  ii. *driving any wild animal for any of the purposes specified in sub clause*
  iii. *injuring, destroying or taking any body part of any such animal, or in the case of wild birds or reptiles, disturbing or damaging the eggs or nests of such birds or reptiles.'*

The Act empowers the Central Government to constitute a Central Zoo Authority. The authority shall comprise a chairperson, member-secretary and members not to exceed 10 in number. The chairperson and other member are appointed for a term, not more than three years. The authority sets the lowest standards for lodging and veterinary care of animals in the zoo. The authority may recognize and de-recognize zoos. They identify the animals for their use of captive breeding. They synchronize training programmes both in and outside India. They organize research and educational programmes. They also co-ordinate the acquisition and exchange of animals.

The Act prohibits teasing, injuring and disturbing of animals. The law prohibits for the establishment of National Board of Wildlife (NBWL) and State Board of Wildlife. The State Wildlife Department is headed by chief wildlife warden.

The Act seeks wildlife protection through the creation of
- national parks;
- sanctuaries;
- conservation reserves; and
- community reserves.

Distinction between national parks and sanctuaries:
- Hunting of animals in Schedule 1 is totally prohibited while that of other in schedule 2, 3 and 4 is permitted only on license.
- Central government is empowered to declare any area a sanctuary. Public entry is barred and hunting without license is prohibited.
- Trade and commerce in animals and their products is forbidden except in specific conditions.
- Major difference between reserved and protected forest on one hand and sanctuaries on the other in at the level of policy. The former are areas for regulated utilization of natural resources as opposed to wildlife conservation, the latter is clearly for protecting wildlife.
- Most of the protected areas are inhabited by local communities. Activists claim that tribal rights are denied and administration is throwing open the habitats to mining and other development projects that is running the habitats.
- Most important aspect is the implementation of existing laws against poachers. Administrators fail due to their well off connections. The department is understaffed and do not have basic facilities to transport, education, housing in the lower rungs. Salaries are low and encounters lead to killing of the officials.

- The NWAP, 2002 focus mainly on the manner in which wildlife should be conserved with more participation of communities. Short term economic goals should not be permitted to undermine ecological security.

- Natural sites are declared as ecologically sensitive zones under EPA where government can regulate developmental activities.

9.9.5 Forest Policy and The Forest Conservation Act, 1980

The Forest Department was formed in 1864. This was followed by the Forest Act of 1865, which was enacted in the year 1878, when the British realized the importance of Indian Forest to be a permanent source of revenue. The Forest Policy of 1894 gave more importance to public benefit and agricultural use. There were many provisions but were never implemented.

Indian Forest Act of 1927, witnessed wide powers given to the state. The state could determine three categories of forest – reserve forest, protected forest and village forest. Forest officers of any rank were given wide range of powers. The Act merely reflects the colonial policy and administration.

The Indian Forest Policy, 1952 focused more on the role of forest as revenue generators. It realized the necessity of balanced and complementary use but even then reiterated the colonial principles.

During this period large tracts of forests were cleared and replaced with monoculture plantations of rapidly growing commercially valuable species.

Forest Conservation Act, 1980: The purpose of this Act is to provide for the conservation of forests. The Act extends to the whole of India except the state of Jammu and Kashmir. The central government may constitute an Advisory Committee to advise the government relating to granting of approvals or any matters related to conservation of forests. Whoever fails to abide by the given rules shall be punished with simple imprisonment for a period that may extend up to 15 days. Without prior approval of the central government the act prevents the state government from ceasing a 'reserved forest' partially or wholly; from giving any forest land for non-forest purpose such as cultivation of cash crops, horticultural or medicinal plants and from clearing naturally growing trees for the purpose of reforestation.

Hence, this act sought to prevent the state from disposing the forest lands in an arbitrary manner. But the administrators were hand in gloves with the offenders or simply closed their eyes and let such activities go on.

The 1989 amendment permitted the forest officers to be prosecuted and to be subjected to fine, similar to other persons. The government adopted and started the concept of social forestry in order to provide fuel wood to the domestic users. The scheme comprised three components:-

- community woodlots on common village lands;

- strip plantations, along the roadsides and canal bank; and

- farm forestry.

9.9.6 Issues involved in enforcement of environmental legislation

Environmental law without enforcement seems to be useless. The governments are generally assigned responsibility to administer statutes and start prosecutions in order to enforce compliance. Often government is found to transgress the law and the entire responsibility falls on the citizens who then pursue court for the enforcement of orders. To keep pace with the rapid environmental degradation more stringent laws are to be brought in and enforced. In spite of all, some remarkable decisions of the upper courts exerted a tough impact on the country's environment law. The principal enforcement instruments in India are the SPCBs and the CPCB. The opportunity of these authorities under the Water act and Air act can be labelled as 'residual' in nature.

Under these Acts, SPCB is allowed to collect samples of any sewage or trade effluent or any emissions from any chimney, etc. The results and reports after sample analysis is however, not acceptable as proof in any lawful procedure, unless the comprehensive process approved by the law has been tracked. The SPCB has the authority to conduct inspection at all places for the reason of investigating – any factory, testimony, inventory, or document. The violation of most of the environmental pollution laws are criminal, unlawful and immoral in nature and entice penalties and feasible incarceration. Infringement of environment zoning notification is civil in nature.

India is an active participant in international forums and has ratified various Multilateral Environmental Agreements. This includes the Ramsar Convention on Wetlands, 1971; CITES, 1973; Montreal Protocol, 1987; Basel Convention, 1989, CBD, 1992; Cartagena Protocol on Biosafety, 1992; the United Nations Framework Convention On Climate Change, 1992 and Kyoto Protocol, 1997; Rotterdam Convention on the Prior Informed Consent Procedure for Certain Hazardous Chemicals and Pesticides in International Trade, 1998 and the Convention on Persistent Organic Pollutants, 2001.

The marine pollution in India is regulated by the Territorial Waters, Continental Shelf (CS), Exclusive Economic Zone (EEZ) and Other Maritime Zones Act, 1976. This Act proclaims India's governance and dominion over the natural resources in the Continental Shelf and the EEZ and deliberates special command to the central government to protect and preserve the oceanic environment. This is accompanied by the Merchant Shipping Act, 1958, which administers the civil and criminal accountability regimes in case of oil spills. The CRZ notification, 1991, governs the development along the coastal stretches. The notice bars 13 categories of actions, together with the setting up of fresh industries and the extension of existing ones. Exceptions being those activities which need water face and foreshore amenities. The notification categorizes the coast into four categories, based on their environmental sensitivity to permit or to restrict different types of activities.

As per the Indian Constitution, water is in the 'state list'. The states have control over the withdrawal of ground water from surface water sources. However, an all-inclusive Act to cover extraction of groundwater is wanting. A replica bill to legalize and control the groundwater management was prepared in 2005, which has been adopted by some states, but, is yet to be passed as a law. Few states such as Maharashtra have enacted distinct laws to regulate the withdrawal of groundwater to the degree that it affects the supply of drinking water.

The Indian Forest Act, 1927 empower authority to the states over both public and private forests, and controls the extraction of lumber for revenue purposes. The forests are grouped into reserve forests, village forests and protected forests. Once an area is declared as a reserve forest, all earlier personal and societal rights over the forest will stand null and void. Admittance to the forest and access to forest produce becomes an issue of state privilege. A Central Empowered Committee has been set up by the Supreme Court to oversee the timber accessibility in India and control all wood-based industries, such as saw-mills, veneer and plywood plants, which require former permission to maneuver.

Twenty five wetland sites were declared by India under the Ramsar Convention on Wetlands, 1971, all of which are well conserved, as per Wildlife (Protection) Act, 1972. Recently, conservation reserves (state-owned land) and community reserves (community or private land) are created to improve the social and financial situation of populace living in ecologically susceptible areas as well as to ensure wildlife conservation. The Government also regulates trade in wildlife as per CITES, 1973.

The Noise Pollution (Regulation and Control) Rules, 2000, were prepared to standardize and manage noise producing and noise generating sources. It lays noise limits and stipulates the existence of silence zones around hospitals, schools, courts, colleges, religious places and any such area declared by the authority.

The Factories Act, 1948, as division of its general environmental health and safety (EHS) has the requirements relating to dangerous processes in factories and ensures safe working conditions. The EPA, 1986, serves as the most comprehensive Act well supported by many other subordinate Acts and Rules. This act is also known as Umbrella Act. The Act illustrates to what degree the Indian judicial system can be concerned in the enforcement of environment legislations.

Box 9.2: Green Bench and National Green Tribunal, 2010

Green Bench

A Supreme Court division bench directed the Chief Justice of Calcutta High Court to create a special division bench called the 'green bench' on 16 April 1996. Green bench, thus, constituted, was to hear environment related petitions. This initiative brought environmental issues to the door steps of the judiciary system. The first green bench in Calcutta was followed by another in Chennai as directed by Justices Kuldip Singh, Falzan-uddin and K. Venkataswami while dealing with the Madras tannery case in the same year on 2 September. Such benches had also been set up in MP, Punjab, Haryana, etc. The green bench had issued several judgements assisting environmental conservation. Public interest litigations such as felling of trees, illegal encroachments and illegal quarrying are the most frequently heard cases. Examples of landmark judgement includes ban on sand quarrying in Tamiraparani river, ban on cutting Karuvelam trees (Acacia nilotica) at Vaagaikulam tank near Alwarkurichi in order to protect the nesting sites of birds. The green bench issued orders to the state of Tamil Nadu to preserve the small hills such as the Yanamalai in Madurai. It also ordered the National Highways Authority to redesign its plan in Kanayakumari in order to preserve the irrigation tank. On 6 October 2007, Chief Justice Dalveer Bhandari and Dr D Y Chandrachud imposed a total freeze over cutting of mangroves in Maharashtra along with ban on dumping of debris and construction within 50 m of the mangrove vegetation. The same year in March the court rejected the petitions made by the builders to encroach and construct townships near Borivili National Park. The bench opined that no trees are to be cut without the government's permission. It ensured all states to collect compensatory afforestation funds where the forestland is to be used for non-forest purpose. These funds, already provided in Forest Conservation Act, 1980, now amounts to ₹ 5,600 crores. The bench instituted Net Present Value (NPV) tax in order to protect the areas not designated as forests as per FCA, 1980, but is controlled by the state forest departments. It prevented government to de-reserve forestland without its permission. But, the clearance to the mining operation by Lafarge in Meghalaya and by Vedanta in Niyamgiri (Orissa) has strongly been criticized. Though mysterious, the Honourable Supreme Court disbanded the green bench.

National Green Tribunal, 2010

The enactment of NGT, 2010 came as an outcome of several observations by the Supreme Court such as:

- the requirement of scientific data evaluation in environment related cases;
- requirement of professional judges and experts in relevant fields;
- establishment of environmental courts with both civil and criminal jurisdiction; and
- expeditious judgments in environment related cases.
- the existence of National Environmental Tribunal under National Environmental Tribunal Act, 1995 and National Environmental Appellate Authority under National Environmental Appellate Authority Act, 1997 only in paper but non-functional in reality.

In view of such findings the Law Commission's 186th report proposed to constitute National Green Tribunal and National Green Tribunal Bill was introduced in the Parliament in 2009. *'An Act to provide for the establishment of a National Green Tribunal for the effective and expeditious disposal of cases relating to environmental protection and conservation of forests and other natural resources including enforcement of any legal right relating to environment and giving relief and compensation for damages to persons and property and for matters connected therewith or incidental thereto'.*

The Tribunal was bestowed ample powers for settlement of environmental disputes, for providing reliefs and compensations; penalties may vary from three years imprisonment or ₹ 10 crores which may be upto ₹ 25 crores for companies. The tribunal comprises a minimum of 10 judicial and expert members to a maximum of 20. Though with many ambiguous words and clauses, yet the Act is remarkable first step towards environmental protection.

9.10 Environment Impact Assessment (EIA)

According to the International Association for Impact Assessment (IAIA), EIA is defined as, *'the process of identifying, predicting, evaluating and mitigating the biophysical, social and other relevant effects of development proposals prior to major decisions being taken and commitments made.'*

It is a systematic process of identifying the future consequences of a current action. In other words, EIA is a tool, seeking to ensure sustainable development through the evaluation and assessment of those impacts, which may arise out of a major activity or project, that are likely to exert environmental impacts. Hence, it is:

- Anticipatory: environmental impacts are presumed.
- Participatory: requires involvement of people from various strata and diverse background.
- Systematic: the entire process follows definite steps and relies on multidisciplinary input.

It evolved in the 1960s in response to increasing environmental awareness. Originally it comes from Section 102 (2) of the National Environmental Policy Act (NEPA), 1969, USA. USA was the first country to enact legislation on EIA.

A. Objectives of environment impact assessment

It developed as an outcome of the failure of the traditional project appraisal methods to account for the impacts on environment. Precisely it aims:

- to assess and present the unquantifiable impacts;
- to provide correct information to the people;
- to formalize the consideration of various alternatives to the project;
- to improve the designs and methodologies to safeguard the environment;
- to shape the projects to suite the local environment; and
- to place the finding with analysis in front of the decision makers.

B. Principles of environment impact assessment

EIA centers on the following principles: transparency, certainty, practicability, flexibility, reliability, accountability, participation and cost-effectiveness.

C. The generic process of environment impact assessment

- *Project conceptualization*
  The detailed information may not be known initially. The proponent in consultation with the consultant shall have to notify in writing by submission of the project proposal with the basic nature of the project, the sites, and the area to be used, etc to the concerned office. The consent from forest department or Airport Authority if required should also be submitted along with the project proposal.

- *Screening*
  Screening is undertaken to determine whether the project proposal requires undergoing the process of environmental impact assessment and getting environmental clearance or it can move ahead without EIA. The process categorize the projects into –

i.   project requiring EIA;

ii.   projects not requiring EIA; and

iii. projects for which the requirement of EIA is not apparent.

Further the initial screening can be screened on the basis of threshold criteria (size, location, infrastructural demands, output, cost, etc) and impact criteria (significant and identifiable impacts, sensitive areas, etc).

■ *Scoping*

Determines the coverage of study of the proposal and establishes the terms of reference according to the guidelines provided for various sectors. It identifies the issues and concerns in the process of EIA, determines the assessment methods and ensures a relevant EIA.

■ *Baseline data collection*

Baseline information serves as an important reference point for the execution of EIA. 'Baseline' refers to the background information on biological, physical, social and economic settings for the proposed location.

The existing environmental status of the study area is to be described and new information is obtained from field samplings. The primary data should be supplemented with secondary data. The task of collecting background information starts right from the inception of the project. It should serve to describe the status and trends of the environmental factors so as to be compared with the project induced changes on environment and thereby prove to be a means of detecting the actual change during monitoring, after the start of the project.

■ *Impact prediction*

This is perhaps the most crucial step where mapping of environmental consequences of the project is done in all aspects. It starts with scoping and ripens with time when more and more data becomes available on biophysical, social and economic environment.

i.   Biophysical impacts on vegetation, wildlife, crops, aquatic life, land forms, soil, propensity of soil erosion, floods, silting and sedimentation, quality of air, water and soil.

ii.   Social impacts like displacement and relocation, authority and leadership, family structures, age, gender, social networks, housing, schools, justice, health, welfare, recreation.

iii. Economic impacts like cost of project construction and operation, material cost, capital investment, price of local goods and services, taxation, income generation opportunities, workforce requirement, employment, income and equity, need of labour peaks, proportion of female labours.

iv.   Cultural impacts like traditional life styles, religious shrines and tribal factors, historic sites, UNESCO heritage sites.

Health impacts like issue of vectors, likelihood of contact between people and vectors, contagious diseases, mental and psychological well being, interaction and impacts of the chemicals should also be considered.

Impacts are generally divided into two broad categories quantifiable and non-quantifiable criteria. The alternatives are to be identified, predicted and compared. Impacts are defined on the basis of its severity of potential impact, whether reversible or irreversible and the potential rate of recovery. The spatial extent or the zone of influence of the impacts should be determined. Impacts that may arise in the various stages should also be considered. The impact that lasts for three to nine years is grouped as Short term; that which lasts for ten to twenty years is Medium term and the one which goes beyond twenty years is Long term. The alternatives are rated and assigned ranks for the choice of the best that yields environmentally optimal financial reimbursement to the society as a whole.

Evaluation of Impacts and mitigation measures

In the subsequent step, the decision makers should have clear information about the environmental scenarios before the start of the project, environmental scenarios with the project and environmental scenarios with the project alternatives. With all other reports and documents in hand, the proponent prepares the final EIA report in a supposedly logical and transparent manner. The key elements for evaluating the impacts are ecological (effects on flora and fauna, carrying capacity, rare, endangered, endemic and keystone species, ecosystem resilience, viability of local species, etc), socio-economical (effect on human life, culture, traditions, displacement, groups and ethnicity, etc), environmental (standards of air, water, soil and limits of discharge). The authority examines whether all the procedures have been followed as per the notifications or not. The executive summary is called the **Environment Impact Statement (EIS)**.

A mitigation plan is to be prepared for chosen alternative to reduce or to avoid the environmental impacts, supplemented with EMP, Risk Assessment Report, disaster management plan, rehabilitation plan etc. The objective of mitigation plan is to maximize the project benefits and minimize the undesirable impacts. Mitigation measures should be integrated in the project design so that they automatically become a part of the construction and operational phase. Such measures should not restrict to one point in the entire EIA process; rather, there should be a strong coordination between mitigation and monitoring.

Public hearing

The present law entails that the public must be informed and allowed to discuss on the proposed project. Public involvement and communication plan is one of the major planning tools in the EIA process. The main purpose is to inform the stakeholders about the project, collect and consider their view points in the decision making process. The stakeholders can be individuals, the community and local people, project beneficiaries, the NGOs, scientists and experts, private sectors etc. There should be a scope for the interested public to get the copy of the project proposal, hold community meetings, workshops, seminars as well as for giving written opinions and/or join public hearings.

Environmental clearance

In the penultimate stage, the assessment authority reviews the report minutely with reference to the guidelines, conducts site visits if necessary and then comes to a decision to grant or reject the EC to the project. Decision is conveyed to the proponent within the stipulated time.

Monitoring

The last step is to monitor the circumstances for clearance, done both during manufacture and operative phases. A meaningful monitoring should determine the indicators to be used in the monitoring process, collect meaningful and accurate data, process the information, draw tangible conclusion and recommend improved corrective measures to be undertaken. This ensures whether commitments made are followed or not and the assumption in the EIA report is correct or not. Corrective actions are to be taken if the impacts exceed the predicted levels.

Figure 9.5: Steps in the process of EIA according to UNEP

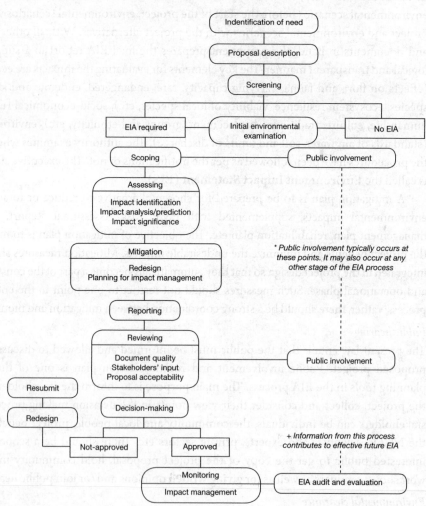

9.10.1 EIA in India

The process of EIA rests on three pillars:

A. **Statutory framework:** In 1972, the National Committee on Environmental Planning and Co-ordination (NCEPC) was formed under the Department of Science and Technology (DST) under the Fourth Five Year Plan. In those days NCEPC was directed to undertake EIA of all major development projects. In 1980, DoE was established, which was upgraded to MoEF in 1985. In 1986, MoEF enacted the EPA, 1986, which made provision for EIA. Environment clearance (EC) is subjected to the stipulated standards under Water Act, Air Act, Noise Pollution Rules, Forest Conservation Act, Hazardous Waste Rules etc, in addition to the stringent regulations of the State government on the local conditions.

B. **Administrative framework:** It involves how the regulatory authority works for Environmental Clearance (EC). Earlier, the process was centralized and conducted by MoEF solely. Presently it is decentralized to some extent between the Central and State government. MoEF serves as the apex body. In the centre, the main body involved in the process is Impact Assessment Agency (IAA) assisted by Regional offices of MoEF and the CPCB.

- IAA with the help of relevant authorities prepares questionnaires, checklists, notifications to amendment with regard to environmental laws. It seeks clarifications and conducts site visits, if necessary. It comprises expert committees called Expert Appraisal Committee (EAC) in the various categories of projects like industry, thermal power, river valley and hydroelectricity, mining, infrastructure and miscellaneous, and atomic power. The EAC comprises professionals and experts with requisite criteria as per Appendix VI; minimum of 15 years of experience or 10 years in case of Doctor of Philosophy (PhD), preferably below 70 years. The membership shall not exceed 15 including a Chairperson and Secretary (representative of MoEF). The Chairperson may choose an external expert as a member. The tenure for the members can be maximum of two terms of three years each. The experts and professionals can be from the field of environmental quality, EIA process experts, risk assessment experts, life science experts, forestry and wildlife experts or expert in appraisal process.

- There are six regional offices of MoEF at Bangalore, Bhopal, Bhubaneswar, Lucknow, Shillong and Chandigarh with the headquarters (HQ) at New Delhi.

- The CPCB, formed in 1974, acts as a research organization by collecting, analyzing and spreading information. Its technical staff and experts comprise the expert committee by IAA. However, it plays no direct role in the EC process.

At state level the process is done with the help of the DoE and the SPCB.

- The DoE is headed by a cabinet minister and formulates guidelines. The states of Maharashtra, Gujarat, West Bengal and Karnataka have State Expert Appraisal Committee (SEAC) under DoE that will issue the EC. In Andhra Pradesh, SEAC is under SPCB, which works under DoE. For the rest of the states, the No Objection Certificate (NOC) is issued by the Member-Secretary or Chairman of PCB.

- The SPCB holds public hearings, the minutes and findings of which needs to be furnished to IAA/SEIAA (State/ Union Territory Environment Impact Assessment Authority) within 30 days.

C. **Procedural framework:** This involves the process or the steps in the process. The cost of undergoing EIA is around 1–5 per cent of the project cost.

Application for Prior Environmental Clearance (PEC) include:

- Any person desiring to undertake a new project or desiring to go for expansion or modernization of any existing industry shall have to submit an application to the Secretary, MoEF, New Delhi.

- The project application will have to be made in the pro-forma prescribed in **Form 1** and the supplementary **Form 1A**, if applicable, accompanied with a Project Report including EIA Report, EMP and details of public hearing.

- The application has to be made after the prospective site has been identified, before any construction activity has begun or land preparation has started.

- The proponent should submit the application to the Central Government/State Government as per its category. The project authorities should intimate the project site location to the Central Government in case of projects like mining, pit-head thermal power stations, hydro power, irrigation or their combination, ports and harbors (except minor ones), mineral prospecting and exploration in $\geq$ 500 hectares. In such case the central government will convey the suitability of site within 30 days with a validity of five years. The submitted reports shall be evaluated and assessed by IAA along with the expert committee. For a thermal power project the application is to given to the State Government Department dealing with environment along with NOC from SPCB.

The environmental clearance for new projects comprises basically of four steps.

A. **Screening:**

- Projects requiring PEC was listed under Schedule I of 1994 Notification.

Box 9.3: Schedule-I, 1994 Notification

List of projects that require environmental clearance from the central government

1. Nuclear power and related projects such as heavy water plants, nuclear fuel complex, rare Earths

2. River valley projects including hydel-power, major irrigation and their combination including flood control

3. Ports, harbours, airports (except minor ports and harbours)

4. Petroleum refineries including crude and product pipelines

5. Chemical fertilizers (nitrogenous and phosphatic other than single superphosphate)

6. Pesticides (technical)

7. Petrochemical complexes (Both Olefinic and Aromatic) and petrochemical intermediates such as N,N-Dimethyltryptamine (DMT), Caprolactam, LAB etc. and production of basic plastics such as Linear low-density polyethylene (LLDPE), HDPE, PP and PVC

8. Bulk drugs and pharmaceuticals

9. Exploration for oil and gas and their production, transportation and storage

10. Synthetic rubber

11. Asbestos and asbestos products

12. Hydrocyanic acid and its derivatives

13. (a) Primary metallurgical industries (such as production of iron and steel, aluminium, copper, zinc, lead and ferro alloys); (b) electric arc furnaces (mini steel plants)

14. Chlor alkali industry

15. Integrated paint complex including manufacture of resins and basic raw materials required in the manufacture of paints

16. Viscose staple fibre and filament yarn

17. Storage batteries integrated with manufacture of oxides of lead and lead antimony alloys

18. All tourism projects between 200–500 m of High Water Line and at locations with an elevation of more than 1,000 m with investment of more than ₹ 5 crores.

19. Thermal power plants

20. Mining projects (major minerals) which leases more than 5 hectares.

21. Highway projects except projects relating to improvement work including widening and strengthening of roads with marginal land acquisition along the existing alignments provided it does not pass through ecologically sensitive areas such as National Parks, Sanctuaries, Tiger Reserves, Reserve Forests

22. Tarred roads in the Himalayas and or forest areas

23. Distilleries

24. Raw skins and hides

25. Pulp, paper and newsprint

26. Dyes

27. Cement

28. Foundries (individual)

29. Electroplating

30. Meta amino phenol

▪ The content of the 1994 notification shall not apply to

i. Item falling under entry nos.

3 {ports, harbours, airports (except minor ports and harbours)},

18 {all tourism projects b/w 200–500 m of High Water Line and at locations with an elevation of more than 1,000 m with an investment of more than ₹ 5 crores}, and

20 {mining projects (major minerals) which leases more than 5 ha of Schedule I}

i. For items under the entry nos. 1, 2, 3, 4, 5, 7, 9, 10,1 3, 14, 17, 19, 21, 25, 27 of Schedule I if the investment for new projects is less than ₹100 crores and for expansion/modernization is less than ₹ 50 crores.

ii. Any item for small scale industrial sector with investment less than 1 crore.

iii. Defence related road construction activities.

iv. Modernization of irrigation projects with an additional command area less than 10,000 ha or project cost less than ₹100 crores.

v. Production of bulk drugs and pharmaceuticals based on Genetically Modified Organism (GMO) covered by Modified Microorganisms Rules.

Box 9.4: Important category of projects as per 2006 notification

Category A projects:

1. Offshore and onshore oil and gas exploration development and production
2. Nuclear power projects and processing of nuclear fuel
3. Petroleum refining
4. Asbestos mining irrespective of mining area
5. Soda ash
6. Chemical fertilizers
7. Petrochemical complexes
8. Oil and gas transportation pipeline passing through ecologically sensitive areas (ESA)
9. Airports
10. Ship breaking yards
11. Mineral mining ≥ 50 ha of lease area
12. River valley projects ≥ 50 MW hydroelectric power generation ≥ 10,000 ha of culturable command area
13. Thermal power plants ≥ 500 MW (coal/lignite/naptha and gas based), ≥ 50 MW (pet coke diesel and all other fuels)
14. Coal washeries ≥ 1 million tone/ year
15. Cement plants ≥ 1 million tone/ year production capacity

16. Chlor-alkali industry ≥ 300 TPD production capacity

17. Petrochemical processing and synthetic organic chemical industries located outside the notified industrial estate

18. Distilleries – all molasses based and non-molasses based ≥ 30 kilolitre per day (KLD)

19. Industrial estates, EPZs, SEZs, biotech parks, leather complexes– if at least one industry falls under category A entire area will be treated as category A, industrial areas > 500 ha

20. Ports and harbours ≥ 5 million terapascal (TPA) cargo handling capacity (excluding fishing harbours)

21. Highways – New national highways, expansion of national highways >30 km involving additional right of way >20 m involving land acquisition through more than one state

Category B projects:

1. Aerial ropeways.

2. Common Effluent Treatment Plants (CETPs)

3. Common Municipal Waste Management Facility (CMSWMF)

4. Mineral mining < 50 ha and ≥ 5 ha of lease area

5. River valley projects < 50 MW and > 25 MW hydroelectric power generation, < 10,000 ha of culturable command area

6. Thermal power Plants < 500 MW (coal/lignite/naptha and gas based), < 50 MW, ≥5 MW (pet coke diesel and all other fuels)

7. Coal washeries < 1 million tone/year

8. Cement plants < 1 million tone/year production capacity

9. Chlor–alkali industry < 300 TPD production capacity

10. Petrochemical processing and synthetic organic chemical industries located inside the notified industrial estate

11. Distilleries– all non-molasses based < 30 KLD

12. Industrial estates at least one industry falls under category B and areas < 500 ha

13. Ports and harbours <5 million TPA cargo handling capacity and/ports and harbours ≥ 10,000 TPA of fishing handling capacity)

14. Highways – New state highways, expansion of national/state highways >30 km involving additional right of way >20 m involving land acquisition through more than one state

15. Building and construction projects ≥20,000 sq. m and < 1,50,000 sq. m of built up area

16. Townships and area development projects covering an area ≥ 50 ha and or built up area of ≥ 1,50,000 sq. m

■ The amendment of 2006 attempted to decentralize the process of EIA. Projects that require getting their PEC from the central government, MoEF, on recommendations of the EAC, are categorized as 'A' in the Schedule and those projects to get their clearance from the state government on recommendations from State /Union territory level Expert Appraisal Committee (SEAC), and are categorized as 'B' in the Schedule. The authority for Central Government is the IAA and that for the State Government is SEIAA. In case of category 'B', the application made in Form 1 will go through scrutiny by the SEAC to determine whether further environmental studies are required for the appraisal and grant of EC. In such case, projects requiring the EIA report shall be designated as category 'B1' while remaining projects will be known as 'B2'.

B. Scoping

■ This refers to the stage where the EAC for category A and SEAC for category B projects determine the details and comprehensive Terms of Reference (TOR) addressing all the relevant environmental issues. The TOR will be determined on the basis of the information furnished in the prescribed application **Form 1/Form 1A** along with the Terms of Reference (TOR) proposed by the applicant/ proponent.

■ All projects listed as category B in Item 8 (Building/Construction projects/ area development projects and townships) will not require scoping. It shall be appraised on the basis of Form1/ Form1A and the conceptual Plan.

■ The TOR shall be communicated to the proponent by the EAC/SEAC within 60 days of the receipt of Form1.

■ In case of Category 'A' hydroelectric projects Item 1c (i) (> 50 MW hydroelectric power generation), the TOR shall be conveyed along with the clearance of the preconstruction activities.

■ In case the TOR were not finalized and conveyed to the proponent within 60 days of the receipt of Form 1, the TOR furnished by the proponent will be the approved TOR for the EIA studies.

■ The application for PEC may also be rejected by the regulatory authority (IAA/SEIAA). The decision along with the reasons for rejection has to be communicated written within 60 days of the receipt of application.

C. Public consultation

■ It is the process whereby the concerns of the plausible stakeholders are taken into account with the purpose of making the design of the project appropriate.

■ All Category A and Category B projects should undergo public consultation except:

    i. Modernization of irrigation projects (item 1c ii of the Schedule)

    ii. All projects located within the industrial estates of parks (item 7c) approved by concerned authorities.

    iii. Expansions of roads and highways (item 7 f) which does not involve land acquisition

   iv.   All building/construction/area development projects and townships (item 8).

   v.   All Category B2 projects and activities.

   vi.   All projects or activities regarding national defence and security or under strategic considerations.

- There are two components under public participation:

   i.   Public hearing to be carried out as sought in Appendix IV.

   ii.   Procuring written responses from concerned people by placing the summary EIA report prepared in the format as per Appendix IV, application, on their website within seven days of the receipt of the written request for holding public hearing.

- Public hearing should be carried out by SPCB/UTPCC (Union Territory Pollution Control Committee) concerned within 45 days of request from the applicant.

- In case the board does not take initiative within the stipulated period, the regulatory authority shall engage any other public agency to complete such a hearing process within a further period of 45 days.

Procedure for conducting public hearing (as per Appendix IV)

a.   The proponent shall make a written request to arrange for public hearing to the member Secretary of the SPCB or UTPCC. In case the project site extends beyond a state, the applicant shall make separate requests to each of the concerned SPCB or UTPCC. The application shall be enclosed with at least 10 hard copies and equivalent number of soft copies of the draft EIA Report with the generic structure given in **Appendix III** including the summary EIA report (as per **Appendix III A**) in English and vernacular in accordance with the TOR communicated after scoping.

b.   At the same time the proponent should make arrangements to forward such reports to MoEF, District Magistrate, District Industry Centre, Office of Municipal Corporation, Zila Parishad office and Regional MoEF offices.

c.   The above mentioned authorities and SPCB except MoEF shall arrange for its full publicity within its jurisdiction and also make the reports available on its website.

d.   The member Secretary will fix the date, time and place of hearing within 7 days of the receipt of the draft EIA report and advertize in at least two newspapers. A minimum of 30 days is provided to the public for giving their responses.

e.   The hearing panel comprises the District Collector, MoEF representative, SPCB representative, State Government representative for relevant sector, not more than 3 representatives of the local bodies such as municipalities and Panchayat and not more than three senior citizens nominated by the District Collector.

f.   A video recording of the entire process should be done and a copy of this is to be enclosed along with the hearing proceedings to the authority.

g.   The proceedings should also be displayed at all the above mentioned offices.

h. The public hearing should be completed within 45 days of the receipt of the request letter from the applicant. It means that SPCB should send the proceedings to the authority within 8 days.

After the completion of the process the proponent shall address all the aspects of environmental concerns expressed during the process, make necessary alterations in the draft EIA and EMP. The **final EIA report** hence prepared shall then be submitted to the authority for appraisal.

The document on EIA will comprise the following (as per Appendix III):

- **Introduction** – with project objective, project proponent, description of the nature of project, size and site along with its significance on the area or the country.

- **Description of project** – project type, its need, location map, size, description of process and the technology, mitigation measures, assessment of novel and untested technology for the possibility of technology failure.

- **Description of environment** – the area under study, components and the methodology, founding of the baseline for all the environmental components and the base maps.

- **Predicted environmental impacts and mitigation** – details of the impacts due to its location, construction, design and operations, measures for minimizing adverse impacts, irreparable commitments of the components, assessments of the importance of impacts and mitigation measures.

- **Analysis of alternatives** – description of each alternative, its adverse impacts, suggested mitigation measures and selection of alternatives.

- **Environmental monitoring**

- **Additional study** – public consultation, assessment of risk, social impact assessment, relocation and rehabilitation plans.

- **Benefits of project** – improvement in physical and social infrastructure, employment opportunities and other material benefits

- **Cost benefit analysis**

- **Environment management plan**

- **Summary and conclusion** – the summary of the EIA report and overall justification of the project. As per **Appendix III A,** the summary of the full EIA report will consist of the project description, environment description, anticipated impacts and its mitigation, programmes on environmental monitoring, any additional studies, the benefits from the project and the EMP within 10 A4 size pages.

- **Disclosure of the consultants**, their names, resume and nature of consultancy.

D. **Appraisal**

- A detailed scrutiny of the application and other documents is to be carried out by EAC/ SEAC in a logical and transparent manner. If necessary, the applicant can also be invited to furnish clarifications.

- After completion, the EAC/SEAC shall make categorical recommendations to the authority for the grant or rejection of PEC on the basis of stipulated terms and conditions as well as reasons.

- Projects not requiring public consultation will undergo appraisal based on the application in Form 1 and Form 1A.

- Appraisal should be completed within 60 days of the receipt of the final EIA report or within 15 days of the receipt of Form 1 and Form 1A where public consultation is not necessary.

Procedure for appraisal (as per Appendix V)

The applicant shall apply with the following documents to the concerned authority

- 20 hard copies of Final EIA Report + 1 soft copy

- A video copy of the public hearing proceedings

- A copy of final layout plan

- 1 copy of project feasibility report

The final EIA report shall be scrutinized within 30 days from the day of receipt in accordance with the TOR and inadequacy to be communicated.

The regulatory authority on recommendations from EAC/SEAC will convey its decision to the proponent within 45 days of the receipt of the appraisal committee's report.

Validity of EC

- Validity of EC is the period from which the PEC is granted to the commencement of the production operations or completion of the construction operations.

- Validity of PEC for river valley project is for a period of 10 years, utmost limit of 30 years for mining projects and 5 years in case of all other projects or activities.

- For construction projects or activities, the validity period may be extended to a maximum of 5 years by the authority if the application is made within the validity period along with updated Form 1 and Form 1A.

Post EC monitoring

- Mandatory submission of bi-annual reports in terms of PEC conditions in hard and soft copy on 1 June and 1 December to the authority.

- All of these reports shall be public documents.

Transference of EC

PEC granted for a particular project to an applicant can be transferred to another legal person during the validity with a written no objection certificate.

Benefits of environment impact assessment

- Undertaking EIA during the design phase provides information about the environmental conditions of the locality; such information affords better design and ensures not to inadvertently put people at risk.

- The categorization of projects into A and B has the potential of being a good move towards decentralization and may also speed up the process of environmental clearance.

- Well defined scoping; the TOR will be determined by the EAC/SEAC.

- In public hearing, a provision is made to arrange for video recording of the entire proceedings to ensure that public hearing is properly carried out.

- The new notification tries to bring quite a number of projects within the purview of the EC and therefore furnishes a revised list. Strikingly important is the removal of categories based on investment; rather project size and capacities determine the mode of clearance.

- Early assessment of the possible impacts of a project ensures appropriate mitigation measures to be undertaken and implemented.

- It helps to reduce the project cost in the long run.

Drawbacks of environment impact assessment

- The constituted team for EIA often lacks expertise in the relevant fields.

- The concept of decentralization can be misused by the state if it actively pursues industrialization; in this case nothing about central government monitoring the state government activities is mentioned.

- Though the TOR is to be provided by the EAC/SEAC, the finalization of the TOR for scoping depends on the information provided by the proponent. In case the EAC does not specify the TOR within 60 days, the proponent may proceed with his/her own TOR.

- There is a lack of extensive ecological and socio-economic indicators for assessment of impacts.

- Public opinion is not considered at the early stage; this often leads to conflicts in the later stage of environmental clearance.

- Public consultation has been exempted for six activities like road expansion, modernization of irrigation projects, etc. These projects often have considerable social and environmental impacts.

- There is a lack of clear guidelines as to who can attend public hearing.

- Most of the reports are written in English and not in local language. The EIA report is too academic, bureaucratic and lengthy. The reports are also incomplete in many cases.

- Keeping in mind the lack of reliable data sources, credibility of primary data is also doubtful. More research and development of improved methodologies are required to overcome difficulties regarding the data.

- The available secondary data in many cases is unreliable.

- The knowledge of the indigenous and local people is not given adequate importance by the data collectors.

- Usually scoping covers the direct impacts only. Scoping is usually done by the proponent or the consultant inclined to meet the pollution control requirements rather that dealing with the range of potential environmental impacts.

- Often the ranges of possible alternatives are small in number.

- The details of prediction and evaluation are not mentioned in the EIA report. Details of the effectiveness of the mitigation measures are not provided.

- Sometimes for strategic projects, such as nuclear power plants, EMP is kept confidential for obvious political and administrative reasons.

- Emergency preparedness plans are not detailed and the information is not properly disseminated to the community people.

- The process of EIA ends immediately after the project clearance; nothing is clearly mentioned about post clearance monitoring in the new notification.

- Earlier the Rapid EIA took 14–19 months and comprehensive EIA took 21–28 months. Under new notification, the EIA for category A will be completed within 10.5–12 months. Squeezing of time can be a compromise with the efficiency and justification of the clearance process.

EIA should be undertaken at policy and planning level rather than at project level. The EIA notification of 1994 has been amended 12 times in the last 11 years. Undoubtedly there have been many improvements in the new notification, but it is also evident that it had failed to meet the expectations of various stakeholders.

9.11 Citizens Actions and Action Groups

Issues related to environment have emerged as the most important distress for people's welbeing since the beginning of the twenty-first century. The idea of environment protection can be realized to begin from the Vedic period.

> O mother earth let thy bosom be free from sickness and decay
> May we through long life
> Be active and vigilant
> And serve thee with
> Devotion
> – Rig Veda

Late Prime Minister Pandit Jawaharlal Nehru and Mrs Indira Gandhi persistently advocated for fortification, conservation and improvement of the environment, since the Stockholm UNCHE, 1972 Conference. Several policies were formulated and legislation enacted. The UN has shown a sweeping transformation in the attitudes, approaches and policies with reference to relations with

NGOs and their participation. Measures have been undertaken to strengthen cooperation with NGOs in virtually all areas of activity such as policy research and examination; policy discourse and discussion; monitoring and advocacy; development activities; humanitarian work like responding to emergency situations, promotion of human rights, democratization, disarmament and peace; and propagation of information and raise public awareness.

Government agencies work with NGOs to:

- augment people's involvement in various programmes;
- to outspread exposure of programmes to the areas that are inadequately served by government staff;
- to examine and repeat inventive approaches; and
- to accomplish greater cost efficiency.

NGOs and Civil Society Organizations (CSOs) can play an increasingly important role in democratizing societies and the challenges. Governments and business may counterattack their advocacy, but the impending roles that NGOs can have in budding and deploying solutions often result marvelous outcome.

The Energy and Research Institute (TERI)

TERI was established in 1974 as a self-governing not-for-profit research organization. It works with the mission of developing and promoting technologies, policies and institutions for proficient and sustainable use of natural resources. It is engaged in imparting environmental edification through various projects, assignments, workshops, audiovisual aids and quizzes.

TERI works in the field of -

- energy and its efficient utilization;
- environment issues;
- sustainable development and sustainable use of natural resources;
- sustainable forestry and biodiversity;
- adoption of renewable energy technologies; and
- reduction of waste generation

Started in 1980, the Centre for Science and Environment (CSE) is an independent, public organization with the objective to increase public awareness in science, technology, environment and development.

Greenpeace is an international organization engaged in campaigning against environmental degradation since 1971. It spreads over 40 countries across Europe, the Americas, Asia and the Pacific. Green Peace's works in the field of marine pollution, oil spills and their impacts on environments, sewage problem, dumping issues, mining, pesticides and fertilizer issues, radioactive discharges, etc.

Case Study 9.5: The Chipko movement

The Chipko movement began in 1973 against illegal felling of trees and rolling them down the slopes of the Himalayas in the upper Alakananda valley village. Such acts led to massive soil erosion, landslides and devastating flood. The villagers and Dasholi Gram Sarajya Sangh, under the leadership of Chandi Prasad Bhatt and Sunderlal Bahuguna came forward protesting against such illegal felling. The main protagonist of the movement were the women folk who protested by embracing and encircling the tress and hence the name 'chipko'. The movement adherents are known as 'tree huggers.' Sunderlal Bahuguna spread the message 'Ecology is permanent economy'. For Bahuguna, shortsighted forest management policy is an indication of a deeper malady, which is the anthropocentric vision. He asserts that man is the 'butcher of Earth.' The protestors appealed to Mrs Indira Gandhi put an end to the felling of trees. Instead the development of local industries were encouraged which was based on conservation and sustainable use of forest wealth to benefit the locals. The movement has been contributory in the social and ecological fragmentation of the hill sects and also to the conceptual clashes between sub-cultures of the movement and it helped redefining the role of gender. The Chipko Movement has saved about 1,00,000 trees from excavation of saplings.

Approximately 250 years ago, a similar movement was started by the Bishnois of Rajasthan. A large group of 24 villages under the leadership of Amrita Devi laid down their lives to protect the trees against the Maharaja of Jodhpur.

Case Study 9.6: Narmada Bachao Andolan (NBA)

Started in 1985, NBA is one of the most powerful mass movements against the construction of huge dam across the Narmada river. There are plans to construct more than 3,000 big and small dams along the river course. It is a multi crore project venture that would generate big revenue.

The proponents claimed that this project will produce about 1,450 MW of electricity and supply potable water to about 40 million people covering thousands of villages and towns. Dams like Tawa and Bargi Dams have already been completed. Opponents protested that this project will rather devastate human lives and biological diversity by destroying thousands of acres of forests and cultivable land. On the other hand, it will by and large rob thousands of people of their livelihood.

The proposed Sardar Sarovar Dam and Narmada Sagar will dislocate more than 2,50,000 people. Resettlement or the rehabilitation of these people becomes the most important issue. These two proposals are already under construction, funded by US$550 million loan by the World Bank. Protestors expressed their grievances through hunger strikes, the mass media, massive marches and rallies and through screening of several documentary films, pressurising the world bank to withdraw its financial support from the project. The movement was led by prominent leaders like Medha Patkar, Baba Amte, Arundhati Roy, Amir Khan etc.

9.12 Public Awareness

We can do our own bit to save our planet and save ourselves.

'The warnings about global warming have been extremely clear for a long time. We are facing a global climate crisis. It is deepening. We are entering a period of consequences.' Al Gore

- Use less fossil fuel.
- Save electricity by switching off lights, fans, air-conditioners, computers etc., when not in use.
- Use energy-efficient products such as CFL bulbs.
- Recycle paper, plastics, glass and minerals.
- Plant of more trees in the process of afforestation, reforestation etc.
- Prevent deforestation.
- Use solar heaters for heating water.
- Harness alternative sources of 'clean' energy such as wave and solar energy.
- Avoid wastage of food and water.

Table 9.5: Environmental calendar of activities

	Events	Dates
1	World Wetlands Day	2-Feb
2	International Day of Action for Rivers	14-Mar
3	World Sparrow Day	20-Mar
4	World Forestry Day	21-Mar
5	World Water Day	22-Mar
6	World Meteorological Day	23-Mar
7	Earth Day	22-Apr
8	Anti Tobacco Day	31-May
9	World Biodiversity Day	22-May
10	World Environment Day	5-Jun
11	World Oceans Day	8-Jun
12	Global Wind Day	15-Jun
13	World Day to Combat Desertification and Drought	17-Jun
14	World Refugee Day	20-Jun
15	World Population Day	11-Jul
16	International Tiger Day	29-Jul
17	International Day for the Preservation of Ozone Layer	16-Sep
18	Zero Emissions Day	21-Sep

Summary

- The industrial revolution fueled by fossil fuels witnessed unimaginable orbits of growth being achieved as a matter of natural consequence. Such development confronts many problems. One such issue is urbanization where more people migrate to the cities in search of livelihood.

- It is a common belief that water is abundant. People utilize lots of water for daily living and for manufacturing articles. *The virtual-water content of a product is the volume of freshwater used to produce the product.* Water footprint is an indicator of both direct and indirect use of water by a consumer or producer. Water can be described in terms of blue water, green water, grey water or black water.

- People are helpless in matters of unequal rainfall that hampers their livelihood and economic well being. This ever increasing scarcity of water makes water conservation even more important. India has a tradition of water harvesting since the Vedic times and rainwater harvesting is an important option. Similarly, various watershed programmes were developed to cater water conservation.

- People migrate or are forced to displace out of their habitats owing to natural and manmade disasters. Even developmental projects like construction, mining displace people of their home and disrupts their livelihood. Wasteland, a degraded land can also be used and brought under vegetative cover with reasonable effort.

- Issues like gender inequality plays a prominent role in access to land and property. Customarily women are prohibited from owning land in many countries and cultures. Humans being the most evolved of all the creatures should also be compassionate about the rights of the animals. Thus, environmental education comprises an ethic that ensures improved behaviour and attitudes among the communities. Morality is an expression of human culture and it is the moral activities that develop moral virtues. A number of philosophies such as anthropocentrism and eco-centrism came up that deals with man's relationship with the environment.

- The disparity in resource consumption and the division between the North and South is basically on the basis of socio-economic and political differences. With industrialization, consumerism has exhibited an exponential increase. Resource consumption patterns thus needs equitable distributions. In this context, rural lifestyles are more sustainable.

- As a solution policies were formulated and enforced through enacting appropriate legislation. The governments are generally assigned responsibility to administer statutes and start prosecutions in order to enforce compliance.

- A very relevant issue in striking a balance between environment protection and development is the Environment Impact Assessment that is based on presuming the impacts of development proposals prior to decisions being taken.

- The UN has shown a sweeping transformation in the attitudes, approaches and policies with reference to relations with NGOs and their participation. Measures have been undertaken to strengthen cooperation with NGOs in virtually all areas of activity such as policy research and analysis; policy dialogue; monitoring and advocacy; development activities; humanitarian work like responding to emergencies, promoting human rights, democratization. Therefore, sustainable development is to be integrated in every activity.

Exercise

MCQs

Encircle the right option:

1. The view that holds that environment should receive moral consideration only as it relates to humans:

    A. anthropocentrism   B. Kant ethics   C. humanism   D. egalitarianism

2. The view that holds that animals should receive moral consideration in addition to humans:

    A. zoocentrism   B. biocentrism   C. anthropocentrism   D. individualism

3. The main unit of moral concern in ecological holism is:

    A. species   B. individuals   C. biotic community   D. none

4. Environmental ethics is a type of _____ ethics:

    A. theoretical   B. subsidiary   C. technological   D. applied

5. Kant's theory is a type of

    A. anthropocentrism   B. zoocentrism
    C. biocentrism   D. egalitarianism

6. As per Kant the reason for granting special moral status to humans is because they are:

    A. moral patients   B. moral agents   C. moral deciders   D. all at times

7. The main purpose of rehabilitation is:

    A. To avert further disease and illness

    B. To reinstate physical, psychological and social functioning

    C. To improve the person's quality of life

    D. All of the above

8. The specific act to declare the regulation of Biosphere Reserves in India is:

    A. WLPA   B. Biodiversity Act   C. EPA   D. None

9. _____ plant cultivation is prohibited from cultivation in India as per WLPA.

    A. Parthenium   B. Dracena (Dragon plant)
    C. Datura   D. Ladies slipper orchids

10. _____ has been kept out of coastal regulation zone (CRZ).

    A. Andaman   B. Nicobar   C. Lakshadweep   D. All

Fill in the blanks.

1. World Water Day is celebrated on_____.
2. The concentration of ozone is measured in _____.
3. Wildlife Protection Act was enacted in _____.
4. The NGO associated with Chipko Movement was _____.
5. The gas to contribute maximally in greenhouse effect is _____.

State whether the statements are true or false.

1. Ozone depletion in the stratosphere increases the incidence of skin cancer. (T/F)
2. The pH of acid rain is less than 5.6. (T/F)
3. Geographical unit to collect, store and release water is a wasteland. (T/F)
4. Appiko movement started in Orissa. (T/F)
5. Smog is a combination of smoke and fog. (T/F)

Short questions.

1. Define sustainable development.
2. State the important criteria's for sustainability.
3. What do you mean by virtual water?
4. What is water footprint?
5. What is gray water?
6. Define watershed.
7. State the utility of watershed management.
8. What are Hariyali guidelines?
9. What are the various factors for displacement?
10. Why is resettlement indispensible?

Essay type questions.

1. 'No water, no life. No blue, no green'. Describe the various practices for water conservation.
2. Compare resettlement and displacement with proper examples.
3. What are the various watershed management practices? State its benefits and limitations.
4. What are the issues in displacement and state the government initiatives towards such process.
5. What are the various impacts that arise out of displacement? Explain with suitable examples.

Human Population and the Environment___10

Learning objectives

- *To develop a comprehensive understanding about population and community.*
- *To know about the population dynamics, such as size, density, age structure, survival patterns, etc.*
- *To know about the various population growth forms.*
- *To know about the various patterns of population distribution all over the world.*
- *To know about the impact of population growth on environment.*
- *To develop a comprehensive understanding about demography and the latest Indian census report.*
- *To know about the various initiatives taken to control population growth.*
- *To know about family welfare programmes.*
- *To know and analyse the relationship between environment and health.*
- *To develop an understanding about the role of information technology in the field of environment.*

10.1 Population and Community

Population consists of a group of organism belonging to a particular species. On the other hand, a community comprises a group of organisms from different species living together. The branch of science that deals with the study of population is called population ecology. Population ecology is an autecology whereas community ecology is synecology.

10.1.1 Population characteristics

A. **Size and density:** simply refers to the number or biomass of individuals. The size of a population depends on birth, death and migration.

Size = number of organisms present at any instant + birth + immigration − death − emigration

This can create ambiguity and so density seems to be a better expression over size. Density is the number of individuals residing per unit area. It may be expressed as **crude density** which is the number or biomass of individuals per total space; say for example all land within organism's range. **Specific density** (ecological density) on the other hand is the number or biomass of organisms per habitat space; say for example, the portion of land for actual living. A population size is governed by

density dependent factors such as competition, predation, and parasitism, disease outbreak etc. and **density independent factors** such as severe weather conditions, disasters, famines, seasonal cycles, sunlight, pesticides and human actions like dam construction, forest clearance, wars, etc.

The population exhibits three main types of **spatial distribution**:

- **Clustered/clumped dispersion**: Individuals usually are aggregated in groups. This can be due to mutual attraction, or availability of a resource in a particular place. For example, herd of antelope, bee hives, ant colony, wolves, geese, sponges, a pack of African wild dogs, herd of gazelle, herd of elephants, pride of lions, chimpanzee. Patches of 'populations' that are usually separated and the individuals in general, stay within a patch, but can and do travel between patches. These interconnected sets of patches make up a **metapopulation**.

- **Random dispersion**: Pattern of spacing varies or is haphazard in nature. This can be due to the absence of strong competition or environmental factors. For example, dandelions, fig trees, oysters, etc.

- **Regular/Even/Uniform dispersion**: Individuals are equidistantly spaced from each other. This can be due to the regularity or evenness of resources or environmental factors. For example, penguins, creosote bushes etc. Such distribution may also result from release of a toxin by the plants to suppress the growth of neighbouring competing plants or **allelopathy**. For example, *Salvia leucophylla*.

B. **Natality:** refers to the number of offspring born per female over unit time. This is also known as the specific natality rate.

The theoretical maximum production of new individuals under ideal environmental condition is known as potential or absolute or **maximum natality**. The production of offspring under existing or actual environmental conditions is known as **ecological natality**. Ecological natality is always lower than the absolute natality due to exerting environmental conditions that put pressure on the population.

$$\Delta N_t = \text{new offsprings produced}$$

$$B(\text{natality rate}) = \frac{\Delta N_t}{\Delta t}$$

$$b \,(\text{natality rate per unit population}) = \frac{\Delta N_t}{N_0 \Delta t}$$

$N_0 =$ initial number of population
$N_t =$ new individuals born
$t =$ time

The maximum reproductive capacity of a population is known as reproductive potential and the maximum reproductive capacity of a population in unlimited resource is its **biotic potential**.

$r = b - d$ where, $r \longrightarrow$ intrinsic rate of growth or biotic potential,

$b \longrightarrow$ specific natality rate, $d \longrightarrow$ specific mortality rate

Natality also depends on the clutch which is the number of organisms produced each time. The **total fertility rate** (TFR) is the total number of offspring that a female would have, on average, if she were to live to the maximum age.

Often the term **crude birth rate (CBR)** is used in demography. This refers to the number of individuals born per thousand in the population per year. For example, CBR 30 means 30 individuals born per 1,000 in a year. **Specific birth rate** refers to the number of offspring born per unit time in various age groups.

■ **Generation time**: is the average age at which a female gives birth to her offspring which is equivalent to the time the population takes to increase by a factor equal to the Net Reproductive Rate (R_0). In other words, it is the average interval between the birth of an individual and the birth of its offspring. The net productive rate is the number of offspring that the individual is expected to produce in its life time.

When growing exponentially,

Generation time or G = t or time (minutes or hours)/n or number of generations

$G = t/n$

If N_0 is the initial population and N_g is the final population, then

$N_g = N_0 \times 2^n$

Or, $\log N_g = \log N_0 + n\log 2$

Or, $n\log 2 = \log N_g - \log N_0$

Or, $n = \dfrac{\log N_g/N_0}{\log 2}$

Or, $n = \dfrac{\log N_g/N_0}{0.301} = 3.3 \log N_g/N_0$

Therefore, $G = \dfrac{t}{3.3 \log N_g/N_0}$

Doubling time is the time period required by the population to attain double its size. If T_d is the doubling time and r is the rate of growth then,

$$T_d = \frac{70}{r}$$

C. Mortality: This refers to the number of death of organisms in a population over time. This is also known as the **specific mortality rate**.

The minimum mortality is the theoretical minimum loss of organisms over time under ideal conditions of environment, whereas the actual loss of organisms under existing environmental conditions is the realized **ecological mortality**. Age-specific mortality rate refers to the fraction of individuals in a population dying during a given age interval as evidenced in many countries due to socio-economic conditions.

The term **crude death rate (CDR)** is used to denote the number of individuals dying per thousand in the population per year.

The difference between birth and death is the intrinsic rate of natural increase denoted by 'r'. This is the instantaneous rate of change assuming the population to be in stable age distribution.

$$\Delta N_t = \text{new offsprings produced}$$

$$D\,(\text{natality rate}) = \frac{\Delta N_t}{\Delta t}$$

$$d\,(\text{natality rate per unit population}) = \frac{\Delta N_t}{N \Delta t}$$

N = initial number of population
N_t = new individuals born
t = time

D. Survivorship and survivorship curve: This refers to the probability of individuals in a population to survive from age zero to a given age. The graphical representation of the variation of mortality with age is called the survivorship curve.

A highly convex curve (Type I) represents that the initial mortality is low until it reaches the end of life span. Examples are humans, elephant, and deer population.

A highly concave curve (Type III) shows high mortality in the initial stages and comparatively very few survive till the end of their life span. Such is seen amongst shell fishes, molluscs and other lower organisms.

On the contrary a diagonal curve (Type II) represents a constant death rate over time. Practically no species exhibit exactly equal death rate and so in nature a slightly sigmoid curve is obtained, commonly found among rabbits, birds etc.

E. Ecological ages

Since organisms have variable life spans it is wise to divide their entire life span into ecological age groups such as young or pre-reproductive, mature or reproductive and old or post-reproductive age groups. In humans the ecological ages are:

- pre-reproductive age refers to population less than 15 years old;

- reproductive age refers to population between 15–49 years; and

- post-reproductive age refers to population more than 49 years.

Figure 10.1: Survivorship curves

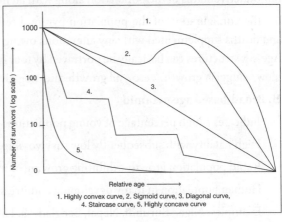

1. Highly convex curve, 2. Sigmoid curve, 3. Diagonal curve, 4. Staircase curve, 5. Highly concave curve

Usually, in a stationary population there is more even distribution of all the age groups. A rapidly expanding population has large proportions of young individuals and children whereas a small proportion of young individuals and a large proportion of old individuals is a characteristic feature of a declining population.

F. Population structure (composition)

Population structure is one of the most important demographic attribute. The structure refers to composition of individuals/organisms in a given population with respect to age, sex, ethnic background, religious background, education, profession, geographic/spatial distribution, socio-economic state, etc. the population structure is also dependent on births, deaths and migration. The study of the change in population size structure in terms of age, sex and other factors is called **population dynamics**.

The age–sex structure is the decisive factor in showing the future trend of growth in all age groups, as well as in the population as a whole. Hence, the age structure is very significant in formulating government policy and its consequences.

The population that has more young people ought to require adequate number of schools and job opportunities for them all. On the other hand, countries with older people should develop retirement and pension systems, old age homes and medical facilities.

G. Age distribution and population pyramid

Age distribution is the proportion of individuals of all ecological age groups of the population, usually displayed in form of a modified bar chart called age pyramid.

So, the age–sex structure of a country at any point of time can best be studied through the pictorial and graphical representation of population pyramids. The age pyramid comprise of a series of horizontal bars for each age group usually taken at five years interval. Conventionally the left half represents the males and the right half represents the females; young individuals are placed at the bottom of the pyramid while the old people are at the top. An age pyramid depicts whether the population is young or old as well as the percentage of various age groups, the present reproductive status of the population along with an indication to future trends.

The configuration of the population pyramid reflects the cardinal factors influencing births and deaths supplemented with any alterations owing to migration over three or four generations. Age-sex structures can basically be portrayed by four fundamental depictions – 'rapid/fast growth', 'slow/ sluggish growth', 'zero/ nil growth' and 'negative growth'.

H. Broad-based age pyramid

- Indicates a high percentage of young population.
- High natality with subsequently low survivorship.
- Rapid population growth exhibiting growing or increasing population.
- Human population in many developing countries.
- Examples: Niger, Burundi, Nepal, Bangladesh, Pakistan and India.

ii. Bell-shaped age pyramid

- The configuration indicates a young to old population in moderate proportions.
- The rate of population growth is slow.
- It indicates a more or less steady or stable population.

▪ It is observed in more developed countries.

▪ Examples: Countries like United States, Sri Lanka, New Zealand, Canada, Iceland, France, and UK.

iii. A rectangular polygon:

▪ All age groups are equally abundant.

▪ It indicates high survivorship and low natality.

▪ It indicates almost zero population growth.

▪ Stable age structure is shown.

▪ Examples: Countries like Bosnia, Norfolk Island and Holy See.

Figure 10.2: Broad-based age pyramid

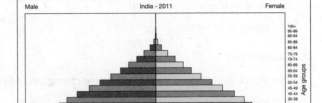

iv. An urn-shaped age pyramid

▪ Indicates a low percentage of young population.

▪ Indicates low natality and high survivorship.

▪ Shows negative population growth.

▪ A declining or dying off population is indicated.

▪ Examples: Countries like Poland, Hungary, Germany, Japan and Russia.

10.1.1.1 Types of population

As mentioned earlier, a given population may be grouped into four main groups based on their expansion status:

A. Expansive or Type I:

Poor countries are traditionally characterized by:

i. high fertility;

ii. high mortality;

iii. high number of children; and

iv. moderate growth rate.

B. Expansive or Type II:

Contemporary, less developed nations are characterized by:

i. high fertility;

ii. declining mortality;

iii. high rate of growth of population; and

iv. very young populace.

C. Stationary

The modern developed nations are characterized by:

  i.   declining or waning fertility;

  ii. declining mortality;

  iii. moderate growth rate; and

  iv. aging populace.

D. Constrictive

The developed nations of the future are characterized by:

  i.   low fertility;

  ii. low mortality;

  iii. very old populace; and

  iv. ceasing growth rate.

10.1.2 Population growth

Population growth over a time period is determined by three variables – number of births (fertility), deaths (mortality) and migration (movement and relocation) or population dispersal.

The equilibrium between these three factors finally deduces the trends in a given population; whether a populace will increase or decrease or the populace shall remain stationary. Thus, Natural Population Increase (Natural Population Growth) is the difference between the number of births and deaths. Upon addition of the net effect of migration to the Natural Population Increase, one obtains Total Increase. Net migration can be calculated as the difference between the number of immigrants and number of emigrants.

So, rate of Natural Population Increase is actually the rate at which

Figure 10.3: Bell-shaped age pyramid

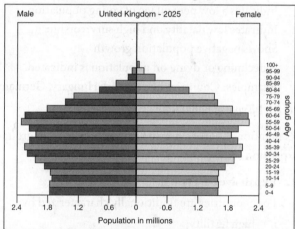

Figure 10.4: Urn-shaped age pyramid

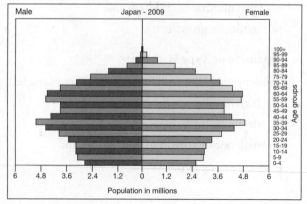

the populace under consideration is rising or declining in a particular year owing to surplus or shortfall of births in excess of deaths; it is usually expressed as a percentage of the base population.

Rate of natural increase or RNI = (births − deaths) X 100

Total population growth rate: The rate at which a population in escalating (or declining) in a particular year because of normal increase and overall migration expressed as a percentage of the base population.

Total growth rate = (natural increase + net migration) X 100

Geometric progression (GP): The ordered numbers are arranged in a series in such a fashion that the proportion of any two adjacent numbers is same; such a series is said to be in geometric progression.

For example, 2, 4, 8, 16 … ….; 3, 9, 27, 81, … …..etc.

Population size over a given period of time is said to display geometric pattern of growth if the change within that specific year is proportional to the size of the population at the commencement or start of that year.

If N_0 is the initial population and N_g is the final population, then

$$N_g = N_0 \times 2^n$$

If the generation value is large then,

$$N_g = N_0 \times e^{rg}$$

Arithmetic Progression (AP): The ordered numbers are arranged in a series in such a fashion that the common difference between any two contiguous number is same; such a series is said to be in arithmetic progression.
For example, 1, 2, 3, 4, 5 … ….; or 3, 5, 7, 9, 11, …..etc.

The doubling time of a given population is the time taken by the population to attain twofold or two times in size, provided the current rate of growth remains same. It represents how quickly the population can grow and double in numbers. But the future size of population cannot be predicted as the growth rate here is assumed as constant. Doubling time changes with any change in fertility or the mortality graphs.

$$\textbf{Doubling Time} = \frac{69.3 \ (or \ approximately \ 70)}{growth \ rate \ in \ percent}$$

10.1.2.1 Population growth forms

A population generally exhibits characteristic pattern of increase known a population growth forms. There are two distinct patterns of population growth forms.

A. Exponential growth form (J- shaped growth curve)

The growth form is expressed by the quick increase in the density of the population in exponential or compound interest fashion. This can only be possible in the absence of environmental restrictions or factors in the initial stages. The growth rate may then stop unexpectedly or abruptly when the factors or resources become limited or effectual more or less all of a sudden. The resistance from the environment, known as carrying capacity is the upper limit beyond no major increase can occur and is represented by K. Alternatively, carrying capacity is the maximum number of individuals that a given environment can sustain over a long period of time.

Figure 10.5: J-shaped growth curve

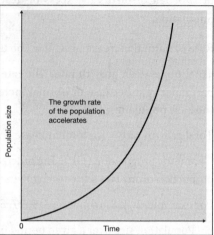

B. Sigmoid growth form (S-shaped growth curve)

In this form of growth the environmental factors or constraints start acting or operating from the very beginning; so the population increases slowly at the beginning, then increases steadily but declines slowly as the environmental resistance or factors build up, till the upper level is reached and maintained, exhibiting a more or less S-shaped curve.

$$N_g = N_0 * e^r\{(K-N)/K\}^g$$

Figure 10.6: J-shaped growth curve

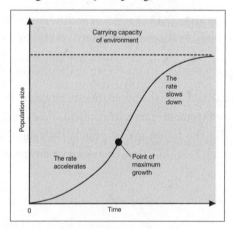

Figure 10.7: S-and J-shaped growth curve together to show the environmental resistance

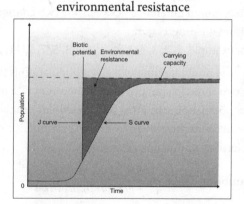

Consequently there is either 'K' selected species or 'R' selected species in a population.

Table 10.1: Comparison between K and R selected species

K selected species	R selected species
1. The size of the population is roughly constant.	1. Population experience rapid growth.
2. Members usually have a low reproductive rate.	2. Offspring produced are numerous. High biotic potential.
3. Offspring require extensive parental care.	3. Offspring require little postnatal care; they mature fast.
4. Individuals are very competitive. For example, humans	4. Population growth fast reproduces rapidly and dies quickly. For example, bacteria

Under practical circumstances, the environment is never constant. The individuals comprising a population vary in terms of age, sex, density etc. No population can start multiplying as soon it is introduced in an environment.

A much more realistic approach is the modified logistic growth curve typically comprising five phases in the growth:

- A lag phase where the organisms acclimatize and establish themselves; they accumulate resources that will be necessary to multiply. The initial phase hardly shows any growth.

- A positive acceleration phase which show very slow initial increase in size in response to the environment.

- A log phase, which is the steady growth phase and where the organisms increase logarithmically under enough supply of food and other factors.

- A negative acceleration phase, where the growth rate slows down or slowly declines as it approaches the carrying capacity.

- Lastly the stationary phase where there is no further growth in size as the population reaches the level of carrying capacity. The natality equals mortality.

Figure 10.8: Modified logistic growth curve

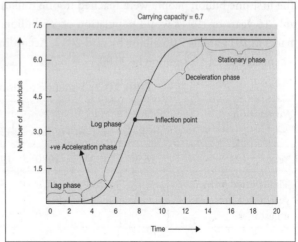

The impacts of population growth can be summarized as follows:

1. shortage of habitable space; land fragmentation;

2. over exploitation of natural resources like water, land, flora, fauna, minerals, energy, etc; coal and petroleum is expected to vanish in the coming few years;

3. increased urbanization and industrialization; problems with housing, increased traffic,

increased waste production; shortage of schools, colleges, hospitals and other basic requirements;

4. deforestation and felling of trees along with shrinking agricultural land;

5. soil erosion and land degradation due to its unsustainable use;

6. shortage of freshwater supply; decreased per capita water availability, water crisis and water conflicts have become local, regional and international issue;

7. shortage of food supply; decreased per capita food availability, increased poverty, malnutrition and famine; decreased availability of clothing; deteriorating healthcare and medical facilities; more spread of contagious and vector-borne diseases;

8. air and water pollution; increased emission of GHGs and rise in global temperatures, acid rain, eutrophication and other forms of ecological degradation;

9. increased incidence of disasters and accidents;

10. low per capita income and low national income; increased inflation, increased government expenses; increase of debt by borrowing from international organizations;

11. unemployment;

12. increase in the number of unproductive consumers – the children and the older group; and

13. increased consumption leads to fall in the rate of savings and investment.

10.1.3 Global population growth and variation among nations

The first one billion people were reached in one million years but the next billion was added in only 10 years to 1994's world population of 5.5 billion. Global population crossed the 7 billion marks in 2011 and is 7.06 billion by mid-2012. According to 2010 revision, the global population is expected to hit 10.1 billion by 2100.

Table 10.2: Timeline of global population

Population	500 million	1 billion	2 billion	3 billion	4 billion	5 billion	6 billion	7 billion
Year	1650	1830	1930	1960	1974	1987	1999	2011
Time period or years taken between milestones		180	123	33	14	13	12	13

More than 90 million individuals are added per year. About 97 per cent of this growth is in the developing countries. On the contrary, the crude birth rate is barely more than the crude death rate in the developed countries; the populace comprises mostly of older individuals. With this, for the first time, the number of deaths will surpass the number of births by 2025.

The non-renewable resources are limited; these resources are under enormous pressures acted on by the rapidly growing population. This, in turn, leads to migration into the regions vulnerable to catastrophe such as cyclones and hurricanes, floods and droughts, earthquakes and landslides, etc. Previously unoccupied and uninhabited, the areas were at all times considered to be unsafe and precarious. Unrestrained growth of population is thus a menace to the safety and security of mankind. In between 1970 and 1985, the number of poor people had risen by one-fifth.

As projected by the UN, Earth's rural population is to reach its upper limit. After this point all of the impending population growth will be mainly focused in the urban areas. By 2015, half the population from developing nations will be residing in the urban areas. Such a speedy urban growth of population along with frantic, hysterical industrial growth will degrade and mortify the urban environment. This will exhaust and drain out natural resources undermining fair and reasonable sustainable development. Estimates say that the urban dwellers have increased by four times from 285 million to 1.4 billion from 1950 to 1987. Additionally, in the last 30 years, 35 million people have migrated to the north from south with nearly one million people joining every year. Of this, people who have migrated illegally are anticipated to be about 15 to 30 million.

The unfortunate, penurious destitutes of the third world nations shall witness the overriding percentage of increase. Sources in United Nation claim that 48 nations have significant squatty income along with high monetary susceptibility. 33 are in sub-Saharan African countries, 14 in Asia, and 1 in the Caribbean, Haiti. The human development indicators such as life expectancy at birth, per capita earning and education levels are also too low. Burundi, Ethiopia, Mozambique, and Zambia belong to Africa while Bangladesh, Nepal, Cambodia and Yemen are such Asian countries. They

Figure 10.9: Proportion of population in developed and developing countries, 2011 as per United Nations Population Division, World Population Prospects: The 2010 Revision

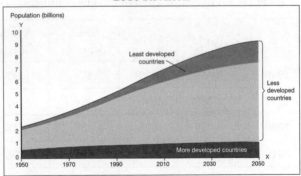

exhibit an annual growth rate of 2.4 per cent and are thus likely to cross 2 billion by 2050. Between 2011 and 2100, the population of 'high-fertility countries' is estimated to triple, from 1.2 billion to 4.2 billion. The population of 'intermediate-fertility countries', for instance the United States, Mexico and India is like to increase by a fair 26 per cent, while that of 'low-fertility countries' like most of Italy, China and Australia, will decline by about 20 per cent.

Trends in developing countries: Most developing nations exhibited population doubling between 1960 and 1990. The population in Africa is projected to nearly double from 1.1 billion to about 2.3 billion by 2050. This depends on the supposition that total fertility rate will decline from 5.1 to nearly 3.0 by 2050. In reality, surveys have indicated either slow TFR decline or are not taking place at all.

With the present population trends, Asian countries are liable to experience a smaller rise in comparison to the African nations; even then one billion people is likely to be added by 2050, most of which is dependent on India and China, the two most populous nations. Asia's TFR is 2.2. Not including China, 47 per cent of Asian women use modern contraception methods. The TFR of Japan, Singapore, South Korea and Taiwan is nearly 1.4 or even less. 24 per cent of the population is in the age group of 65 and older group in Japan.

The smallest proportional growth to be likely by 2050 from 599 million to 740 million, is in Latin America and the Caribbean. Such a reduction owes largely to the declining fertility in few of the biggest nations like Brazil and Mexico. With the present TFR of 2.2 and application of modern methods of contraception and there will be some balance in this regard.

Trends in developed countries: A very steep reduction in fertility is prominently seen in the first world nations. Europe is amongst the first and foremost to experience long term decline in population. TFRs of 1.4 and even lower are a surprise. TFR close to 2.0 is prevalent in the countries such as France and Norway. As projected, the populace of Europe is expected to decline from 740 million to 732 million. A load of elderly people, shortages of pension system and high healthcare costs are all responsible for low TFRs.

In USA, Australia, Canada and New Zealand, growth from higher births or immigration is likely to continue. In 2010, USA had a TFR of 1.9 births/woman.

Table 10.3: Population (in millions) by continents: 1980 to 2050 as per US Census Bureau

Year	World	Africa	North America	South America	Asia	Europe	Oceania
1980	4,453	479	371	242	2,644	695	23
1990	5,289	630	424	297	3,189	723	26
2000	6,089	803	486	348	3,691	730	30
2010	6,853	1,015	539	396	4,133	734	35
2020	7,597	1,261	595	440	4,531	731	39
2030	8,259	1,532	648	477	4,841	718	43
2040	8,820	1,827	695	504	5,049	698	46
2050	9,284	2,138	739	520	5,167	671	49
Percentage distribution							
1980	100	10.7	8.3	5.4	59.4	15.6	0.5
2000	100	13.2	8	5.7	60.6	12	0.5
2050	100	23	8	5.6	55.7	7.2	0.5

10.1.4 Population trends in India (Census 2011)

The first population census was done in 1872 with census usually being taken after 10 years. The process of census in India is governed by the Census Act, 1948 as amended in 1994, the Census Rules, 1990 as amended in 1994 and the Citizenship Act, 1955.

India has 28 states and 7 union territories with around 30 languages spoken all over. The current population of India as of 2013 is 1.27 billion as compared to 1.21 billion of 2011 census and the figure is likely to overtake China by 2020. The number of males and females are 623,724,248 and 586,469,174, respectively. The decadal growth rate is 17.64 per cent for 2001–11 as compared to 21.15 per cent for 1991–2001.

The most populous states in descending order are UP (1,99,581,477), Maharashtra (1,12,372,972), Bihar (1,03,804,637), West Bengal (91,347,736) and Andhra Pradesh (84,655,533), whereas Sikkim (607,688), Mizoram (1,091,014) and Arunachal Pradesh (1,382,611) have the lowest population. Considering the union territories, Lakshadweep (64,429), Daman and Diu (242,911), and Dadra and Nagar Haveli (342,853) stand out as the least populous amongst all. The city of Mumbai is the highest populous followed by Delhi, Bangalore and Hyderabad while Nagada in UP is the least populous with 100,036 people. 53 Indian cities are the home to at least 1 million people.

The population density is 382 persons per km^2. Bihar has beaten West Bengal and has the highest population density of 1,106/km^2. This is followed by West Bengal and Kerala with population density of 1,029 /km^2 and 1,084/km^2 respectively. Considering the union territories Delhi is the highest with 11,297/km^2, Chandigarh with 9,252/km^2 and Puducherry with 2,598/km^2. The state of Arunachal Pradesh is with the least population density with 17/km^2.

The female is to male sex ratio is 940:1000. Kerala has the highest sex ratio of 1,084 females per 1,000 males while Haryana has the least with 877 females per 1,000 males. The average life expectancy is 64.4 years for males and 67.6 years for females.

The literacy rate is 74.04 per cent with 82.14 per cent in males and 65.46 per cent in females. The country's highest literacy is 94 per cent in Kerala, 91.85 per cent in Lakshadweep, 91.33 per cent in Mizoram, 88.7 per cent in Goa and 87.22 per cent in Tripura. Bihar has the lowest literacy rate of 63.82 per cent. Amongst the cities Aizwal (Mizoram) has the highest literacy rate and Rampur in UP has the lowest of 60.74 per cent.

10.1.5 Population explosion and demographic transition

Population explosion: Population explosion can be referred to as 'the geometric expansion of a biological population, especially the unchecked growth in human population resulting from a decrease in infant mortality and an increase in longevity'.

'When the population increases in such a way that increases in the fixed proportion of its own size at any time, its growth is said to be exponential'.

In simple words it is a quick increase in the size of a population due to sudden decline in infant mortality, increase in natality or an increase in life expectancy in abundant supply of resources. With increase in size, the increase also becomes bigger. Thus, the actual increase depends on (a) rate of increase in proportion to its size; and (b) its own size at that instant.

Table 10.4: Countries with highest population as on 31. 07. 2014

Country	Population	Share of world population (per cent)
China	1,394,457,758	19.24
India	1,268,657,757	17.5
United States	322,790,888	4.45
Indonesia	2,531,700,942	3.49
Brazil	202,009,622	2.79
Pakistan	185,379,534	2.56
Nigeria	178,923,659	2.46
Bangladesh	158,670,365	2.19
Russia	142,437,770	1.97
Japan	126,988,062	1.75

Table 10.5: Cities with highest population

Rank	Countries	City	Population (millions)	Remarks if any
1	Japan	Tokyo	37.83	Population likely to decline to 37 million by 2030 yet it will remain at the top.
2	India	New Delhi	24.95	Likely to increase 36 million by 2030
3	China	Shanghai	22.99	
4	Mexico	Mexico city	20.84	
5	Brazil	Sao Paolo	20.83	
6	India	Mumbai	20.74	To become the fourth largest by 2030 with a population of 28 million.
7	Japan	Osaka	20.12	
8	China	Beijing	19.52	
9	USA	New York	18.59	
10	Egypt	Cairo	18.41	

Demographic transition: The model pursues to explicate the change of population over time. This model is rested on the survey started by Warren Thompson in 1929. This American demographer began analysing the observable changes or shift in birth and death rates in the societies of developed nations that started over a period of 200 years or so. The model also elucidates the transitions of the nations from high birth and death rates towards low birth and death rates. Such shift or change started in the developed nations since the eighteenth century and persists till today. On the other side, such change in the LDCs began much later and is in the middle of the earlier stages even now. This model also takes into account the CBR and crude death rate (CDR) over a period of time.

A. Stage I: In the pre-modern times there existed a steadiness between the birth rates and death rates. Prior to the Industrial Revolution, this balance was disrupted in Western Europe which exhibited a high crude birth rate and high death rate. At least the population growth was extremely slow, ever since the 'Agricultural Revolution', almost 10,000 years ago. The growth was less than 0.05 per cent which emanated in extended doubling times of 1 to 5,000 years.

More children implies more recruits or more employees for working and as a result the natality is high. Also, with high rate of death, families were in want of more and more children so as to ensure existence or survival and perpetuation of their family. Death rates were on the high because of outbreak of diseases like influenza, scarlet fever, plague, cholera, typhoid, typhus, dysentery, diarrhea, tuberculosis (TB), measles, diphtheria, whooping cough, lack of adequate knowledge of disease prevention and cure, lack of clean drinking water, occasional food shortage, lack of sewage disposal and proper hygienic conditions.

The state of equilibrium between CBR and CDR reflected as a slow growth of the population. Sporadically, outbreak of epidemics would raise the CDR radically for a few years (shown by the 'waves' in Stage I of the model). Their balance consequently results in only very slow population growth and is also known as 'High Stationary Stage'. Stage I illustrates all world regions until the seventeenth century. Some demographers summarizes its character as a 'Malthusian stalemate'.

B. Stage II: In the mid-eighteenth century, in western European nations, there was a considerable slump in death rate due to better conditions of sanitation and medicine. But traditionally, owing to practice, the natality prevailed and kept high. The decline in the mortality in Europe that initiated in northwest Europe disseminated to the south and east in the next century. It mainly happened due to advancement and upgradation in the areas of food supply, high produce and agricultural practices like crop rotation, selective breeding, and seed drill technology. Significant improvements in public health reduced mortality, mostly in childhood. Improvements in water supply, sewage system, food handling, and general personal hygiene developed from growing scientific knowledge helped greatly.

The falling mortality along with steady natality in the start of Stage II resulted in shooting up of the rate of population growth. The chasm or gulf between number of births and deaths grew wider and wider. With time children turned into supplementary encumber since they were unable to put in to the wealth of the family. Population that expanded rapidly began to slow down. Many of the least developed countries are still in Stage II of the demographic transition model. Kenya's high growth rate is driven by a high crude birth rate of 32/1000 and a low crude death rate of 14/1000. The age structure of the population is yet another characteristic. Most of the death in Stage I was between 5 to 10 years of life. The age structure of Stage II becomes progressively youthful as this stage calls for increased survival of children.

C. Stage III: In the latter half of twentieth century, both the CBR and CDR in the first world nations started falling. In many cases, the crude birth rate is little more than the crude death rate (as evidenced in USA, 14 versus 9). In many other nations, CDR is more than CBR (as in Germany, 11 versus 9). Immigration from LDC contributes to a great deal of the population growth in industrialized countries. China, Singapore, South Korea and Cuba are speedily heading towards Stage III.

Population moves towards stability through a reduction in natality. Such transition is contradictory of Malthus's belief where alterations in death rate were the prime cause of change in population. In general, the decline in birth rate in the first world nations started towards the end of nineteenth century that followed the drop in death rates.

This ultimate decline may be the contribution of various factors, such as

- Childhood death continues and parents came to comprehend that they do not require many children.

- Increasing urbanization changes the traditional values. Residing in urban areas also increases the cost of living. Joint families reorganize into small nuclear families. People started to think and validate rationally so as to how many children they could bear to live comfortably.

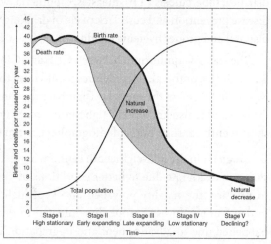

Figure 10.10: Demographic transition

- Growing female literacy and their subsequent employment brought changes in attitude towards accepting and willing to bear children and motherhood. Women valued themselves beyond childbearing and motherhood. They entered workplace, mix with a good number of other people, changed their attitudes and became vocal about their own decisions.

- Improvements in contraceptive technology also leads to drop in death rates.

D. Stage IV: This period is characterized by stability where both the birth rate and death rate have fallen drastically, mostly seen in very highly developed countries. Such changes have happened in places where the females are more knowledgeable. This is not merely due to the solicitation of modern discovery of drugs, but rather due to changes in attitude and perceptions linked to sanitation or hygiene that have bettered the endurance and favoured life over death.

10.2 Family Welfare Programme

National Family Welfare Programme: The National Family Welfare Programme was launched in India in 1951. It is a centrally-sponsored programme. The programme initiated with the objective of reducing birth rate that was indispensible to bringing stability to the nation's populace to an extent that was steady with Indian economy. Under the First and Second Five Year Plans, the approach was largely clinical. Based on the 1961 census report, the clinical approach was replaced by 'Extension and Education Approach'. High importance was rendered to the programme with the proposal of reducing natality from 35 per thousand to 32 per thousand by the end of Fourth Five Year Plan. This continued in the Fifth Five Year Plan, from 1974 to 1979 in order to bring down the natality rate furthermore to 30 per thousand at the end of 1978–79. This was coupled with an amalgamation of family planning services with matters of health, maternal and child health (MCH) and nutrition. Striking increase in sterilization was observed during 1975–77. Government changed the name of the programme from Family Planning to Family Welfare. Between 1985 and 1990, the Seventh Five Year plan went on an unintended basis with emphasis on support for spacing methods, achieving greatest community participation and supporting mother and child healthcare. Mahila Swasthya Sanghs (MSSs) at village level were constituted comprising 15 persons like adult education instructor, anganwadi worker, primary school teacher, mahila mukhya sevika and the dai. The Member-Convenor is the Auxiliary Nurse Midwife (ANM). In the Eighth Plan (1992–97), greater emphasis was put on the of NGOs involvement to complement the government initiatives. During Ninth and Tenth Five Year planning period decrease or bringing down the rate of growth of population was accepted as one of the most important objectives. The National Family Welfare Programme recommended the various contraceptive devices like condoms, oral contraceptive pills and intra uterine device (IUD) such as Copper-T for spacing births. Public awareness about the contraceptive measures is done by the government agencies, education and extension workers. Emergency contraceptive pill (ECP) was also introduced during 2002–03 that was to be used for checking or preventing unwanted pregnancy following an unprotected act of sexual intercourse. Above all the decisive factor in restricting family size depends was the couple's education and background.

10.2.1 Methods of sterilization

Under National Family Welfare Programme the common sterilization or terminal methods are:

- Tubectomy: It is a method of female sterilization by ligating the fallopian tube. It can be Mini Lap Tubectomy and Lapro Tubectomy.
- Vasectomy: It is a method of male sterilization by ligating the vas deferens. It can be Conventional Vasectomy or No-Scalpel Vasectomy.

10.2.2 Urbanization

The world is progressively becoming more urban, as more people move to cities and towns. This may be in search of employment opportunities, educational facilities and higher living standards. More than 75 per cent of inhabitants in Australia, Canada, New Zealand and the United States are urban. In Africa and Asia, about a third of the total populations represent urban dwellers. 261 cities in developing countries will have populations over 1 million by the end of the century. This was

213 in the mid-1990s. A megacity refers to a place with more than 10 million inhabitants. There were 14 so-called 'mega-cities', in 1994 with at least 10 million people. Their number is predicted to double by 2015. But rapid urban growth pressurizes the local and national governments to provide basic services such as water supply, electricity supply and sewerage system. Most of the new growth is likely to occur in small towns and cities. Cities provide more promising backdrops for the resolving social and environmental problems than the rural areas. They generate employment and income rate is higher than the rural areas. The governance is good, cities offer congenial atmosphere for work, communication, education, healthcare and other services more efficiently. Opportunities for social mobilization and women's empowerment are great. People leaving native towns in pursuit of better lives in the city often end up in shanty towns and slums; a place where they lack access to decent housing, proper sanitation, proper healthcare and education, i.e., trading in rural for urban poverty. Over 1 billion people reside in urban slums, which are highly overcrowded, polluted and dangerous.

International migration is also increasing day by day. In 1990, there were 12 million refugees. Approximately, half of the over 125 million people live outside their national boundaries. The emigration is mainly driven by income disparities between nations, political conflicts, loss of homeland, environmental degradation etc.

10.3 Environmental and Human Health

The environmental changes catered by anthropogenic activities have an immense impact on the environment. Urbanization, industrialization, technological advancements, economic growth and GDP can be indicators of development and prosperity. Similarly the incidence and outbreak of several diseases are also indicators of the morbid state of the environment. The different forms of air-borne, water-borne and other diseases do indicate the state of air we breathe, water we drink and food we eat, all of which are integral part of environment. The developmental strategies without ecological safeguards culminate in bad health. Therefore, environmental management should incorporate measures to improve health.

10.3.1 Environmental health

Environmental health comprises the facets of human health and the class or quality of life that are governed by physical, chemical, biological and social factors in the nature. The environment surrounding does affect our health. Better health leads to longer life span but also results in population inflation. On an average, the life span has almost doubled over the past century primarily because we now have safe and clean drinking water. We can take in fresh oxygen and are immunized against deadly diseases. But many of the commonly used drugs have severe side effects. Curing of one disease often leads to damage to another organ. If parts of the environment, like air and water or soil degrade, it can culminate in health problems. Hence, environmental health refers to all the physical, chemical, biological factors outside an individual. It takes into account all allied factors that affect the behaviour of an individual. It involves the assessment and regulation of the total factors of the environment that has the potential of affecting the state of well-being. It is intended to prevent ill health and diseases so as to create health-supportive environments. Behaviour is well correlated to the social and cultural environment and genetics.

Some environmental risks are natural, such as, radon in the soil. Others are the result of anthropogenic activities, like lead poisoning from paint, mercury from industrial use. Cigarette smoke comprises nearly 4,000 various chemicals that are able to damage the tissue systems and cells. At least 80 chemicals out of these are carcinogenic such as tar, arsenic, C_6H_6, cadmium and formaldehyde (HCHO), nicotine etc. Various diseases can be correlated with the environment. Asthma causes inflammation of the nasal passage and thus blocks the normal air flow. The symptoms are wheezing, coughing, chest tightness and troubled breathing. Asthma is now very frequent among children and affects millions. The genetic connection is obvious but the increase in incidence does direct towards the increase in allergens, aerosols, particulate matter etc. Other lung diseases like emphysema, pulmonary fibrosis, pneumonia, lung cancer show a vigorous relation to the subjection of ozone, particulates and oxides of sulphur. There is an increasing tendency for reduced lung capacity, risk for early death and other respiratory symptoms. Researches show that pesticide contact might prompt genetic disorders like Parkinson's disease. Reproductive diseases and disorders including congenital defects, development disorders, reduced weight at birth, reduced fertility, impotency and menstrual disorders have strong correlations to environmental pollutants like mercury, lead, thallium etc. People who are exposed to crystalline silica dust, have double or quadrupled threat for culminating chronic Lupus disease. Asbestosis is very common among people working with asbestos as the fibre is easily inhaled along with air, though the time lag between significant inhalation and any adverse health manifestations can extend to 30 or more years. Electrical and magnetic fields (EMFs) surround electrical devices such as power lines, electrical wiring, and appliances. In adults, EMFs may reduce heart rate and interfere with brain activity during sleep.

10.3.2 Climate and health

Climate change does affect the weather conditions. Air and water pollutant concentration may occur due to average warmer temperatures leading to longer heat waves. The spread of diseases could be augmented by alterations in temperature, precipitation and other extreme events. Such weather extremes can be storms, flood, higher velocity wind current, etc., that pose direct intimidation to people and their property.

The effects of climate change on health are dependent on the effectual public health and safety systems of the society. The consequences may vary with areas, reactivity and vulnerability of populations, the extent and time span of contact and society's ability to adapt to cope with the change.

Climate change upsets human health and well-being: Global climate change may affect the human well-being in the following way:

- The poor and impoverished people, the aged and elderly group, the disabled, infants and children and the uninsured section are most likely to be affected in terms of health impacts.

- Adaptation should initiate from now, beginning with public health infrastructure. Steps are to be taken by individuals, society, governmental agencies in order to abate or lessen the climate change effects on human well-being.

Impact from heat waves and extreme weather events:

Heat waves can lead to dehydration and heat stroke. The people residing in northern and upper latitudes will be affected more since they are less equipped to withstand high temperature extremes. Unsurprisingly warmer urban areas in relation to rural surroundings would seek more electric supply to run the air-conditioning machines in turn boosting up the GHG emissions and air pollution.

The frequency and intensity of tropical storms and flooding is likely to increase. These could cause injuries and, in several cases, death. This may have the following impacts:

- paucity in drinking water and fresh food availability;
- impaired communication and interruption of crucial health care services;
- vulnerability to CO poisoning from the portable generators in use during storms and following storms;
- increased incidence of gastric and intestinal disorders among evacuees; and
- susceptibility to different types of mental states such as depression and post–traumatic stress.

Impact from reduced air quality:

- Ozone: Researches project the increase in ground level ozone with increase in temperatures leading to smog formation, for example, smog in Los Angeles. Tropospheric ozone damages the pulmonary tissue, causes a reduction in functioning of lungs, induces inflammation of the airpipes and aggravates asthma. Tropospheric ozone is formed when primary air pollutants like CO, NOx and VOCs are exposed to each other.

- Fine Particulate Matter: Particulate matter less than 2.5 microns may be formed by chemical reactions between sulphur dioxide, nitrogen dioxide, and VOCs, inhalation of which aggravate heart and lung diseases, chronic pulmonary ailments, aggravation of asthma. Other sources include thermal plants, petroleum and diesel engines, wood combustion, smelters and steel mills and forest fires.

- Allergens: Climate change may affect the length of the season, flowering and pollination patterns. Experiments show that both the amount and time of ragweed pollen grains production will increase with the increasing concentrations of atmospheric carbon dioxide.

Impact on climate-sensitive diseases: More diseases will spread with the change in climate as more pathogens get transmitted through water, edibles, birds, insects, etc. Climate change would also affect all the vectors and carriers probably.

A. Food-borne diseases

- Higher air temperatures and hot environments may facilitate the bacterial multiplication like that of Salmonella and other bacteria linked to food poisoning. These diseases disturb the gastrointestinal system and in extreme cases they can also be fatal. Salmonellosis outbreak indicates its relation to poultry egg consumption.

- Listeriosis by *Listeria monocystogenes* associated with the consumption of processed meat, soft cheese may cause septicemia. It causes abortion and still birth in pregnant women. Bovine spongiform encephalopathy is linked to contaminated bovine carcasses used as cattle feed. It causes fatal and infectious neurodegenerative disease. Liver disease is caused by fish products contaminated with trematodes.

- Flooding and heavy rainfall can overflow sewage. Overflows into freshwater and soil could contaminate certain food crops with parasite and pathogen-infested faecal matter.

B. Water-borne diseases

In the case of heavy precipitation and deluge water-borne pathogens like *Giardia, Cryptosporidium, and Entamoeba* are increased in number. Such pathogens infect the intestinal tract manifesting in diarrhoea and blood stool. Gastroenteritis is the most frequent infection from contamination of water. Its symptoms are vomiting and headache accompanied by fever and minor auditory, ophthalmic, nasal and throat ailments.

C. Animal-borne diseases

- Mosquitoes are favoured by warm and humid climatic conditions concerned with the spreading of malaria, dengue, West Nile virus, yellow fever, filaria etc.

- The geographic range of ticks carrying Lyme disease with symptoms such as fever, headache, exhaustion, and a typical skin rash is limited by temperature. With rise in temperatures, ticks are likely to migrate northward.

10.3.3 Diseases

10.3.3.1 Infectious diseases and noninfectious diseases

A non-communicable or non-infectious disease is one which non-transmissible among people. These are not contagious and cannot be transmitted from one person to another. They include autoimmune diseases, myasthenia gravis, cardio-vascular disease, chronic obstructive pulmonary disease (COPD), stroke, cancer, asthma, malnutrition, diabetes, chronic kidney disease, osteoporosis, Down's syndrome, cystic fibrosis, Tay-Sach, Huntington, Alzheimer's disease, cataracts, psychosis and more.

- Lung cancer: may be caused by excessive exposure to cigarette smoke or other pollutants.

- Leukemia: cancer of the white blood cells (WBC) – malignant cells are released and carried through blood. The diseases penetrate body tissues like hepatic organ, the dermis and the CNS.

- Skin cancer: Caused primarily by exposure to UV rays that damage DNA and the DNA fails to repair. This is triggered by stratospheric ozone depletion.

- Thyroid cancer: may be caused by previous radiation therapy.

- Heart Disease: the heart gets partially cut off from the supply of oxygen as a result of the blockages of the coronary arteries that supply oxygen to the heart. Such cases are mostly triggered by smoking of cigarettes, hypertension, high cholesterol levels, diet comprising saturated fat, obesity, lack of exercise and diabetes mellitus.

- Diabetes mellitus: is a metabolic disorder characterized by high blood glucose level due to inability of the pancreas to secrete insulin. Patients often suffer from renal damage, peripheral nerve degeneration and disorders in vision.

- Seizures or epilepsy: are changes or modifications in neurological functions caused due to outburst of abnormal electrical impulse in the brain. Seizures occurring partially may elicit isolated movements such as chewing, lip smacking, swallowing or sensations like numbness. Damage to brain, tumour or stroke may also cause seizures.

10.3.3.2 Water-related diseases

Water, sanitation and hygiene have important impacts on both health and disease.

Water-related diseases may include:

- diseases due to presence of microbes and chemical substances in potable water;
- diseases where the pathogen spent part of their lifecycle in water, like schistosomiasis; and
- diseases like malaria where the vector life cycle is related to water.

Arsenic is found in soil and minerals. Arsenic contamination is a common concern in countries including Argentina, India, Bangladesh, Chile, Mexico, China, Thailand and US. Compounds of arsenic are often used in making pesticides. Arsenic can get into water, air and ground from wind-blown dust. The runoff containing arsenic may contaminate water. People can have arsenic exposure by:

- ingestion of arsenic contaminated foodstuff, water and even inhaling air;
- breathing saw dust or smoke from wood combustion that is treated with arsenic;
- living in areas having high levels of pyrites in the rock; and
- occupational exposure to arsenic.

Arsenicosis is a chronic illness that occurs from drinking arsenic-contaminated water with high levels of arsenic over 5 to 20 years. It shows various health effects including skin problems, corns and small warts, skin cancer, bladder cancer, kidney cancer and lung cancer, affects peripheral circulation, leads to high blood pressure and reproductive disorders. (Refer to Chapter 6)

Excessive intake of **fluorine** in drinking water causes fluorosis, characterized chiefly by mottling of the teeth. Syria, Jordan, Egypt, Libya, Algeria, Sudan, Kenya, Turkey, Iraq, Iran, Afghanistan, India, northern Thailand and China are rich in fluorine in water. Acute high-level exposure causes abdominal pain, excessive salivation, nausea, vomiting, seizures and muscular spasm. Ingestion of fluorides for a long period results in skeletal fluorosis symptomized by rigidity and joint pain. In severe cases there is deformation of osseous structures, calcification of ligaments, muscle impairment and severe pain. (Refer to Chapter 6)

Hepatitis is a disease highlighting liver. Communication of Hepatitis A and E disease is generally through water, foodstuff and from person to person. Hepatitis A is repeated in unsanitary and filthy conditions of African, Asian and Latin American nations. The symptoms begin with a sudden onset of pyrexia, body weakness, loss of appetite, queasiness, abdominal discomfort followed by jaundice within a couple of days.

Excess **nitrate** forms methaemoglobin after reacting with haemoglobin of blood causing methaemoglobinemia. The disease is characterized by the reduced ability of the blood to transport oxygen because of reduced levels of normal free haemoglobin. The symptoms are blueness around the mouth, hands and feet, troubled breathing, vomiting and diarrhoea. Extreme cases are marked by lethargy, salivation, loss of consciousness. (Refer to Chapter 6)

Polio spreads through human-to-human contact, usually routing the body through the mouth by faecally contaminated water or food. The virus attacks the nerve cells in the brain causing paralysis of muscles as those controlling swallowing, heartbeat, and respiration. The disease is fatal if left untreated. Polio-endemic countries are northern India, Egypt, Pakistan, Afghanistan, Somalia and Nigeria.

Cholera is caused by *Vibrio cholera*. Symptoms are very watery diarrhoea, queasiness, spasms, nose-bleeding, fast pulse and vomiting.

Fasciolopsiasis is caused by *Fasciola hepatica*. Symptoms include GIT disorder, diarrhea, enlargement of liver, cholecystitis and obstructive jaundice.

Giardiasis is caused by *Giardia lamblia*, resulting in diarrhoea, abdominal uneasiness, bloating and flatulence.

Amoebiasis is caused by *Entamoeba histolytica* that leads to abdominal distress, tiredness, weight loss, diarrhea and dysentery.

Dysentery is caused by *Shigella* and *Salmonella* and manifests by frequent passage of blood-tinged feces with mucus and blood vomiting.

Typhoid fever caused by *Salmonella typhi* is characterized by sustained fever up to 104°F, profuse sweating, diarrhea, delirium and liver and splenic enlargement.

SARS caused by coronavirus has the symptoms like fever, myalgia, lethargy, gastrointestinal symptoms, cough and sore throat.

Mercury is used in the thermometers, barometers as it is liquid and in fluorescent light bulbs. Metallic mercury is extremely dangerous with skin contact and results in the absorption of mercury into the blood stream and potential health problems. Mercury poisoning may include skin rashes, muscular weakness, mental disturbance, speech impairment, defective of hearing and vision and 'pins and needles' sensation in the hands and feet. Mercury in soil and water undergoes bioconcentration and biomagnification as evidenced in the Minamata incident. (Refer to Chapter 6)

Lead is has been used in paint, ceramics, lead pipes, solders, petroleum, batteries, and cosmetics, etc. The most common sources of lead exposure are lead-based paints, contaminated soil and drinking water, lead crystal, and glazed pottery. Lead exposure can cause a range of neurological disorders. The symptoms can be lack of muscular coordination, convulsions and coma. Adult individuals exposed to lead continually suffer from hypertension, reduced fertility, neuronal disorders, severe pain in muscles and joints and problems related to memory. (Refer to Chapter 6)

Infectious diseases, or **transmissible diseases** or **communicable diseases** exhibit characteristic signs and/or symptoms of disease. It happens due to the presence and growth of pathogens that causes infection in the body of the host. In specific cases like HIV/AIDS, the disease is usually asymptomatic. The pathogens can be viruses, bacteria, fungi, protozoa, multi-cellular parasites, or prions. The disease incidence can be sporadic, epidemics or pandemic. Infectivity is the ability of an organism to inoculate, incubate and multiply in the host. Infectiousness specifies the ease with which the disease gets transmitted to other hosts. The disease can spread from one infected individual to other through transmission of pathogen by physical contact, contaminated food and water, body fluids transfusion, fomites, inhalation, or through carriers and vectors.

The disease is called contagious when they are transmitted by contact with an affected person or through their secretions.

H1N1 (Swine) Flu cases have been reported in 168 countries, including U.S. Novel Influenza A H1N1 Virus is a respiratory disease in pigs that can also infect humans. It spreads from human to human, and causes illness. Antiviral drugs like oseltamivir (Tamiflu) and zanamivir (Relenza) are given to lessen the severity of signs.

Measles, also called red measles, is a highly contagious respiratory illness caused by a *Rubeola* virus. The air-borne virus usually spreads through coughs, sneezes, food or drinks. Measles, Mumps, Rubella and Varicella (MMRV) vaccines are available against measles, mumps, rubella and varicella.

Hantavirus Pulmonary Syndrome (HPS) is an uncommon respiratory disease caused by Hantaviruses or the Sin Nombre virus. HPS can be fatal. Deer, mice and other rodents are the primary carriers who shed the virus in the droppings, urine, and saliva and can then be released in air. Humans can become infected through inhalation. Vaccination given is Andes virus M genome segment-based DNA vaccine.

Pertussis (whooping cough) disease is a very contagious disease of the lungs and respiratory system. It is caused by bacteria, *Bordetella pertussis*. Children are very prone. Pertussis or 'whooping cough' can spread through coughs, sneezes or during talks. DPT vaccine is used to immunize the children against diptheria, pertussis and tetanus.

TB is a contagious disease that is caused by *Mycobacterium tuberculosis* that primarily attacks the lungs and even kidney, spine or brain. Primary symptoms of infection are chronic cough with blood-tinged sputum, pyrexia, night perspirations and loss of weight. If not treated suitably TB is fatal. About one-third of global TB patients are from India and more than 1,000 people die each day. Nearly 2,000,000 Indians suffer from TB, of which about 3.3×10^5 Indians die due to TB each year. The GOI has been executing WHO-recommended Directly Observed Therapy Scheme (DOTS) via the Revised National Tuberculosis Control Programme (RNTCP). Phase II of this programme, in progress from October 2005, is a footstep towards accomplishing the Millennium Development Goals objectives. Bacille Calmette Guerin (BCG) is the vaccine used for tuberculosis.

10.3.3.3 Risks due to chemicals in food

The human body and all foods are made up of chemicals. Naturally found chemical substances are carbohydrates, proteins, lipids, fibres and a magnitude of other minerals and compounds. The foodstuff that is consumed may contain a host of potentially injurious or toxic chemicals either as natural ingredients, as contaminants, or as pesticides and fertilizers, or as food additives.

Plants may contain some natural substances that are toxic. Glycoalkaloids are present in potatoes, aubergines, tomatoes of the Solanaceae family. Glycoalkaloid toxins are responsible for queasiness, vomiting, watery stool or diarrhoea, gastric cramps, headache with possibly more rigorous cases having problems in nervous systems including hallucinations and paralytic attack.

Contaminants are the synthetic/ natural, unwanted matters that can be in particular food owing to the farming practices, manufacturing process or techniques, food preparation techniques, etc. If contaminants are present above permitted limits, they cause a severe health threat. For instance, excess intake of acrylamide can cause DNA damage, cancer etc. Aflatoxins are known to cause liver damage and dioxins are carcinogenic.

Food additives are materials concerned with imparting flavour or enhancing taste of the food as well as its appearance, when supplemented with edibles. Flavours and colours are the top class additives of food.

The substances added for preventing the growth of bacteria and fungi, to arrest further oxidation, to hold back natural ripening of fruits and vegetables are known as preservatives. Packaged foods are always added with preservatives like sodium benzoate and benzoic acid. Preservatives occurring usually in sauces, fruit juices, jams, pickled products can behave as allergens manifesting allergic reactions. Bakery products are often supplemented with potassium bromate to boost their volume. Residual amount of potassium bromate left in the product is detrimental if eaten. Often bacon and hotdogs employ the use of sodium nitrates and nitrites. Consumption of a large amount of sodium nitrate increases the propensity of asthma and reduces pulmonary functioning.

Colour additives are added to enhance natural colours and to provide colour to colourless and 'fun' foods. These artificial colours are mostly made out from petro-chemicals. The colours used may have certain health impacts such as –

Cancer may be caused by Citrus Red 2, Red 3, Red 40, Blue 1, Blue 2 Yellow 5, Yellow 6, and Green 3.

Allergies may be often caused by Blue 1, Red 40, Yellow 5, and Yellow 6.

Hyperactivity: most food dyes are linked to hyperactivity, attention deficit disorder and related health issues.

The sweet recipes, syrups and deserts are often added with orange colour. They cause face behaviour problems and growth problems. Red colour used, causes allergies and thyroid problems. Blue 1 and Blue 2 colours may trigger tumour formation along with red rashes on the skin. Green colour is used while cooking green peas, gel, kulfis, paneer recipes, cakes and jams. Excess use of this colour causes asthma and cancer. The orange flavour is due to octyl acetate while the banana flavour in food is due to isoamyl acetate. Comparatively flavours cause harm to a lesser extent.

Substitutes – Natural dyes are substituted. For example, McDonald's strawberry sundae is coloured with strawberries and not red 40. Fanta orange soda gets its colour from pumpkin and carrot extracts, instead of Red 40 and Yellow 6.

Frequently, pesticides include herbicides, fungicides, insecticides and heavy metals. They are mostly carcinogenic and also reported to trigger Parkinson's disease. Organochlorines are by far more hazardous than others. Exposure to DDT reduces the avian population since the egg shells become thin, break easily and its viability is decreased. In man, such exposure affects the new born. Pesticides damage the reproductive system, nervous system, induces behavioural and developmental abnormalities interfere with hormone function and affect the immune system. They are deposited and concentrate in the adipose tissue and can easily be passed from mothers to breast-fed children.

10.3.3.4 HIV/AIDS

HIV or the human immunodeficiency virus damages or kills the cells of the body's immune system. The most advanced stage of HIV infection is AIDS. There may not be signs and symptoms. A flu-like weakness often lasts for few weeks. Other symptoms may be fever, headache, fatigue, and enlarged lymph nodes. The person may feel normal and the asymptomatic phase may last for years. The virus in the incubation period multiplies rapidly and destroys the primary infection fighters, the CD4 cells, a type of white blood cell. The infected person is highly contagious and the person loses his immunity. The infections in AIDS are called opportunistic infections as they take advantage of the opportunity to infect a debilitated host. The infections may be pneumonia, MAC (mycobacterium avium complex), histoplasmosis, lymphoma, Kaposi's sarcoma with various symptoms.

Most commonly, HIV infection spreads by having unprotected sex with an infected partner. The virus finds its route through vaginal lining, vulva, penis, rectum, or oral sex. HIV is frequently among injection-drug users sharing contaminated needles or syringes. Infected women can transmit HIV though placenta to their babies during pregnancy or through breastfeeding. HIV may be transmitted if tissues or organs or blood from an infected donor is transplanted or transfused to the recipient. People suffering from sexually transmitted infections, such as syphilis, genital herpes, human papilloma virus (HPV), gonorrhea, etc. are more prone to HIV infections.

Globally 34 million plus people are living with HIV/AIDS of which 3.3 million are children under the age group 15. About 7000 contract HIV daily of which 300 is per hour.

The (HIV/AIDS) has assumed pandemic status and have far-reaching impacts in the society. The infection reduces the ability to work, increases the poverty rates, reduces the agricultural output and completely transforms the socio-economic arrangement of many countryside households. The increasing number of death in various countries alters the population dynamics. This has decreased the proportion of prime-age working adults, resulting in gender disparity, whereby, females have to bear added burden of household responsibilities. According to FAO, decline in agricultural productivity is noticed in AIDS-affected households in Kenya. People lose their basic access to land. In a patriarchal societies, the death of a male head often causes women and children to lose ownership of land rights. HIV/AIDS distresses the society through the loss of vital human capital. Resource management organizations suffer because both the knowledge and manual effort are lost with the demise of prime-age adults. Environmental scarcity can intensify and increase HIV/AIDS susceptibility. Actually scarcity of resource deepens poverty and robs the households of feasible livelihood choices. Poverty is found to be the driving force behind HIV/AIDS and unsustainable use of resources. So the efforts to reduce poverty could yield considerable gains both in terms of public health and environmental protection.

10.4 Human Rights

Human rights are defined as the common standards achievement for all people of all nations.

Human rights, environment and sustainable development are interdependent and inseparable. Everyone has the right to secure a healthy and ecologically sound environment. The environment should be adequate to meet the needs of both the present and future generation. Everyone has the right to healthy food and safe drinking water. Everyone has the right to work in a safe and healthy environment. It entails freedom from pollution that threatens health. Everyone has the right to information regarding environment as well as to express opinions regarding environment. Everyone has the right not to be driven out or evicted from their homes except during emergency which will overall be beneficial to the society. Everyone has the right to timely assistance in case of natural disaster. Everyone has the right to benefit impartially and equitably from conservation and sustainable use of nature and natural resources.

10.4.1 Equity

The constitution of India seeks to provide all Indian citizens with justice, liberty of thought, expression, belief, faith and worship; equality of status and of opportunity and to promote among them all, fraternity, assuring the dignity of individual and the integrity of the nation. To ensure liberty and equality for all the parliament have enacted several legislations such as, adoption of Untouchability Act, Suppression of Immoral Traffic act, 1956, Dowry Prohibition act, 1961, National Commission for Women At, 1990, National Commission for Minorities Act, 1992, Protection of Human Rights Act, 1993 along with separate commissions on minorities, SC and ST, Language etc.

Economic disparities are a fact and this is evident when a comparison between the global north and south is made on regional, national, family and individual basis. There is an eclectic gap between the rich and the poor, the men and the women and the present and the future generations. Everyone has the right to basic amenities of life – food, shelter, water etc. We cannot utilize and exhaust the available resources to leave our future generations with empty hand. Therefore, our basic concern should be the equitable distribution of resources, wealth and energy.

10.4.2 Nutrition, health and human rights

All citizens have the right to health which is 'economic, social and cultural right to the highest attainable standard of health', as accepted by the Universal Declaration of Human Rights, International Covenant on Economic, Social and Cultural Rights and the Convention on the Rights of Persons with Disabilities. According to Article 25 of Universal Declaration of Human Rights, 1948, everyone has a right to live which should be ample for health and well-being, not only for themselves but also for their families. The constitution of WHO also affirms that health is one of the fundamental rights of human beings. Every country being a party to human rights treaty must promote and protect health to respect and fulfill the human rights. Poverty, hunger and malnutrition affect the health of the people and restrict the socio-economic progress of the country.

10.4.3 Intellectual property rights and community biodiversity registers

Intellectual property rights: The TRIPs under World Health organization (WTO) was negotiated during the Uruguay Round and India being a party to WTO has to abide by the general provisions of WTO. The intellectual property rules and regulations are thus introduced in the multilateral trading system. These are private rights and ensure the minimum standards of protection that each government has to afford to the intellectual property right in all the WTO member parties. TRIPS cover copyrights, trademarks, patents, geographical indications, trade secrets, industrial designs and layout designs of integrated circuits.

Indigenous people living in and in harmony with nature have used the local flora and fauna since long time. This knowledge is often exploited by the multinational pharma companies to produce modern day drugs with various brand names. They earn in millions and not a penny is given to the native people whose knowledge has been used by those companies. Establishment of community biodiversity registers (CBR) is one such measure to protect the rights of these tribal people. For instance, an extract of pigeon pea was patented by an USA company, which was long been used by the Indian people for treatment of diabetes and arterial blockages. This was challenged by CSIR as an infringement of India's knowledge.

Community biodiversity registers:

The history of Keoladeo Ghana Bird Sanctuary at Bharatpur illustrates an outstanding model of the significance and application of people's knowledge. After years and years of research, conclusion was drawn to exclude buffalo grazing in that habitat of aquatic birds and well accepted by the government with its up gradation to national park in 1982. In the absence of grazing, Paspalum, a grass, grew unrestrained and suffocated the wetland; making it a far inferior habitat. The wetland is famous for Siberian crane, one of the flagship species. The numbers of these migrant birds have been diminishing even after the constitution of the National Park.

The inhabitants of Aghapur opined that the National Park regulations barred the people from digging out roots of Khas or Vetiver grass. This in turn has made the soil compact, making it difficult and harder for the Siberian Cranes to procure underground tubers and corms as food. This is one of the earliest attempts in forming People's Biodiversity Registers during 1996–97 involving the village of Aghapur.

Scientists, therefore, now support and campaign the flexibility of ecosystem management that must always be equipped to make modifications on the basis of continuous watching of ongoing changes. This is what is known as knowledge-based, flexible system of 'adaptive management'.

Many Indians love to live in nature as ecosystem people, closely bonded to the nature to accomplish several of their daily necessities. They farm various indigenous plant species, eat wild fruits, corm bulbs and fish, and graze livestock for milk and dairy products, use fuel-wood to prepare their meals, use grass and mud to thatch their huts and cowsheds. They broadly make use of herbal remedial measures and remedy to cure their ailments and worship nature objects like peepal trees and hanuman.

The traditional knowledge of local people regarding neighbouring biological resources, their therapeutic value or any other utility can be enlisted to provide comprehensive information to all, known as community biodiversity registers or people biodiversity registers.

The intensive, fertilizer-aided cultivation is afflicted from an intensification of pest and disease outbreaks, with numerous pests having acquired resistance. Both mosquito and malarial pathogens have acquired resistance sabotaging their eradication programme. Croplands are becoming unproductive due to mineral deficiency, erosion, water logging, salinity etc. Many medicinal trees and herbs are on the verge of extinction. Forests are being depleted resulting in shortage of fuelwood and village economy is in deep trouble. Under such circumstances, a CBR acts an important instrument that can be used by local people to record the resources in terms of biodiversity, traditional knowledge and good practices. A CBR helps in creating a sense of ownership and also helps to safeguard indigenous knowledge, local crops and livestock resources. It helps to strike a balance between development and conservation to eliminate the unwanted consequences leading to the loss of important genetic resources and diversity.

Agriculture has been the main means of livelihood of communities in rural Sarawak for generations. They grow rubber, rice and pepper for household income. Many other crops supply food for a varied diet, construction supplies, medicines, animal fodder. The area has experienced development in terms of supply of electricity, clean water and good roads. But yet they face challenges such as unpredictable climatic effects, market uncertainties and unplanned development. Increased human encroachment for increased farming is replacing traditional agriculture with monoculture using oil palm. But a balance between development and conservation approach is the utmost need and this can be achieved by CBR.

10.5 Value Education

Values refer to the handling of one's own moral, ethics and standards which helps in distinguishing between right and wrong deeds. It is the values within us that helps in decision making that lead us to our justified course of action. Values are time-specific and do change with time. Hunting, poaching, killing of wildlife which was regarded manly and royal activities once upon a time is now banned by law. Western approach is utilitarian stressing on the value of resources in nature. But Eastern approach is spiritual and traditional. Education can be formal or informal. But its main aim is to inculcate moral values in cultural, social aspect so that one can develop, progress and justify their course of actions. Environmental values are inherent and augment a feeling of sensitivity for preserving nature. Environmental values should enable one to comprehend the core aspects of nature, their relationship to cultural assets. Human heritage will help us properly use and consume resources, ensure social justice and prevent environmental degradation. Though materialistic, we have an inborn desire to explore and unravel the mysteries of nature. We should develop an appreciation of our natural treasures but also move towards a sustainable mode of lifestyle. Once we explore the wonders; we can develop a close bond with nature. We cannot ignore the aesthetic beauty of our wilderness, the serenity of forests, the magnificent personality and grandeur of lions, tigers and elephants. We still worship God in the form of trees and animals. Emphasis should be laid on the significance of preserving antique structures. The distinctive buildings, monument, artworks are all treasures of early cultures and priceless environmental assets. They provide valuable information on our origin, our status at present and where we are heading. The problems created by technological advance and financial growth are mainly due to lack of proper planning, lack of thinking process, inadequate policy and a lack of urge for living in healthy state of mind and health. It is not technology that causes pollution but it is our lack of awareness of the consequences of limitless and uncontrolled encroaching and anti-environmental activities. This mindset should be changed and a deep sense of preserving nature should develop which should be based on spiritual and traditional values. We should learn to appreciate and value the resources we are using, the food we eat, the house we live in, the water that we drink and all other products that forms the basis of ecosystem services that we must appreciate. We should broaden our horizon when we look and perceive our environment. People from all sectors – policymakers, administrators, planners, industrialists, farmers, educationalists, scientists, students, technocrats, etc., from all background should join hands in making this planet a habitable place.

During the rule of emperor and kings there were no environmental concerns. Colonial rule was only motivated towards the exploitation of natural resources. The primary objective post-independence was centered on poverty eradication, increase in literacy rate and equity and growth. The 1972 Stockholm UNCHE was the eye opener to all of us. An entire chapter on environment was incorporated in the Fourth Five Year Plan with emphasis on environmental and economic principles in the use of resources. Environment was included in the Constitution of India in the 42nd Amendment of which Article 48(A) and 51A (g) are most important.

10.5.1 Common property resources

Resources are generally congestible. That means if a resource is used for one purpose the same cannot not be used for another purpose. For example, the fish that we eat cannot be eaten by others. But let it be considered as a congestible resource that everybody can use. Everybody can use as much resource as they want without paying for it. That means it has a zero value. These resources are termed common property resources. The use or access to it is not controlled. Nobody effectively owns these resources. It comprises a core resource that is often subjected to overuse. Environmental resources are largely uncontrolled resources. A common property resource can be the air we breathe, the water with which we sustain the forests, the pasture, minerals etc. Take for example, a pasture where limited amount of grazing is allowed. This will keep it unharmed. Overgrazing the same may deteriorate its condition, the soil may become loose and susceptible to erosion and fail to benefit subsequently. In the beginning, resources were abundant and there was no need to control. All of the resources started off as common property resource but there is now a need to control access over them due to their scarcity.

10.5.2 Ecological degradation

Environment is relentlessly changing over time. This may be catered by human activities or natural. Sometimes people do not have control on some changes like the natural disaster. Landslides, earthquakes, tsunamis, hurricanes, and wildfires can completely devastate plant and animal lives. But they cannot ignore their contribution to environmental degradation. An ecosystem is a distinctive unit that comprises all the living and non-living elements interacting within it. Some area sensitive species may require large stretches of land to meet all of their requirements for food and habitat. Habitats fragmentation breaks up such stretches. Construction of roads across forests for tourism purpose often disrupts the normal lifestyle of wildlife. Invasion by exotic species can displace the native species resulting in food shortage of the organisms that depend on these native ones. Environment degrades when we pollute the air, water and soil with its emissions, discharges or effluents. Agricultural runoff is a lethal source of pollutants. Excess use of fertilizers and pesticides leads to bio-concentration, biomanification and eutrophication. Plants, fish, and other organisms start dying slowly and turn the seas and lakes into dead zone. Nowadays, urban development is one of the principal causes of environmental degradation. With more people pouring in, more towns and cities to be built, more roads to be constructed, more homes needed, there has been an increase in felling of tress and use of cropland and thus more ecological imbalance. City planners and resource managers must consider and predict the long term effects of so called development on the environment. This can be achieved with sound policy, good planning and effective execution.

It may not always be the fault of humans, but we sitting in the seat of highest cerebral activity need to recognize the extent to which nature could provide resources and be responsible enough to utilize it prudently. As a matter of fact, environmental stewardship should form an integral part of healthy resource management practice.

10.6 Women and Child Welfare

Every year, throughout the world, several million children die of malnutrition and diseases. Majority of these cases are in the developing countries. In most of the LDCs, child mortality is more frequent at infancy; one in five children death are before five years old. Several factors relate environment with the well-being of mother and child. Children mainly die of diseases like marasmus, kwashiorkor, pneumonia, diarrhea, measles, and malaria. A close link exists between poverty, malnutrition and environmental degradation. The problem is further intensified by a lack of consciousness on how the children become malnourished. The kind of food offered to children is a very important factor during the transition from breast feeding to external food based diet. A slight raise in breastfeeding could save up to 10 per cent of all children below the age five. Continual breastfeeding possibly maintains good nutritional status of the children. Mothers individually should be very careful and selective about home care and their children's diet that should be a complete diet. The mothers should make sure that a child gets sufficient food and should be able to select better and healthy substitutes for food. Women especially of the low income groups spend long hours in working. Women serve as cheaper sources of labour than men. Their working pattern is also variable making them susceptible to various health hazards. Some are garbage pickers, some work in tobacco factories, in plastic factories or in drug companies, work as construction labourers etc. Handling such disposed, hazardous items they get exposed to various infections. Many living in the slum areas stay in unhygienic contions and this gives rise to respiratory ailments. Frequently they have to walk few kilometers to get water and fuelwood. They serve both within their family and outside it. They are the last persons in the family to eat after feeding other members. Often they eat the residual food which is inadequate and this leaves them undernourished. This socio-environmental division and gender discrimination is a key concern that warrants rectification. Maternal and child healthcare is an important international issue stated in the Millennium Development Goals and is in the provisions of national policy.

10.7 Role of Information Technology in Environment and Human Health

Information technology has incredible prospective in the arena of environmental educational and health as it does have in the field of commerce, money matters, political affairs or culture. It includes development of World Wide Web, GIS and remote sensing, satellite imaging and internet facilities so as to keep us updated and well acquainted with the unknown world. It enables us to keep a database of forests, animals, diseases like malaria, dengue cancer, HIV/AIDS, population, resources, etc. A database is the assortment of interrelated data on various topics in a computerized form in such a manner that it can be recovered and processed whenever required. The information in the database is organized in a methodical and logical manner and can be easily manageable. A number of user-friendly software have been developed in the field of environment studies. Information technology increases our efficiency as it saves both time and human endeavour.

National Management Information System (NMIS) under the DST has assembled a database on Research and Development projects along with complete information about research scientists and persons involved.

The MoEF, GOI, has formed an information system called Environmental Information System (ENVIS). Headquartered in Delhi, it functions in 25 centres throughout the country. ENVIS is engaged in generating a database in Himalayan ecosystem, mangroves, coastal ecosystems, environmental media, environmental management, biodiversity, clean technologies, pollution etc.

The National Institute of Occupational Health provides electronic information on the health aspects of people working with hazardous and non-hazardous substances, their safety measures etc.

Remote sensing and geographical information system, satellite images provide real information about different topographical features and biological resources, their state of dilapidation in a digital form through remote sensing. GIS has become a very important and operational instrument in environmental management. It is a method of overlaying various thematic maps gathered from digital data on a large number of interrelated aspects. They maps can be physical, chemical, biological, social, economic etc. Such maps provide information about ground water, surface water, soil types, farmland, wasteland, human settlements, industries as so; this can then help us in effective and wise planning of subsequent development projects, roadways, railways, town planning etc It provides an easy means of monitoring the present status and also assist us in predicting our future. GIS can even interpret polluted zones and degraded lands. GIS can be used to prepare Zonal Atlas. Planning for location of suitable areas for industries is now being accomplished using GIS by preparing Zoning Atlas. They help solve global environmental issues such as, smog formation, ozone depletion, oil and mineral reserves, weather changes, algal bloom etc.

The World Wide Web with resource material on all aspects, power-point lecture presentations, digital photos, animations, web-exercises and quiz have proved to be enormously useful to both students and the teachers of environmental studies.

Summary

- Population is a group of organism belonging to a particular species. The branch of science that deals with the study of population is called population ecology. The characteristics of a population are governed by size and density, natality, generation time, mortality, survivorship pattern, ecological age structures, etc.

- Population structure refers to the composition of individuals in the population in terms of age, sex, ethnicity, religion, educational, occupation, geographical distribution, socioeconomic condition, births, deaths and migration. The study of the change in population size and structure is called population dynamics.

- The balance among births, deaths and migration concludes whether a population will increase, remains stationary or decrease in number. Population may increase in geometric progression or arithmetic progression and thus exhibit exponential growth form (J-shaped growth curve) and sigmoid growth form (S–shaped).

- The impacts of population growth are environmental, social and economical leading to misery and impoverished condition. World population crossed 7 billion marks. However, trends in developing countries are alarming in comparison to the developed countries. On account of population explosion, India was the first country to launch a national family planning programme. Its first population policy came in 1976. Maternal and child healthcare is an important international issue stated in the Millennium Development Goals.

- Urbanization, industrialization, technological advancements, economic growth and GDP can be indicators of development and prosperity. Similarly the incidence and outbreak of several diseases are also indicators of the morbid state of environment. Human rights, environment and sustainable development are interdependent and inseparable. Everyone has the right to secure a healthy and ecologically sound environment.

- All citizens have the right to health which is economic, social and cultural right to the highest attainable standard of health. The constitution of India seeks to provide all Indian citizens with justice, liberty of thought, expression, belief, faith and worship; equality of status and of opportunity and to promote among them all, fraternity, assuring the dignity of individual and the integrity of the nation.

- The intellectual property rights ensure minimum standards of protection that each government has to give to the intellectual property rights in all the WTO member parties. The knowledge of the indigenous people is often exploited by the multinational pharma companies to produce modern day drugs with brand names. Establishment of community biodiversity registers (CBR) is one such measure to protect the rights of these tribal people.

- Resources are generally congestible. But a resource that is usable to everybody without paying for it has zero value. Such resources are common property resources.

- Information technology has incredible prospective in the field of environmental educational and health through the development of World Wide Web, GIS and remote sensing and internet facilities.

- It may not always be the fault of humans, but we sitting in the seat of highest cerebral activity. As a matter of fact, environmental stewardship should form an integral part of healthy resource management practice.

Exercise

MCQs

Encircle the right option:

1. The current size of population of human beings worldwide is:
   A. More than 12 billion   B. More than 7 billion
   C. Less than 1 billion   D. Immediately under 7million

2. The numerical study relating to populations is referred to as:
   A. Biotic potential   B. Demography   C. Dispersal
   D. Natality   E. Fecundity

3. In a population, the number of organisms inhabiting per unit area is known as:
   A. Survival   B. Dispersion   C. Natality
   D. Density   E. Hypervolume niche

4. What type of effect is said to have a rising impact with the increase in population size?

    A. Density-independent effect   B. Age effect

    C. Survivorship effect   D. Density-dependent effect

5. The mathematical expression -

    [(number of births + number of immigration) – (number of deaths + number of emigration)]

    X 100/original size of population is utilized to evaluate:

    A. Population change   B. Exponential growth rate

    C. Growth rate   D. Geometric expansion

6. Determine the density of the orange tree population in an orchard having 40 oranges trees in an area of 2 hectare:

    A. 20 trees /hectare   B. 80 trees/hectare   C. 40 trees/hectare

    D. 38 trees/hectare   E. 0.10 trees/ hectare

7. The number of individuals belonging to same species inhabiting the total area of their entire habitat is its:

    A. Populace dimension   B. Ecological density   C. Crude density

    D. None of these

8. A population increases provided:

    A. Birth rate decreases   B. Death rate increases   C. Biotic potential increases

    D. Ecological resistance increases   E. All of the above

9. _____ is not a consequence of soaring populace density:

    A. Accumulation of noxious wastes   B. Increasing mortality

    C. Predators likely to ignore over abundant prey   D. Decrease in reproduction

10. Which of the following is used to measure population density?

    A. Individuals/m   B. Individuals/ha   C. Individuals/cm

    D. All   E. A and B only

Fill in the blanks.

1. Number of organisms living per habitat space is _____ density.

2. The time required by the population to attain double its size is known as _____ time.

3. The biotic potential of r selected species is _____ .

4. Cholera is caused _____ .

5. BCG vaccine refers to _____ .

State whether the statements are true or false.

1. Community ecology is an example of autecology. (T/F)

2. The age pyramid of Bangladesh, at present, is broad based. (T/F)

3. In arithmetic progression the common difference between two contagious number is same. (T/F)

4. Tubectomy is a process of male sterilization. (T/F)

5. Bisphenol A is a type of endocrine disruptor.(T/F)

Short questions.

1. Define population.

2. State the difference between population and community.

3. What is population density?

4. What is the difference between crude density and ecological density?

5. Differentiate between density and dispersion of a population with examples.

6. Define the demographic transition.

7. What are migration, emigration and immigration?

8. Define biotic potential.

9. What is environmental resistance?

10. Explain the difference between population dynamics and demography.

Essay type questions.

1. What conditions may result in clumped, uniform and random dispersion of individuals in a population?

2. Compare the exponential form of population growth with the logistic form.

3. Distinguish between r-selected populations and K-selected populations. Give examples.

4. Explain how density-dependent factors may affect growth of a population.

5. Explain with examples, how biotic and abiotic factors may influence population's growth.

Bibliography

Agrawal, K.M., P.K. Sikdar and S.C. Deb. 2002. *A Text book of Environment*. Macmillan Publication.

Ambinakudige, S. and K. Joshi. 2012. 'Remote Sensing of Cryosphere'. *Journal of Photogrammetry and Remote Sensing* 57: 5–6.

Amirthalingam, M. 2004. 'The Sacred Groves of Tamilnadu'. *The Indian Forester* 130 (11): 1279–85

Basu, R.N. (ed). *Environment*. Kolkata: University of Calcutta.

Bharucha, E. 2005. *Text Book of Environmental Studies for UG Course*. University Press (India) Pvt. Ltd.

Blumenthal, D.S. and J. Ruttenber. 1995. *Introduction to Environmental Health*, Second Edition. New York: Springer.

Botanical Survey of India. 1983. *Flora and Vegetation of India - An Outline*. Howrah: Botanical Survey of India.

Botkin, D.B. and A.E. Keller. *Environmental Science*, Eighth Edition. Wiley India Pvt. Ltd.

Bustard, H.R. 1982. 'Crocodile Breeding Project'. In *Wildlife in India*, edited by V. B. Saharia. Dehradun: Natraj Publishers.

Centre for Science and Environment. 1994. 'The Spirit of Sanctuary'. *Down to Earth*: 21–37.

Central Water Commission. 2010–11. *Annual Report, 2010–2011*. GOI.

Chandran, M.D.S. and D. Mesta. 2001. 'On the Conservation of the Myristica Swamps of the Western Ghats'. In *Forest Genetic Resources: Status, Threats, and Conservation Strategies*, edited by U. Shaanker, R.K.N. Ganeshaiah and K.S. Bawa, 1–19. Oxford & IBH Publishing.

Chary, S.N., and B. Vyasulu. 2001. *Environmental Management: An Indian Perspective*. MacMillan India Ltd.

Chiras, D.D. 2001. *Environmental Science: Creating a Sustainable Future*, Sixth Edition. Jones & Bartlett Publishers.

Costanza, R. 1997. 'The Value of the World's Ecosystem Services and Natural Capital'. *Nature* 387: 253–59.

Cunningham, W., and M.A. Cunningham. 2003. *Principles of Environmental Science; Enquiry and Applications*, Second Edition. New Delhi: Tata McGraw Hill Publication.

De, A.K. 1996. *Environmental Chemistry*, Third Edition. New Delhi: New Age International (P) Ltd. Publishers.

Deevey, E.S. 1960. 'The Human Population'. *Scientific American* 203: 195–204.

Elton, C.S. 1927. *Animal Ecology*. New York: Macmillan.

Enger, E. and B. Smith. 2010. *Environmental Science: A Study of Interrelationships*. Twelfth Edition. McGraw-Hill Higher Education.

EPA, Ardcavan, Wexford, Ireland, Wastewater Treatment Manuals primary, secondary and tertiary treatment.

Esakku, S., A. Swaminathan, O. Parthiba Karhtikeyan, J. Kurian and K. Palanivelu. 2007. 'Municipal Solid Waste Management in Chennai City, India', Proceedings Sardinia 2007, Eleventh International Waste Management and Landfill Symposium, S. Margherita di Pula, Cagliari, Italy.

Fernandes, W. 2004. 'Rehabilitation Policy for the Displaced'. *Economic and Political Weekly* 39 (12): 1191–93.

Foster, P. 1992. *The World Food Problem: Tackling the Causes of Undernutrition in the Third World*. London: Lynne Rienner Publishers.

Gadgil, M. 1989. 'Heritage of a Conservation Ethic'. In *Conservation of an Indian Heritage*, edited by B. Allchin, F.R. Allchin and B.K. Thapar, 13–22. New Delhi: Cosmo Publications.

Ghosh Roy, M.K. 2011. *Sustainable Development (Environment, Energy and Water Resources)*. Ane Books Pvt. Ltd.

Glanntz, M.H. (ed). 1987. *Droughts and Hunger in Africa: Denying Famine a Future*. Cambridge: Cambridge University Press.

Goyal, S., B. S. Sergi and M. Esposito. *Social Entrepreneurship in Developing Economies – Understanding the Constraining Factors and Key Focus Areas from the Literature Review*.

Hardin, G. 1968. 'The Tragedy of the Commons'. *Science* 162: 1243–48.

Hawkins, R.E. 1986. *Encyclopedia of Indian Natural History*. Centenary Publication of B.N.H.S. Mumbai: OUP.

Hoekstra, A.Y. and A.K. Chapagain. 2008. *Globalization of Water*. Malden MA: Blackwell Publishing.

International Energy Agency. 2013. *World Energy Outlook*.

Intergovernmental Panel on Climate Change (IPCC). 2007. *Climate Change 2007: The Physical Science Basis, Contribution of Working Group I to the Fourth Assessment Report of the Intergovernmental Panel on Climate Change*, edited by S. Solomon, D. Qin, M. Manning, Z. Chen, M. Marquis, K.B. Averyt, M. Tignor, and H.L. Miller, 996. Cambridge United Kingdom and New York, NY, USA: Cambridge University Press.

Ison, S., S. Peake, and S. Wall. 2002. *Environmental Issues and Policies*. Prentice Hall.

Khaitan, B. and N. Priya. 2009. 'Rehabilitation of the Displaced Persons in India'. *NUJS Law Review* 2 (111).

Kumar, H.D. 2001. *Forest Resources; Conservation and Management*. Affiliated East West Press Pvt. Ltd.

Kumar, R. 2013. 'Green Marketing - A Brief Reference to India'. *Asian J. Multidis Stu.* 1(4).

Lasgorceix, A. and Kothari. 2009. 'A Displacement and Relocation of Protected Areas: A Synthesis and Analysis of Case Studies'. *Economic and Political Weekly* 44 (49): 37–47.

Lippmann, M. and R.B. Schlesinger. 1979. *Chemical Contamination in the Human Environment*. New York: Oxford University Press.

Lovelock, J. 1995. *The Ages of Gaia: A Biography of our Living Earth*. New York: Norton.

Lutgens, F.K. and E.J. Tarbuck. 2001. *The Atmosphere*. Upper Saddle River, NJ: Prentice-Hall.

Metcalf, L. and H.P. Eddy. 2002. *Wastewater Engineering, Treatment, Disposal and Reuse*, Fourth Edition. New Delhi: Tata McGraw-Hill Publishing Company Limited.

Micklin, P. and N.V. Aladin. 2008. 'Reclaiming the Aral Sea'. *Scientific American*.

Miller, G.T. 2001 *Environmental Science*, Eighth Edition. USA: Brooks/Cole, Thomas Learning, Inc.

Miller, T.G. Jr. *Environmental Science*. Wadsworth Publishing Co.

Mishra, P. and P. Sharma. 2010. 'Green Marketing in India: Emerging Opportunities and Challenges'. *J. Engin. Sci. Man. Edu.* 3: 9–14.

Misra, S.P. and S.N. Pande. 2011. *Essential Environmental Studies*, Third Edition. Ane Books Pvt. Ltd.

Mohan, D. and P.M.K. Bhamawat. 2008. 'E-waste Management Global Scenario: A Review'. *J. Env. Res. Dev.* 2 (4).

National Conservation Strategy and Policy Statement on Environment and Development, Govt. of India, Ministry of Environment and Forest, June, 1992.

Odum, E.P. 1971. *Fundamentals of Ecology*, Third Edition. Japan: WB Saunders Company.

Peavy, H.P. and D.R. Rowe. 1985. *Environmental Engineering*. McGraw-Hill International Editions.

Pimentel, D., R. Harman, M. Pacenza, J. Pecarsky and M. Pimentel. 1994. 'Natural Resources and an Optimum Human Population'. *Population and Environment: J. Interdis. Stud.* 15 (5): 347–69.

Population Action International. 1997. *Why Population Matters*. Washington, DC: Population Action International.

Ramachandra, T.V. and V. Kulkarni. *Environmental Management*. Capital Publishing Company.

Rodgers, W.A. and H.S. Panwar. 1988. *Planning a Wildlife Protected Area Network in India*, 2 Volumes, 339, 267. Project FO: IND/82/003. Dehradun: FAO.

Sankar, U. 1998. Laws and Institutions Relating to Environmental Protection in India, Conference on The Role of Law and Legal Institutions in Asian Economic Development, 1–4 November.

Sao, A. 2013. 'Research Paper on Green Marketing'. *J. Sci. App. Res* 3 (1): 1–6.

Singh, G., D. Laurence, and L.D. Kauntala. 2006. *Managing the Social and Environmental Consequences of Coal Mining in India*. Dhanbad, India: The Indian School of Mines University.

Sinha, R.K., M. Dubey, R.D. Tripathi, A. Kumar, P. Tripathy and S. Dwivedi. 2010. 'India As A Megadiversity Nation'. *International Society of Environmental Botanists* 16 (4).

Sodhi, G.S. 2001. *Fundamental Concepts of Environmental Chemistry*. New Delhi: Narosa Publishing House.

Terminski, B. 2012. *Mining-Induced Displacement and Resettlement: Social Problem and Human Rights Issue (A Global Perspective)*. Available at SSRN 2028490.

Quammen, D. 1997. *The Song of the Dodo: Island Biogeography in an Age of Extinctions*. New York: Simon & Schuster.

Wani, S.P. and K.K. Garg. 2009. 'Watershed Management Concept and Principles'. In *Best-bet Options for Integrated Watershed Management*, Proceedings of the Comprehensive Assessment of Watershed Programs in India, 25–27 July 2007, ICRISAT Patancheru, Andhra Pradesh, India.

Wilson, E.O. 1992. *The Diversity of Life*. New York: W.W. Norton & Co.

World Bank. 1984. *World Development Report 1984*. New York: Oxford University Press.

_____. 1996. *Poverty Reduction and the World Bank Progress and Challenges in the 1990s*. Washington, DC.

World Resources Institute. 1999. *Urban Growth*. Washington DC.

Wright, R.T. 2008. *Environmental Science: Towards a Sustainable Future*. Prentice-Hall Inc.

Information collected from different websites

http://www.ipcc.ch/pdf/assessment-report/ar4/wg3/ar4-wg3-chapter1.pdf

Aldwell, C.R., D.J. Burdon and M. Sherwood. 1983. 'Impact of Agriculture on Groundwater in Ireland'. *Environmental Geology* 5: 39–48.

Alfoldi, L. 1983. 'Topic 2: Movement and Interaction of Nitrates and Pesticides in the Vegetation Cover-Soil Groundwater-Rock System'. *Environmental Geology* 5:19–25.

Alford, H.G., and M.P. Ferguson (eds). 1982. 'Pesticides in Soil and Groundwater', Proceedings of a Conference, Division of Agricultural Sciences, University of California, Berkeley.

Bidwai, Praful. 9 September 2013. 'Food Bill is a First Step'. *The Daily Star*.

Case Studies Water Conflict Management and Transformation at OSU.htm. Available at: http://www.transboundarywaters.orst.edu.

Central Water Commission. 1989. *Major River Basins of India-An Overview*. New Delhi: Ministry of Water Resources, Government of India.

Vedanta and Posco: A Tale of Two Projects. Available at: http://infochangeindia.org/environment/features/vedanta-and-posco-a-tale-of-two-projects.html

Conservation India. TRAFFIC's latest study. *'Illuminating the Blind Spot' A Study on Illegal Trade in Leopard Parts in India*. Available at: www.conservationindia.org/tag/hunting.

Conservation International. *The Biodiversity Hotspots*. Available at: http://www.conservation.org/where/priority_areas/hotspots/Pages/hotspots_defined.aspx

_____. *Endemic Plant Species*. Available at: www.conservation.org/.../key_findings/Pages/endemic_plant_species.aspx.

'Conservation, Relocation and The Social Consequences of Conservation Policies in Protected Areas'. *Case Study of the Sariska Tiger Reserve, India*. Available at: http://conservationandsociety.org/article.asp?issn=09724923;year=2011;volume=9;issue=1;spage=54;epage=64;aulast=Torri;type=3.

Despite Ban, Mines Thrive in Sariska Reserve Down To Earth.mht. Available at: http://www.downtoearth.org.in/.

Earthworks Fact Sheet: Hardrock Mining and Acid Mine Drainage. Available at: http://www.earthworksaction.org/pubs/FS_AMD.pdf.

'Tiger Population of India Facing "Total Disaster" Due to Tourism Ban'. *The Guardian*. Available at: http://www.theguardian.com/environment/2012/aug/20/tiger-population-india-tourism-ban.

Environment Australia. 2002. *Overview of Best Practice Environmental Management in Mining*. Available at: http://www.ret.gov.au/resources/Documents/LPSDP/BPEMOverview.pdf.

Environmental Movements. Available at: http://environmentalmovemnts.blogspot.com/.

Flood: The River Kosi - All Systems Need A Little Disorder. Available at: http://alittledisorder.com/case–studies–in–control–failure/flood–the–river–kosi/.

Global Wind Energy Council. February 2013. 'Wind Energy Statistics 2012'. *Report*.

'Goverment Defers Promulgation of Ordinance on Food Security Bill'. *Times of India*. 13 June 2013.

GSA Today - Land Transformation by Humans A Review.htm. Available at: http://www.geosociety.org/.

GWEC. Global Wind Report Annual Market Update. Available at: Gwec.net.

How Many Species on Earth 8.7 Million, says New Study. Available at: http://www.unep-wcmc.org/how-many-species-on-earth-87-million-says-new-study-_704.html.

http://earthobservatory.nasa.gov/Features/Ozone/.

http://education.nationalgeographic.co.in.

http://energy.gov/eere/energybasics/energy-basics.

http://indiagovernance.gov.in/.

http://indianpowersector.com/.

http://india-wris.nrsc.gov.in/wrpinfo/index.php?title=Basins.

http://india-wris.nrsc.gov.in/wrpinfo/index.php?title=Basins.

http://nexusnovus.com/biogas-opportunities-india.

http://pib.nic.in.

http://saarc-sdmc.nic.in/pdf/Earthquake3.pdf.

http://www.alternative-energy-news.info/.

http://www.alternative-energy-tutorials.com/energy-articles/ocean-thermal-energy-conversion.html.

http://www.bene.ie/.

http://www.cea.nic.in.

http://www.energy.eu/stats/energy-coal-proved-reserves-total.html.

http://www.epa.gov/acidrain/what/.

http://www.ethanolindia.net/.

http://www.iea.org/topics/nuclearfissionandfusion/.

http://www.ifrc.org/Global/Publications/disasters/dref/cs-India.pdf.

http://www.ifrc.org/Global/Publications/disasters/dref/cs-India.pdf.

http://www.internationalrivers.org/environmental-impacts-of-dams.

http://www.inwea.org/.

http://www.mnre.gov.in/.

http://www.nrc.gov/reading-rm/basic-ref/students/reactors.html.

http://www.seai.ie.

http://www.sea-technology.com.

http://www.uclm.es/area/amf/Antoine/Energias/Worldwide%20Renewable%20Electricity%20Production.pdf.

http://www.udayindia.org/english/content_07may2011/cover_story.htmlUDAY INDIA17040.

http://www.weforum.org/content/global-agenda-council-food-nutrition-security-2012–2014.

http://www.windpowertv.com/.

International Institute for Environment and Development. 2002. 'Breaking New Ground: Mining, Minerals and Sustainable Development', Chapter 9 of *Local Communities and Mines. Breaking New Grounds.* Available at: http://www.iied.org/pubs/pdfs/G00901.pdf.

IFC/World Bank. December 2007. *Environmental, Health and Safety Guidelines for Mining.* Available. at: http://www.ifc.org/ifcext/sustainability.nsf/AttachmentsByTitle/gui_EHSGuidelines2007_Mining/$FILE/Final+–+Mining.pdf.

IUCN - Facts and Figures on Biodiversity. Available at: http://www.iucn.org/what/biodiversity/about/

Journal of Defence Studies. Available at: www.idsa.in/?q=journals.

Martí, José A., Manuel A.J. Laboy and Dr Orlando E. Ruiz. *Commercial Implementation Of Ocean Thermal Energy Conversion: Using the Ocean for Commercial Generation of Baseload Renewable Energy and Potable Water.* Available at : http://www.seatechnology.com/features/2010/0410/thermal_energy_conversion.html.

MINEO Consortium. 2000. *Review of Potential Environmental and Social Impact of Mining.* Available at: http://www2.brgm.fr/mineo/UserNeed/IMPACTS.pdf.

NDTV.com. 2011. *Tiger Census 295 tigers Added, Population Estimated at 1706.* Available at: www.ndtv.com/article/india/tiger-census-295-tigers-added

Osmanian: Water Resources. Available at: http://www.osmanian.com/2011/01/water–resources.html.

Planning Commission. Available at: http://planningcommission.nic.in/reports/genrep/bkpap2020/16_bg2020.doc5.

Press Information Bureau English releases Tiger Census 2011. Available at: http://pib.nic.in/newsite/erelease.aspx?relid=71310.

The Bougainville Conflict.htm. Available at: www.speedysnail.com/pacific.

'The Global Implications of India's Food Security Law'. *Business Standard.html*. Available at: http://www.business-standard.com.

'The National Food Security Bill, 2013 Receives the Assent of the President', published in the *Gazette of India* as Act No. 20 of 2013' (Press release). Press Information Bureau.

The Uranium Mining Project in Andhra Pradesh A Dossier.htm. Available at: http://www.minesandcommunities.org/.

The Selenium Scourge in Punjab | Down To Earth. Available at: http://www.downtoearth.org.in/node/10900.

Threats To Global Biodiversity. Available at: http://www.greenfacts.org/en/global-biodiversity-outlook/l-3/6-threat-biodiversity.htm

Vij, N.C. and N.V.C. Menon. *The Earthquake and Tsunami in Japan of 11th March 2011: A Wake Up Call for India*. Available at: http://ndma.gov.in/ndma/pressrelease/ncvij&nvcmprmessage.pdfhttp://ozonewatch.gsfc.nasa.gov/facts/hole.html.

Wildlife Protection Society of India (WPSI). Current Status of Tiger in India. Available at: wpsi–india.org/tiger/tiger_status.php.

World Health Organization. 1946. 'Preamble to the Constitution of the World Health Organization'. *Official Records of the World Health Organization* 2: 100.

www.griequity.com/resources/Environment/Geo3/Chapter2urban.pdf.

www.fao.org/docrep/016/i3027e/i3027e.pdf.

www.unep.org/vitalforest/report/vfg-01-forest-definition-and-extent.pdf.

www.indiaenvironmentportal.org.in/files/book_WaterSecurity.pdf.

The World Conservation Union. 2010. *IUCN Red List of Threatened Species. Summary Statistics for Globally Threatened Species*. [Table 1: Numbers of threatened species by major groups of organisms (1996–2010)].

World Wind Energy Association. February 2011. World Wind Energy Report 2010. (PDF). Report.

www.moef.nic.in/report/0405/Chap-06.pdf.

www.mse.ac.in/pub/op_sankar.

www.iso.org/iso/home/standards/management-standards/iso14000.htm.

www.iso.org/iso/theiso14000family_2009.pdf.

www.cpcb.nic.in/Eco_Label.php.

www.ceeraindia.org/documents/ecomarkindia.htm.cercenvis.nic.in/.

www.ecolabelindex.com/ecolabel/ecomark-india.

www.physicalgeography.net/fundamentals/10h.html.